世界技能大赛 3D 数字游戏艺术项目创新规划教材

Maya 2018 英汉速查手册

伍福军　张巧玲　朱海波　主　编

石志花　肖红梅　邓玉莲　沈振凯　副主编

彭　放　主　审

电子工业出版社
Publishing House of Electronics Industry
北京·BEIJING

内 容 简 介

本书是"世界技能大赛 3D 数字游戏艺术项目创新规划教材（Maya 2018 三维动画系列）"的工具书，按 Maya 2018 的模块结构和读者使用习惯进行编写，方便读者快速查找进行三维动画制作所需命令的中英文对照，以及命令参数的中英文对照、功能和使用方法。读者还可以通过第 13 章 Maya 中英文命令对照表，查找 Maya 英文版任意英文单词的中文对照。本书可以让读者轻松运用 Maya 英文版进行学习、工作和教学。

本书内容包括公共菜单、建模菜单、装备菜单组、动画模块菜单组、FX 模块菜单组、渲染模块菜单、视图模块菜单组、工具属性及工具架图标、界面与快捷菜单、对象属性、渲染设置属性和 Maya 中英文命令速查，最后还附有 Maya 常用命令操作快捷键（附录 A）。本书免费提供教学视频（含综合案例），请登录百度云盘下载，详细链接网址和密码请参考附录 B。

本书适合高职高专、职业院学校学生和三维动画爱好者自学参考，也可以作为相关机构短期培训的参考教材。

未经许可，不得以任何方式复制或抄袭本书之部分或全部内容。
版权所有，侵权必究。

图书在版编目（CIP）数据

Maya 2018 英汉速查手册 / 伍福军等主编．—北京：电子工业出版社，2018.3
世界技能大赛 3D 数字游戏艺术项目创新规划教材
ISBN 978-7-121-33634-8

I. ①M⋯ II. ①伍⋯ III. ①三维动画软件—高等学校—教材 IV. ①TP391.414

中国版本图书馆 CIP 数据核字(2018)第 022023 号

责任编辑：郭穗娟
印　　刷：三河市鑫金马印装有限公司
装　　订：三河市鑫金马印装有限公司
出版发行：电子工业出版社
　　　　　北京市海淀区万寿路 173 信箱　邮编　100036
开　　本：787×1092　1/16　印张：33.25　字数：851 千字
版　　次：2018 年 3 月第 1 版
印　　次：2018 年 3 月第 1 次印刷
定　　价：69.80 元

凡所购买电子工业出版社图书有缺损问题，请向购买书店调换。若书店售缺，请与本社发行部联系，联系及邮购电话：(010)88254888，88258888。
质量投诉请发邮件至 zlts@phei.com.cn，盗版侵权举报请发邮件至 dbqq@phei.com.cn。
本书咨询联系方式：(010)88254502，guosj@phei.com.cn

前 言

本书是根据 2018 年编者所在学院的专业建设和参加世界技能大赛的培训计划，结合技师学院教学实际和第 44 届世界技能大赛 3D 数字游戏艺术项目技术文件规范，组织开发的世界技能大赛 3D 数字游戏艺术项目创新规划教材之一。

本书对编写体系进行了精心的设计，按照"公共菜单→模块菜单→视图菜单→界面与快捷菜单→对象属性面板→Render Settings【渲染设置】面板→Maya 中英文命令对照"这一思路进行编排，读者可以快速查找命令的中英文对照。本书可以作为教学和培训参考教材。

本书知识结构如下：

第 1 章 Public【公共】模块菜单组，主要介绍 File【文件】、Edit【编辑】、Create【创建】、Select【选择】、Modify【修改】、Display【显示】、Window【窗口】、Cache【缓存】、【Arnold】和 Help【帮助】10 个菜单组中命令的中英文对照、命令作用和使用方法。

第 2 章 Modeling【建模】模块菜单组，主要介绍 Mesh【网络】、Edit Mesh【编辑网络】、Mesh Tools【网络工具】、Mesh Display【网络显示】、Curves【曲线】、Surfaces【曲面】、Deform【变形】、【UV】和 Generate【生成】9 个菜单组中命令的中英文对照、命令作用和使用方法。

第 3 章 Rigging【装备】模块菜单组，主要介绍 Skeleton【骨架】、Skin【蒙皮】、Deform【变形】、Constrain【约束】和 Control【控制】5 个菜单组中命令的中英文对照、命令作用和使用方法。

第 4 章 Animation【动画】模块菜单组，主要介绍 Key【关键帧】、Playback【播放】、Visualize【可视化】和【MASH】4 个菜单组中命令的中英文对照、命令作用和使用方法。

第 5 章 FX 模块菜单组，主要介绍 nParticle【n 粒子】、Fluids【流体】、nCloth【n 布料】、nHair【n 头发】、nConstraint【n 约束】、nCache【n 缓存】、Fields/Solvers【字段/结算器】、Effects【效果】、【Bifrost】和 Boss【凸柱】10 个菜单组中命令的中英文对照、命令作用和使用方法。

第 6 章 Rendering【渲染】模块菜单组，主要介绍 Lighting/Shading【照明/着色】、Texturing【纹理】、Render【渲染】、Toon【卡通】和 Stereo【立体】5 个菜单组中命令的中英文对照、命令作用及其使用方法。

第 7 章 View【视图】模块菜单组，主要介绍 View【视图】、Shading【着色】、Lighting【照明】、Show【显示】、Renderer【渲染器】和 Panels【面板】6 个菜单组中命令的中英文对照、命令作用和使用方法。

第 8 章工具属性及工具架图标，主要介绍 Maya 中的工具图标和工具架图标的中英文对照、作用和使用方法。

 Maya 2018 英汉速查手册

第 9 章界面与快捷菜单，主要介绍 Maya 2018 中的界面布局和快捷菜单中命令的中英文对照。

第 10 章创建对象的属性，主要介绍 Maya 2018 中创建对象的属性，主要包括多边形对象、NURBS、体积对象、摄影机和类型等对象属性。

第 11 章 Rigging【装备】和 FX 中的对象属性，主要介绍 Maya 2018 中 Rigging【装备】和 FX 中的对象属性，主要包括 Joint【关节】、Partice【粒子】、Emitter【发射器】、3D Container【3D 容器】和 2D Container【2D 容器】、Ocean【海洋】、Pond【池塘】、Air【空气】、Drag【阻力】、Gravity【重力】、Newton【牛顿】、Radial【径向】、Turbulence【湍流】、Uniform【统一】和 Vortex【漩涡】15 个对象。

第 12 章 Render Settings【渲染设置】面板，主要介绍 Maya Software（软件）渲染器、Maya Hardware（硬件）渲染器、Maya Hardware 2.0（硬件 2.0）渲染器和 Maya Vector（向量）渲染器的相关属性。

第 13 章 Maya 中英文命令对照速查表，主要按命令字母的顺序排列 Maya 中的命令，英汉对照。读者可以快速查找每个命令中英文对照和命令参数英汉对照。

拥有该手册，读者可以轻松使用 Maya 英文版本进行学习、工作和教学。该手册特别适合高职、高专及职业学校教师和学生，也可作为培训机构参考手册。

本书由伍福军、张巧玲和朱海波担任主编，石志花、肖红梅、邓玉莲和沈振凯担任副主编。彭放对本书进行了主审，在此表示感谢！

由于编者水平有限，本书可能存在疏漏之处，敬请广大读者批评指正！联系电子邮箱：763787922@qq.com。

编　者
2018 年 1 月 1 日

目　　录

第1章　Public【公共】模块菜单组 ································· 1

1.1　File【文件】菜单组 ·· 1
1.2　Edit【编辑】菜单组 ·· 11
1.3　Create【创建】菜单组 ·· 18
1.4　Select【选择】菜单组 ·· 32
1.5　Modify【修改】菜单组 ·· 36
1.6　3Display【显示】菜单组 ·· 53
1.7　Window【窗口】菜单组 ·· 59
1.8　Cache【缓存】菜单组 ··· 62
1.9　Help【帮助】菜单组 ·· 68

第2章　Modeling【建模】模块菜单组 ······························· 71

2.1　Mesh【网络】菜单命令组 ·· 71
2.2　Edit Mesh【编辑网络】菜单组 ····································· 77
2.3　Mesh Tools【网络工具】菜单组 ···································· 82
2.4　Mesh Display【网络显示】菜单组 ·································· 103
2.5　Curves【曲线】菜单组 ·· 108
2.6　Surfaces【曲面】菜单组 ·· 114
2.7　Deform【变形】菜单组 ·· 123
2.8　【UV】菜单组 ·· 142
2.9　Generate【生成】菜单组 ·· 152

第3章　Rigging【装备】模块菜单组 ································ 175

3.1　Skeleton【骨架】菜单组 ·· 175
3.2　Skin【蒙皮】菜单组 ·· 179
3.3　Constrain【约束】菜单组 ··· 186
3.4　Control【控制】菜单组 ··· 190

第4章　Animation【动画】模块菜单组 ······························ 193

4.1　Key【关键帧】菜单组 ··· 193

4.2 Play back【播放】菜单组 …… 198
4.3 Visualize【可视化】菜单组 …… 201
4.4 【MASH】菜单组 …… 203

第5章 FX 模块菜单组 …… 204

5.1 nParticle【n 粒子】菜单组 …… 204
5.2 Fluids【流体】菜单组 …… 212
5.3 nCloth【n 布料】菜单组 …… 219
5.4 nHair【n 头发】菜单组 …… 223
5.5 nConstraint【n 约束】菜单组 …… 227
5.6 nCache【n 缓存】菜单组 …… 231
5.7 Fields/Solvers【字段/结算器】菜单组 …… 236
5.8 Effects【效果】菜单组 …… 245
5.9 Bifrost【Bifrost】菜单组 …… 248
5.10 Boss【凸柱】菜单组 …… 251

第6章 Rendering【渲染】模块菜单组 …… 252

6.1 Lighting/Shading【照明/着色】菜单组 …… 252
6.2 Texturing【纹理】菜单组 …… 253
6.3 Render【渲染】菜单组 …… 255
6.4 Toon【卡通】菜单组 …… 257
6.5 Stereo【立体】菜单组 …… 259

第7章 View【视图】模块菜单组 …… 260

7.1 View【视图】菜单组 …… 260
7.2 Shading【着色】菜单组 …… 262
7.3 Lighting【照明】 …… 264
7.4 Show【显示】菜单组 …… 264
7.5 Renderer【渲染器】菜单组 …… 269
7.6 Panels【面板】菜单组 …… 269

第8章 工具属性及工具架图标 …… 271

8.1 Maya 工具图标 …… 271
8.2 Shelf Tabs【工具架】图标 …… 274

第 9 章 界面与快捷菜单 ………………………………………………………… 293
9.1 Maya 2018 界面布局模式的分类 ……………………………………… 293
9.2 快捷菜单介绍 ……………………………………………………………… 298

第 10 章 创建对象的属性 ……………………………………………………… 302
10.1 Polygon【多边形】对象公共属性面板 ……………………………… 302
10.2 Polygon【多边形】对象的 Shape【形状】公共属性面板 ………… 304
10.3 Polygon Primitives【多边形基本体】对象属性面板 ……………… 309
10.4 NURBS Primitives【NURBS 基本体】对象的 Shape【形状】公共属性面板 …… 313
10.5 NURBS Curve【NURBS 曲线】对象的 Shape【形状】公共属性面板 …… 316
10.6 NURBS Primitives【NURBS 基本体】对象属性面板 ……………… 317
10.7 Volume Primitives【体积基本体】对象的 Shape【形状】公共属性面板 …… 319
10.8 Volume Primitives【体积基本体】对象属性面板 ………………… 321
10.9 Lights【灯光】对象的 Shape【形状】属性面板 …………………… 324
10.10 Cameras【摄影机】对象属性面板 ………………………………… 335
10.11 Type【类型】对象属性面板 ………………………………………… 343

第 11 章 Rigging【装备】和 FX 中的对象属性 …………………………… 348
11.1 Joint【关节】对象属性面板 ………………………………………… 348
11.2 Particle【粒子】对象的 Shape【形状】属性面板 ………………… 350
11.3 Emitter【发射器】属性面板 ………………………………………… 357
11.4 3D Container【3D 容器】的 Shape【形状】属性面板 …………… 359
11.5 2D Container【2D 容器】的 Shape【形状】属性面板 …………… 368
11.6 Ocean【海洋】对象属性面板 ………………………………………… 368
11.7 Pond【池塘】对象属性面板 ………………………………………… 374
11.8 Air【空气】属性面板 ………………………………………………… 383
11.9 Drag【阻力】属性面板 ……………………………………………… 385
11.10 Gravity【重力】属性面板 …………………………………………… 387
11.11 Newton【牛顿】属性面板 …………………………………………… 389
11.12 Radial【径向】属性面板 …………………………………………… 390
11.13 Turbulence【湍流】属性面板 ……………………………………… 392
11.14 Uniform【统一】属性面板 ………………………………………… 394
11.15 Vortex【漩涡】属性面板 …………………………………………… 396

第 12 章 Render Settings【渲染设置】面板 ……………………………… 399
12.1 Render Settings【渲染设置】面板中的 Common【公用】属性面板 …… 399
12.2 Render Settings【渲染设置】面板中的 Maya Software【Maya 软件】渲染器属性面板 ……………………………………………………… 401

  12.3 Render Settings【渲染设置】面板中的 Maya Hardware【Maya 硬件】
     渲染器属性面板 ·········· 404

  12.4 Render Settings【渲染设置】面板中的 Maya Hardware 2.0
     【Maya 硬件 2.0】渲染器属性面板 ·········· 405

  12.5 Render Settings【渲染设置】面板中的 Maya Vector 【Maya 向量】
     渲染器属性面板 ·········· 408

第 13 章 Maya 命令中英文对照速查表 ·········· 411

  A ·········· 411
  B ·········· 418
  C ·········· 422
  D ·········· 434
  E ·········· 440
  F ·········· 445
  G ·········· 450
  H ·········· 452
  I ·········· 454
  J ·········· 458
  K ·········· 459
  L ·········· 459
  M ·········· 463
  N ·········· 470
  O ·········· 472
  P ·········· 475
  Q ·········· 482
  R ·········· 483
  S ·········· 491
  T ·········· 508
  U ·········· 513
  V ·········· 516
  W ·········· 518
  X ·········· 520
  Z ·········· 520
  数字 ·········· 521

附录 A Maya 常用命令操作快捷键 ·········· 522

附录 B 本书教学视频链接网址和密码 ·········· 524

第 1 章　Public【公共】模块菜单组

Public【公共】模块菜单组主要包括 File【文件】菜单组、Edit【编辑】、Create【创建】、Select【选择】、Modify【修改】、3Display【显示】、Window【窗口】、Cache【缓存】和 Help【帮助】，共 9 个菜单命令组。

1.1　File【文件】菜单组

1. New Scene【新建场景】命令

New Scene【新建场景】命令的快捷键为 Ctrl+N，New Scene Options【新建场景选项】面板，如图 1.1 所示。

（1）Linear【线性】参数组：在该参数组中提供了 millimeter【毫米】、centimeter【厘米】、meter【米】、inch【英寸】、foot【英尺】和 yard【码】6 个选项，默认 centimeter【厘米】选项。

（2）Angular【角度】参数组：在该参数组中提供了 degrees【度】和 radians【弧度】2 个选项，默认 degrees【度】选项。

（3）Time【时间】参数组：在该参数组中提供了【15fps】、Film（24fps）【电影（24fps）】、【PAL（25fps）】、【NTSC（30fps）】、Show（48fps）【演示（48fps）】、PAL Field（50fps）【PAL 场（50fps）】、NTSC Field（60fps）【NTSC 场（60fps）】、milliseconds【毫秒】、seconds【秒】、minutes【分钟】、hours【小时】、【2fps】、【3fps】、【4fps】、【5fps】、【6fps】、【8fps】、【10fps】、【12fps】、【16fps】、【20fps】、【40fps】、【75fps】、【80fps】、【100fps】、【120fps】、【125fps】、【150fps】、【200fps】、【240fps】、【250fps】、【300fps】、【375fps】、【400fps】、【500fps】、【600fps】、【750fps】、【1200fps】、【1500fps】、【2000fps】、【3000fps】和【6000fps】42 个选项，默认 Film（24fps）【电影（24fps）】选项。

2. Open Scene…【打开场景…】命令

Open Scene…【打开场景…】命令的快捷键为 Ctrl+O，Open Options【打开选项】面板如图 1.2 所示。

（1）File type【文件类型】参数组：在该参数组中提供了 Best Guess【最佳推测】、【mayaAscii】、【mayaBinary】、【mel】、【OBJ】、【audio】、【move】、【EPS】、【Adobe（R）Illustrator（R）】、【image】、【fluidCache】、【editMB】、【IGES_ATF】、【JT_ATF】、【SAT_ATF】、【STEP_ATF】、【SEL_ATF】、【INVENTOR_ATF】、【CATIAV4_ATF】、【CATIV5_ATF】、【NXZ_ATF】、【PROE_ATF】、【FBX】、【DAF_FBX】、【ASS】、【Alembic】和【BIF】，共 28 个选项，默认 Best Guess【最佳推测】选项。

图1.1　New Scene Options【新建场景选项】面板　　图1.2　Open Options【打开选项】面板

（2）Load Settings【加载设置】参数组：在该参数组中提供了 Load default references【加载保存的引用加载状态】、Load all references【加载所有引用】、Load top-level references only【仅加载顶层级引用】和 Load no references【不加载引用】4个选项，默认 Load default references【加载保存的引用加载状态】选项。

3. Save Scene【保存场景】命令

Save Scene【保存场景】命令的快捷键为 Ctrl+S，Save Scene Options【保存场景选项】面板如图1.3所示。

4. Save Scene As…【场景另存为…】命令

Save Scene As…【场景另存为…】命令的快捷键为 Ctrl+Shift+S，Save Scene As Options【场景另存为选项】面板如图1.4所示。

【文件类型】参数组：在该参数组中提供了【mayaAscii】和【mayaBinary】2个选项，默认【mayaBinary】选项。

图1.3　Save Scene Options【保存场景选项】面板　　图1.4　Save Scene As Options【场景另存为选项】面板

第1章 Public【公共】模块菜单组

5. Increment and Save【递增并保存】命令

Increment and Save【递增并保存】命令的快捷键为"Ctrl+Alt+S",该命令没有选项面板。

6. Archive Scene【归档场景】命令

Archive Scene【归档场景】命令的 Archive Scene Options【归档场景选项】面板如图 1.5 所示。

7. Save Preferences【保存首选项】命令

Save Preferences【保存首选项】命令没有选项面板。

8. Optimize Scene Size【优化场景大小】命令

Optimize Scene Size【优化场景大小】命令的 Optimize Scene Size Options【优化场景大小选项】面板,如图 1.6 所示。

图 1.5　Archive Scene Options
【归档场景选项】面板

图 1.6　Optimize Scene Size Options
【优化场景大小选项】面板

──────────── Import/Export【导入/导出】────────────

9. Import…【导入…】命令

Import…【导入…】命令的 Import Options【导入选项】面板,如图 1.7 所示。

(1) File type【文件类型】参数组:在该参数组中提供了 Best Guess【最佳推测】、【MayaAscii】、【MayaBinary】、【Mel】、【OBJ】、【Audio】、【Move】、【EPS】、【Adobe(R)

Illustrator（R）】、【Image】、【FluidCache】、【EditMB】、【IGES_ATF】、【JT_ATF】、【SAT_ATF】、【STEP_ATF】、【SEL_ATF】、【INVENTOR_ATF】、【CATIAV4_ATF】、【CATIV5_ATF】、【NXZ_ATF】、【PROE_ATF】、【FBX】、【DAF_FBX】、【ASS】、【Alembic】和【BIF】28个选项，默认 Best Guess【最佳推测】选项。

（2）Load Settings【加载设置】参数组：在该参数组中提供了 Load default references【加载保存的引用加载状态】、Load all references【加载所有引用】、Load top-level references only【仅加载顶层级引用】和 Load no references【不加载引用】4个选项，默认 Load default references【加载保存的引用加载状态】选项。

10. Export All…【导出全部…】命令

Export All…【导出全部…】命令的 Export All Options【导出全部选项】面板，如图1.8所示。

图1.7 Import Options【导入选项】面板

图1.8 Export All Options【导出全部选项】面板

【文件类型】参数组：在该参数组中提供了【MayaAscii】、【MayaBinary】、【Mel】、【Move】、【Edit MA】、【Edit MB】、【FBX Export】、【DAE_FBX Export】、【ASS Export】和【OBJ Export】10个选项，默认【MayaBinary】选项。

11. Export Selection…【导出当前选择】命令

Export Selection…【导出当前选择】命令的 Export Selection Options【导出当前选择选项】面板如图1.9所示。

File type【文件类型】参数组：在该参数组中提供了【MayaAscii】、【MayaBinary】、【Mel】、【Move】、【Edit MA】、【Edit MB】、【FBX Export】、【DAE_FBX Export】、【ASS Export】和

【OBJ export】10 个选项，默认【mayaBinary】选项。

12. Game Exporter【游戏导出器】命令

Game Exporter【游戏导出器】命令 Game Exporter【游戏导出器】面板，如图 1.10 所示。

图 1.9　Export Selection Options
【导出当前选择选项】面板

图 1.10　Game Exporter
【游戏导出器】面板

（1）Exporter【导出】方式：主要提供了 Export All【导出全部】、Export Selection【导出当前选择】和 Export Object set【导出对象集】3 种方式，默认 Export All【导出全部】方式。

（2）Settings【设置】卷展栏参数中提供了 Export to Single File【导出到单个文件】和 Export to Multiple Files【导出到多个文件】2 个导出选项。默认 Export to Single File【导出到单个文件】导出选项。

（3）Up Axis【上方向轴】参数组：在该参数组中提供了【Y】和【Z】2 个选项。默认【Y】选项。

（4）File Type【文件类型】参数组：在该参数组中提供了 Binary【二进制】和 ASCII【ASCII】2 个选项，默认 Binary【二进制】选项。

（5）FBX Version【FBX 版本】参数组：在该参数组中提供了【FBX 2016/2017】、【FBX 2014/2015】、【FBX 2013】、【FBX 2012】、【FBX 2011】、【FBX 2010】、【FBX 2009】和【FBX 2006】8 个选项，默认【FBX 2016/2017】选项。

13. Cloud Import/Export…【云导入/导出…】命令

Cloud Import/Export【云导入/导出】面板如图 1.11 所示。

14. Send To Unity【发送到 Unity】命令组

Send To Unity【发送到 Unity】命令组主要包括如图 1.12 所示的 3 个命令。

15. Send To Unreal【发送到 Unreal】命令组

Send To Unreal【发送到 Unreal】命令组主要包括如图 1.13 所示的 3 个命令。

图 1.11　Cloud Import/Export
【云导入/导出】面板

图 1.12　Send To Unity
【发送到 Unity】命令组面板

图 1.13　Send To Unreal
【发送到 Unreal】命令组面板

16. Send to Adobe（R）After Effects（R）【发送到 AE】命令组

Send to Adobe（R）After Effects（R）【发送到 AE】命令组主要包括如图 1.14 所示的 2 个参数组。

17. Send To Print Studio【发送到 Print Studio】命令（新增功能）

Send To Print Studio【发送到 Print Studio】命令没有选项面板。

18. Export to Offline File…【导出到脱机文件…】命令

Export to Offline File…【导出到脱机文件…】命令的 Export to Offline File Options【导出到脱机文件选项】面板如图 1.15 所示。

【文件类型】参数组：在该参数组中主要提供了【editMA】和【editMB】2 个选项，默认【editMA】选项。

19. Assign Offline File…【指定脱机文件…】命令

Assign Offline File…【指定脱机文件…】命令的 Assign Offline File Options【指定脱机文件选项】面板，如图 1.16 所示。

第 1 章　Public【公共】模块菜单组

图 1.14　Send to Adobe（R）After Effects（R）
【发送到 AE】命令组面板

图 1.15　Export to Offline File Options
【导出到脱机文件选项】面板

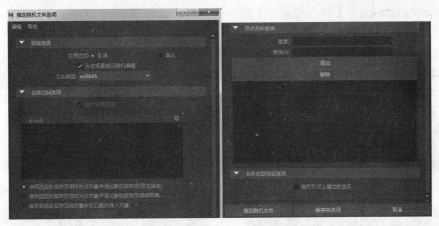

图 1.16　Assign Offline File Options【指定脱机文件选项】面板

File Type【文件类型】参数组：在该参数组中提供了【MayaAscii】、【MayaBinary】、【Mel】、【Move】、【Edit MA】、【Edit MB】、【FBX export】、【DAE_FBX Export】、【ASS Export】和【OBJ Export】10 个选项，默认【MayaBinary】选项。

20.【ATOM】命令组

【ATOM】命令组主要包括如图 1.17 所示的 3 个命令。

1）Export Animation…【导出动画…】命令

Export Animation…【导出动画…】命令的 Maya Atom Exporter Options【Maya Atom 导出器选项】面板，如图 1.18 所示。

图 1.17 ATOM【ATOM】
命令组面板

图 1.18 Maya Atom Exporter Options
【Maya Atom 导出器选项】面板

2）Import Animation…【导入动画…】命令

Import Animation…【导入动画…】命令的 Maya Atom Importer Options【Maya Atom 导入器选项】面板，如图 1.19 所示。

3）Create ATOM Template…【创建 ATOM 模板…】命令

【创建 ATOM 模板…】命令的 Maya Atom Template Options【Maya Atom 模板选项】面板，如图 1.20 所示。

图 1.19 Maya Atom Template Options
【Maya Atom 导入器选项】面板

图 1.20 Maya Atom Template Options
【Maya Atom 模板选项】面板

━━━━━━━━━━━━━━【引用】━━━━━━━━━━━━━━

21.【创建引用…】命令

【创建引用…】命令的快捷键为 Ctrl+R，Reference Options【引用选项】面板，如图 1.21 所示。

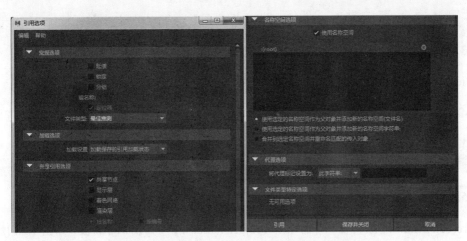

图 1.21　Reference Options【引用选项】面板

（1）File type【文件类型】参数组：在该参数组中提供了 Best Guess【最佳推测】、【mayaAscii】、【mayaBinary】、【mel】、【OBJ】、【audio】、【move】、【EPS】、【Adobe（R）Illustrator（R）】、【image】、【fluidCache】、【editMB】、【IGES_ATF】、【JT_ATF】、【SAT_ATF】、【STEP_ATF】、【SEL_ATF】、【INVENTOR_ATF】、【CATIAV4_ATF】、【CATIV5_ATF】、【NXZ_ATF】、【PROE_ATF】、【FBX】、【DAF_FBX】、【ASS】、【Alembic】和【BIF】28 个选项，默认 Best Guess【最佳推测】选项。

（2）Load Settings【加载设置】参数组：在该参数组中提供了 Load default references【加载保存的引用加载状态】、Load all references【加载所有引用】、Load top-level references only【仅加载顶层级引用】和 Load no references【不加载引用】4 个选项，默认 Load default references【加载保存的引用加载状态】选项。

22. Reference Editor【引用编辑器】命令

Reference Editor【引用编辑器】命令的 Reference Editor【引用编辑器】面板，如图 1.22 所示。

图 1.22　Reference Editor【引用编辑器】面板

————————————————View【视图】————————————————

23. View Image…【查看图像…】命令

View Image…【查看图像…】命令没有选项面板。

24. View Sequence…【查看序列…】命令

View Sequence…【查看序列…】命令没有选项面板。

————————————————Project【投影】————————————————

25. Project Window【项目窗口】命令

Project Window【项目窗口】命令的 Project Window【项目窗口】面板，如图 1.23 所示。

图 1.23　Project Window【项目窗口】面板

26. Set Project…【设置项目…】命令

Set Project…【设置项目…】命令的 Set Project【设置项目】面板，如图 1.24 所示。

————————————————Recent Files【最近】————————————————

27. Recent Files【最近文件】命令

Recent Files【最近文件】命令显示最近打开的文件，没有选项面板。

28. Recent Increments【最近的递增文件】命令

Recent Increments【最近的递增文件】命令显示最近使用过的递增文件，没有选项面板。

29. Recent Projects【最近的项目】命令

Recent Projects【最近的项目】命令显示最近使用过的项目，没有选项面板。

30. Exit【退出】命令

Exit【退出】命令的快捷键为 Ctrl+Q，没有选项面板。

1.2　Edit【编辑】菜单组

1. Undo【撤销】命令

Undo【撤销】命令的快捷键为 Ctrl+Z，没有选项面板。

2. Redo【重做】命令

Redo【重做】命令的快捷键为 Ctrl+Y，没有选项面板。

3. Repeat【重复】命令

Repeat【重复】命令的快捷键为 G。

4. Recent Commands List【最近命令列表】命令

单击该命令以列表的方式显示最近使用过的命令。

5. Cut【剪切】命令

Cut【剪切】命令的快捷键为"Ctrl+X"，没有选项面板。

6. Copy【复制】命令

Copy【复制】命令的快捷键为"Ctrl+C"，没有选项面板。

7. Paste【粘贴】命令

Paste【粘贴】命令的快捷键为"Ctrl+V"，没有选项面板。

8. Keys【关键帧】命令组

Keys【关键帧】命令组主要包括如图 1.25 所示的 7 个命令。

图 1.24　Set Project【设置项目】面板

图 1.25　Keys【关键帧】命令组面板

1）Cut Keys【剪切关键帧】命令

Cut Keys【剪切关键帧】命令的 Cut Keys Options【剪切关键帧选项】面板，如图 1.26 所示。

2）Copy Keys【复制关键帧】命令

Copy Keys【复制关键帧】命令的 Copy Keys Options【复制关键帧选项】面板，如图 1.27 所示。

图 1.26　Cut Keys Options
【剪切关键帧选项】面板

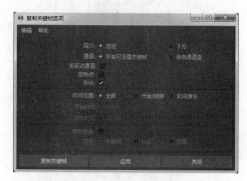
图 1.27　Copy Keys Options
【复制关键帧选项】面板

3）Paste Keys【粘贴关键帧】命令

Paste Keys【粘贴关键帧】命令的 Paste Keys Options【粘贴关键帧选项】面板，如图 1.28 所示。

4）Delete Keys【删除关键帧】命令

Delete Keys【删除关键帧】命令的 Delete Keys Options【删除关键帧选项】面板，如图 1.29 所示。

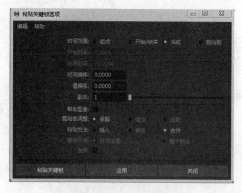
图 1.28　Paste Keys Options
【粘贴关键帧选项】面板

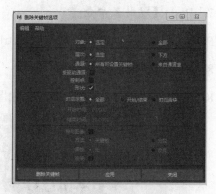
图 1.29　Delete Keys Options
【删除关键帧选项】面板

5）Scale Keys【缩放关键帧】命令

Scale Keys【缩放关键帧】命令的 Scale Keys Options【缩放关键帧选项】面板，如图 1.30 所示。

6）Snap Keys【捕捉关键帧】命令

Snap Keys【捕捉关键帧】命令的 Snap Keys Options【捕捉关键帧选项】面板，如图1.31所示。

图1.30　Scale Keys Options
【缩放关键帧选项】面板

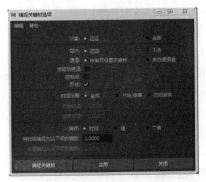

图1.31　Snap Keys Options
【捕捉关键帧选项】面板

7）Bake Simulation【烘焙模拟】命令

Bake Simulation【烘焙模拟】命令的 Bake Simulation Options【烘焙模拟选项】面板，如图1.32所示。

Baked layers【烘焙层】参数组：在该参数组中提供了 Keep【保持】、Remove Attributes【移除属性】和 Clear Animation【清除动画】3个选项，默认 Keep【保持】选项。

————————Delete【删除】————————

9. Delete【删除】命令

Delete【删除】命令没有选项面板。

10. Delete by Type【按类型删除】命令组

Delete by Type【按类型删除】命令组主要包括如图1.33所示的10个命令。

图1.32　Bake Simulation Options
【烘焙模拟选项】面板

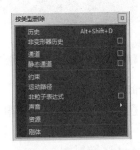

图1.33　Delete by Type
【按类型删除】命令组面板

1）History【历史】命令

History【历史】命令的快捷键为"Alr+Shift+D",没有选项面板。

2）Non-Deformer History【非变形器历史】命令

Non-Deformer History【非变形器历史】命令的 Delete Non-Deformer History Options【删除非变形器历史选项】面板,如图 1.34 所示。

3）Channels【通道】命令

Channels【通道】命令的 Delete Channels Options【删除通道选项】面板,如图 1.35 所示。

图 1.34　Delete Non-Deformer History Options
【删除非变形器历史选项】面板

图 1.35　Delete Channels Options
【删除通道选项】面板

4）Static Channels【静态通道】命令

Static Channels【静态通道】命令的 Delete Static Channels Options【删除静态通道选项】面板,如图 1.36 所示。

5）Constraints【约束】命令

Constraints【约束】命令没有选项面板。

6）Motion Paths【运动路经】命令

Motion Paths【运动路经】命令没有选项面板。

7）Non-Particle Expressions【非粒子表达式】命令

Non-Particle Expressions【非粒子表达式】命令的 Delete Expressions Options【删除表达式选项】面板,如图 1.37 所示。

图 1.36　Delete Static Channels Options
【删除静态通道选项】面板

图 1.37　Delete Expressions Options
【删除表达式选项】面板

8）Sounds【声音】命令

Sounds【声音】命令在默认情况下没有任何选项,只有在添加了声音之后,才会出现下级选项。

9）Assets【资源】命令

Assets【资源】命令没有选项面板。

第1章 Public【公共】模块菜单组

10）Rigid Bodies【刚体】命令

Rigid Bodies【刚体】命令没有选项面板。

11. Delete All by Type【按类型删除全部】命令组

Delete All by Type【按类型删除全部】命令组主要包括如图 1.38 所示的 31 个命令。

12. Duplicate【复制】命令

Duplicate【复制】命令的快捷图标为 Ctrl+D，该命令没有选项面板。

13. Duplicate Special【特殊复制】命令

Duplicate Special【特殊复制】命令的快捷键为 Ctrl+Shift+D，Duplicate Special Options【特殊复制选项】面板，如图 1.39 所示。

图 1.38 Delete All by Type
【按类型删除全部】命令组面板

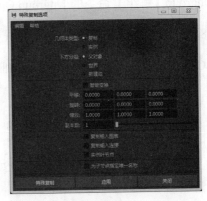

图 1.39 Duplicate Special Options
【特殊复制选项】面板

14. Duplicate With Transform【复制并变换】命令

Duplicate With Transform【复制并变换】命令的快捷键为"Shift+D"，没有选项面板。

15. Transfer Attribute Values【传递属性值】命令

Transfer Attribute Values【传递属性值】命令的 Transfer Attribute Values Options【传递属性值选项】面板如图 1.40 所示。

──────────── Hierarchy【层次】────────────

16. Group【分组】命令

Group【分组】命令的快捷图标为 Ctrl+G，Group Options【分组选项】面板如图 1.41 所示。

图 1.40　Transfer Attribute Values Options
【传递属性值选项】面板

图 1.41　Group Options
【分组选项】面板

17. Ungroup【解组】命令

Ungroup【解组】命令的 Ungroup Options【解组选项】面板如图 1.42 所示。

18. LOD（Level of Detail）【LOD（详细级别）】命令组

LOD（Level of Detail）【LOD（详细级别）】命令组主要包括如图 1.43 所示的 3 个命令组。

图 1.42　Ungroup Options
【解组选项】面板

图 1.43　LOD（Level of Detail）
【LOD（详细级别）】命令组面板

1）Create LOD Group【创建 LOD 组】命令

Create LOD Group【创建 LOD 组】命令的图标为 ▦，Setup LOD Options【创建 LOD 组】面板如图 1.44 所示。

2）Generate LOD Group Mesh【生成 LOD 网络】命令

Generate LOD Group Mesh【生成 LOD 网络】命令的图标为 ▦，Generate LOD Meshes Options【生成 LOD 网络选项】面板，如图 1.45 所示。

（1）Reduction method【减少方法】参数组：在该参数组中提供了 Percentage【百分比】、Vertex limit【顶点限制】和 Triangle Limit【三角形限制】3 个选项，默认 Percentage【百分比】选项。

（2）Number of LOD Levels【LOD 级别的数量】参数组：在该参数组中提供了【1】、【2】、【3】、【4】、【5】、【6】、【7】、【8】和【9】9 个选项，默认【1】选项。

（3）Symmetry tolerance【对称类型】参数组：在该参数组中提供了 None【无】、Automatic【自动】和 Plane【平面】3 个选项，默认 None【无】选项。

图 1.44 Setup LOD Options【创建 LOD 组】面板   图 1.45【生成 LOD 网络选项】面板

（4）Symmetry Plane【对称平面】参数组：在该参数组中提供了【XZ】、【XY】和【YZ】3 个选项，默认【XZ】选项。

3）Ungroup LOD Group【解组 LOD 组】命令

Ungroup LOD Group【解组 LOD 组】命令的图标为 ，该命令没有选项面板。

19. Parent【父对象】命令

Parent【父对象】命令的快捷键为 ，Parent Options【父对象选项】面板如图 1.46 所示。

20. Unparent【断开父子关系】命令

Unparent【断开父子关系】命令的快捷键为 Shift+P，Unparent Options【断开父子关系选项】面板如图 1.47 所示。

图 1.46 Parent Options
【父对象选项】面板

图 1.47 Unparent Options
【断开父子关系选项】面板

1.3 Create【创建】菜单组

―Objects【对象】―

1.【NURBS 基本体】命令组

【NURBS 基本体】命令组主要包括如图 1.48 所示的 10 个命令。

1）Sphere【球体】命令

Sphere【球体】命令的图标为◎，Tool Settings【工具设置】面板如图 1.49 所示。

图 1.48　【NURBS 基本体】命令组面板

图 1.49　Tool Settings【工具设置】面板

2）Cube【立方体】命令

Cube【立方体】命令的图标为◎，Tool Settings【工具设置】面板如图 1.50 所示。

3）Cylinder【圆柱体】命令

Cylinder【圆柱体】命令的图标为◎，Tool Settings【工具设置】面板如图 1.51 所示。

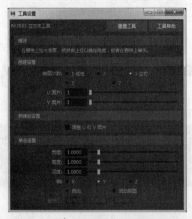

图 1.50　Tool Settings【工具设置】面板

图 1.51　Tool Settings【工具设置】面板

4）Cone【圆锥体】命令

Cone【圆锥体】命令的图标为 ，Tool Settings【工具设置】面板如图1.52所示。

5）Plane【平面】命令

Plane【平面】命令的图标为 ，Tool Settings【工具设置】面板如图1.53所示。

图1.52　Tool Settings【工具设置】面板

图1.53　Tool Settings【工具设置】面板

6）Torus【圆环】命令

Torus【圆环】命令的图标为 ，Tool Settings【工具设置】面板如图1.54所示。

7）Circle【圆形】命令

Circle【圆形】命令的图标为 ，Tool Settings【工具设置】面板如图1.55所示。

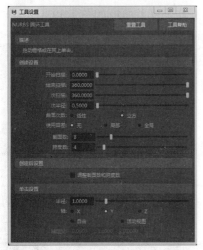

图1.54　Tool Settings【工具设置】面板

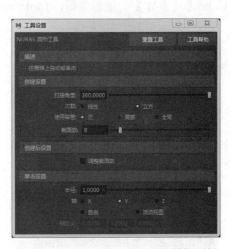

图1.55　Tool Settings【工具设置】面板

8）Square【方形】命令

Square【方形】命令的图标为■，Tool Settings【工具设置】面板如图 1.56 所示。

9）Interactive Creation【交互式创建】命令

Interactive Creation【交互式创建】命令没有选项面板。

10）Exit On Completion【完成时退出】命令

Exit On Completion【完成时退出】命令没有选项面板。

2. Polygon Primitives【多边形基本体】命令组

Polygon Primitives【多边形基本体】命令组主要包括如图 1.57 所示的 14 个命令。

图 1.56　Tool Settings【工具设置】面板　　图 1.57　Polygon Primitives【多边形基本体】命令组面板

1）Sphere【球体】命令

Sphere【球体】命令的图标为■，Tool Settings【工具设置】面板如图 1.58 所示。

2）Cube【立方体】命令

Cube【立方体】的图标为■，Tool Settings【工具设置】面板如图 1.59 所示。

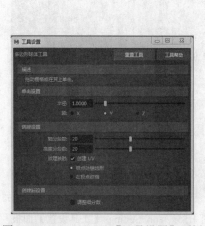
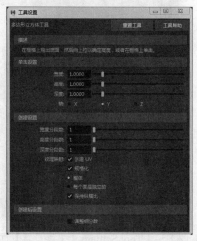

图 1.58　Tool Settings【工具设置】面板　　图 1.59　Tool Settings【工具设置】面板

3）Cylinder【圆柱体】命令

Cylinder【圆柱体】命令的图标为▇，Tool Settings【工具设置】面板如图 1.60 所示。

4）Cone【圆锥体】命令

Cone【圆锥体】命令的图标为▇，Tool Settings【工具设置】面板如图 1.61 所示。

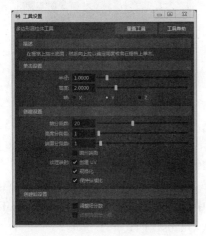
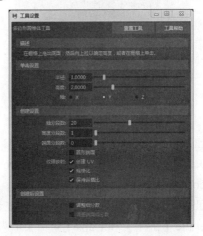

图 1.60　Tool Settings【工具设置】面板　　图 1.61　Tool Settings【工具设置】面板

5）Torus【圆环】命令

Torus【圆环】命令的图标为▇，Tool Settings【工具设置】面板如图 1.62 所示。

6）Plane【平面】命令

Plane【平面】命令的图标为▇，Tool Settings【工具设置】面板如图 1.63 所示。

图 1.62　Tool Settings【工具设置】面板　　图 1.63　Tool Settings【工具设置】面板

7）Disc【圆盘】命令

Disc【圆盘】命令的图标为▇，Polygon Disc Options【多边形圆盘选项】面板如图 1.64 所示。

Subdivision【细分模式】参数组：在该参数组中提供了 Quads【四边形】、Triangles【三角形】、Pie【饼图】、Caps【封口】和 Circle【圆形】5 选项，默认 Quads【四边形】选项。

8）Platonic Solid【柏拉图多面体】命令

Platonic Solid【柏拉图多面体】命令的图标为■，Polygon Platonic Options【多边形柏拉图多面体选项】面板如图 1.65 所示。

图 1.64　Polygon Disc Options
【多边形圆盘选项】面板

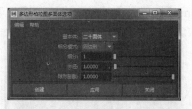
图 1.65　Polygon Platonic Options
【多边形柏拉图多面体选项】面板

Primitive【基本体】参数组：在该参数组中提供了 Tetrahedron【四面体】、Cube【立方体】、Octahedron【八面体】、Dodecahedron【十二面体】和 Icosahedron【二十面体】5 选项，默认 Icosahedron【二十面体】选项。

Subdivision【细分模式】参数组：在该参数组中提供了 Quads【四边形】、Triangles【三角形】、Pie【饼图】和 Caps【封口】4 选项，默认 Quads【四边形】选项。

9）Pyramid【棱锥】命令

Pyramid【棱锥】命令的图标为■，Tool Settings【工具设置】面板如图 1.66 所示。

10）Prism【棱柱】命令

Prism【棱柱】命令的图标为■，Tool Settings【工具设置】面板如图 1.67 所示。

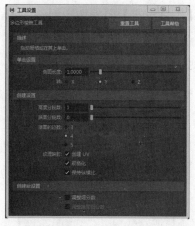
图 1.66　Tool Settings【工具设置】面板

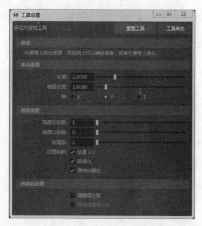
图 1.67　Tool Settings【工具设置】面板

11）Pipe【管道】命令

Pipe【管道】命令的图标为■，Tool Settings【工具设置】面板如图 1.68 所示。

12）Helix【螺旋线】命令

Helix【螺旋线】命令的图标为■，Tool Settings【工具设置】面板如图 1.69 所示。

第 1 章　Public【公共】模块菜单组

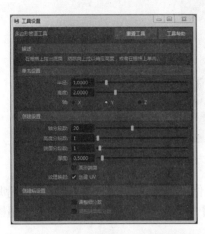

图 1.68　Tool Settings【工具设置】面板

图 1.69　Tool Settings【工具设置】面板

13）Gear【齿轮】命令

Gear【齿轮】命令的图标为 ，Polygon Gear Options【多边形齿轮选项】面板如图 1.70 所示。

14）Soccer Ball【足球】命令

Soccer Ball【足球】命令的图标为 ，Tool Settings【工具设置】面板如图 1.71 所示。

图 1.70　Polygon Gear Options【多边形齿轮选项】面板

图 1.71　Tool Settings【工具设置】面板

──────Super Shapes 超形状──────

15）Super Ellipse【超椭圆】、Spherical Harmonics【球形谐波】和 Ultra Shape【Ultra 形状】命令

Super Ellipse【超椭圆】命令的图标为 ，Spherical Harmonics【球形谐波】命令的图标为 ，Ultra Shape【Ultra 形状】命令的图标为 ，这 3 个命令的面板完全相同，如图 1.72 所示。

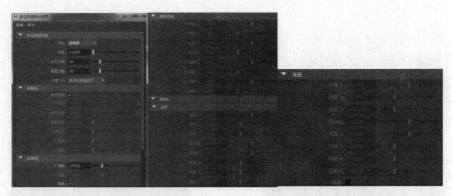

图1.72　Polygon Shape Options【多边形超形状选项】面板

（1）Shape【形状】参数组：在该参数组中提供了 Super Ellipse【超椭圆】、Spherical Harmonics【球形谐波】和 Ultra Shape【Ultra 形状】3 个选项，默认 Super Ellipse【超椭圆】选项。

（2）Create UV【创建 UV】参数组：在该参数组中提供了 None【无】、Pinched at Pole【已收缩到极点】和 Sawtooth at pole【极点处的锯齿形】3 个选项，默认【极点处的锯齿形】选项。

16）Interactive Creation【交互式创建】命令

Interactive Creation【交互式创建】命令没有选项面板。

17）Exit On Completion【完成时退出】命令

Exit On Completion【完成时退出】命令没有选项面板。

3．Volume Primitives【体积基本体】命令组

Volume Primitives【体积基本体】命令组主要包括如图 1.73 所示的 3 个命令。

4．Lights【灯光】命令组

Lights【灯光】命令组主要包括如图 1.74 所示的 6 个命令。

1）Ambient Light【环境光】命令

Ambient Light【环境光】命令的图标为 ，Create Ambient Light Options【创建环境光选项】面板如图 1.75 所示。

图1.73　Volume Primitives　　图1.74　Lights　　图1.75　Create Ambient Light Options
【体积基本体】命令组面板　　【灯光】命令组面板　　【创建环境光选项】面板

2）Directional Light【平行光】命令

Directional Light【平行光】命令的图标为■，Create Directional Light Options【创建平行光选项】面板如图 1.76 所示。

3）Point Light【点光源】命令

Point Light【点光源】命令的图标为■，Create Point Light Options【创建点光源选项】面板如图 1.77 所示。

图 1.76 Create Directional Light Options
【创建平行光选项】面板

图 1.77 Create Point Light Options
【创建点光源选项】面板

Decay rate【衰退速率】参数组：在该参数组中提供了 None【无】、Linear【线性】、Quadratic【二次方】和 Cubic【立方】4 个选项，默认 None【无】选项。

4）Spot Light【聚光灯】命令

Spot Light【聚光灯】命令的图标为■，Create Spot Light Options【创建聚光灯选项】面板如图 1.78 所示。

Decay rate【衰退速率】参数组：在该参数组中提供了 None【无】、Linear【线性】、Quadratic【二次方】和 Cubic【立方】4 个选项，默认 None【无】选项。

5）Area Light【区域光】命令

Area Light【区域光】命令的图标为■，Create Area Light Options【创建区域光选项】面板如图 1.79 所示。

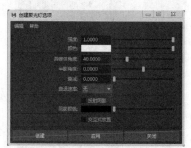
图 1.78 Create Spot Light Options
【创建聚光灯选项】面板

图 1.79 Create Area Light Options
【创建区域光选项】面板

Decay rate【衰退速率】参数组：在该参数组中提供了 None【无】、Linear【线性】、Quadratic【二次方】和 Cubic【立方】4 个选项，默认 None【无】选项。

6）Volume Light【体积光】命令

Volume Light【体积光】命令的图标为■，Create Volume Light Options【创建体积光选

项】面板如图 1.80 所示。

5. Cameras【摄影机】命令组

Cameras【摄影机】命令组主要包括如图 1.81 所示的 5 个命令。

图 1.80　Create Volume Light Options
【创建体积光选项】面板

图 1.81　Cameras
【摄影机】命令组面板

1）Camera【摄影机】命令

Camera【摄影机】命令的图标为 ，Create Camera Options【创建摄影机选项】面板如图 1.82 所示。

Film fit【胶片适配】参数组：在该参数组中提供了 Horizontal【水平】、Vertical【垂直】、Fill【填充】和 Overscan【过扫描】4 个选项，默认 Fill【填充】选项。

2）Camera and Aim【摄影机和目标】命令

Camera and Aim【摄影机和目标】命令的图标为 ，Create Camera and Aim Options【创建摄影机和目标选项】面板，如图 1.83 所示。

图 1.82　Create Camera Options
【创建摄影机选项】面板

图 1.83　Create Camera and Aim Options
【创建摄影机和目标选项】面板

Film fit【胶片适配】参数组：在该参数组中提供了 Horizontal【水平】、Vertical【垂直】、Fill【填充】和 Overscan【过扫描】4 个选项，默认 Fill【填充】选项。

3）【摄影机、目标和上方向】命令

Camera，Aim and Up【摄影机、目标和上方向】命令的图标为，Create Camera 和 Aim and Up Options【创建摄影机、目标和上方向选项】面板如图 1.84 所示。

Film fit【胶片适配】参数组：在该参数组中提供了 Horizontal【水平】、Vertical【垂直】、Fill【填充】和 Overscan【过扫描】4 个选项，默认 Fill【填充】选项。

4）Stereo Camera【立体摄影机】命令

Stereo Camera【立体摄影机】命令没有选项面板。

5）Multi Stereo Rig【多重立体设想机】命令

Multi Stereo Rig【多重立体设想机】命令没有选项面板。

6. Curve Tools【曲线工具】命令组

Curve Tools【曲线工具】命令组主要包括如图 1.85 所示的 6 个命令。

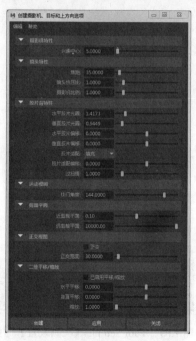

图 1.84　Create Camera，Aim and Up Options 【创建摄影机、目标和上方向选项】面板

图 1.85　Curve Tools 【曲线工具】命令组面板

1）CV Curve Tool【CV 曲线工具】命令

CV Curve Tool【CV 曲线工具】命令的图标为，Tool Settings【工具设置】面板如图 1.86 所示。

2）EP Curve Tool【EP 曲线工具】命令

EP Curve Tool【EP 曲线工具】命令的图标为■，Tool Settings【工具设置】面板如图 1.87 所示。

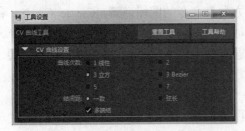

图 1.86　Tool Settings【工具设置】面板

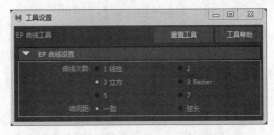

图 1.87　Tool Settings【工具设置】面板

3）Bezier Curve Tool【Bezier 曲线工具】命令

Bezier Curve Tool【Bezier 曲线工具】命令的图标为■，Tool Settings【工具设置】面板如图 1.88 所示。

4）Pencil Curve Tool【铅笔曲线工具】命令

Pencil Curve Tool【铅笔曲线工具】命令的图标为■，Tool Settings【工具设置】面板如图 1.89 所示。

图 1.88　Tool Settings【工具设置】面板

图 1.89　Tool Settings【工具设置】面板

5）Three Point Circular Arc【三点圆弧】命令

Three Point Circular Arc【三点圆弧】命令的图标为■，Tool Settings【工具设置】面板如图 1.90 所示。

6）Two Pont Circular Arc【两点圆弧】命令

Two Pont Circular Arc【两点圆弧】命令的图标为■，Tool Settings【工具设置】面板如图 1.91 所示。

第 1 章　Public【公共】模块菜单组

图 1.90　Tool Settings【工具设置】面板　　　　图 1.91　Tool Settings【工具设置】面板

7. Type【类型】命令（新增功能）

Type【类型】命令的图标为，该命令没有选项面板。

8.【SVG】命令

【SVG】命令的图标为▨，该命令没有选项面板。

9. Adobe（R）Illustrator（R）Object…【Adobe（R）Illustrator（R）对象…】命令

Adobe（R）Illustrator（R）Object…【Adobe（R）Illustrator（R）对象…】命令的图标为▨，Adobe（R）Illustrator（R）Object【Adobe（R）Illustrator（R）对象】面板如图 1.92 所示。

―――――――――――― Construction Aids【构建辅助工具】――――――――――――

10. Construction Plane【构造平面】命令

Construction Plane【构造平面】命令的图标为▨，Construction Plane Options【构造平面选项】面板如图 1.93 所示。

图 1.92　Adobe（R）Illustrator（R）Object　　　图 1.93　Construction Plane Options
　　　【Adobe（R）Illustrator（R）对象】面板　　　　　　【构造平面选项】面板

11. Free Image Plane【自由图像平面】命令

Free Image Plane【自由图像平面】命令的 Create Image Plane Options【创建图像平面选项】面板，如图 1.94 所示。

12. Locator【定位器】命令

Locator【定位器】命令的图标为，该命令没有选项面板。

13. Annotation…【注释…】命令

Annotation…【注释…】命令的图标为，该命令没有选项面板。

14. Measure Tools【测量工具】命令组

Measure Tools【测量工具】命令组主要包括如图 1.95 所示的 3 个命令。

图 1.94　Create Image Plane Options
【创建图像平面选项】面板

图 1.95　Measure Tools
【测量工具】命令组面板

1）Distance Tool【距离工具】命令

Distance Tool【距离工具】命令的图标为，该命令没有选项面板。

2）Parameter Tool【参数工具】命令

Parameter Tool【参数工具】命令的图标为，该命令没有选项面板。

3）Arc Length Tool【弧长工具】命令

Arc Length Tool【弧长工具】命令的图标为，该命令没有选项面板。

──────────────── Scene Management【场景管理】────────────────

15. Scene Assembly【场景集合】命令组

Scene Assembly【场景集合】命令组主要包括如图 1.96 所示的 2 个命令。

1）Assembly Definition…【集合定义…】命令

Assembly Definition…【集合定义…】命令的 Assembly Definition Options【集合定义选项】面板，如图 1.97 所示。

图 1.96　Scene Assembly
【场景集合】命令组面板

图 1.97　Assembly Definition Options
【集合定义选项】面板

2）Assembly Reference…【集合引用…】命令

Assembly Reference…【集合引用…】命令没有选项面板。单击 Assembly Reference…【集合引用…】命令，弹出如图 1.98 所示的 Set Definition File【设置定义文件】对话框。

16．Empty Group【空组】命令组

Empty Group【空组】命令组在默认情况下没有任何命令。

17．Sets【集】命令组

Sets【集】命令组主要包括如图 1.99 所示的 3 个命令。

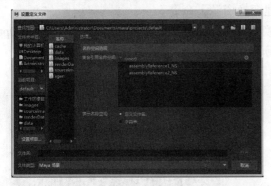

图 1.98　Set Definition File【设置定义文件】对话框

图 1.99　Sets【集】命令组面板

1）Set【集】命令

Set【集】命令的 Create Set Options【创建集选项】面板如图 1.100 所示。

2）Partition【划分】命令

Partition【划分】命令 Partition Options【划分选项】面板如图 1.101 所示。

图 1.100　Create Set Options
【创建集选项】面板

图 1.101　Partition Options
【划分选项】面板

3）Quick Select Set…【快速选择集…】

Quick Select Set…【快速选择集…】的 Create Quick Select Set【创建快速选择集】面板，如图 1.102 所示。

18．Asset【资源】命令组

Asset【资源】命令组主要包括如图 1.103 所示的 2 个命令。

图 1.102　Create Quick Select Set
【创建快速选择集】面板

图 1.103　Asset
【资源】命令组面板

1）Create Asset【创建资源】命令

Create Asset【创建资源】命令的 Create Advanced Asset Options【创建高级资源选项】面板，如图 1.104 所示。

2）Create Asset with Transform【创建变换资源】命令

Create Asset with Transform【创建变换资源】命令的 Create Asset with Transform Options【创建变换资源选项】面板，如图 1.105 所示。

图 1.104　Create Advanced Asset Options
【创建高级资源选项】面板

图 1.105　Create Asset with Transform Options
【创建变换资源选项】面板

1.4　Select【选择】菜单组

1. All【全部】命令

All【全部】命令的快捷键为"Ctrl+Shift+A"，该命令没有选项面板。

2. All by Type【全部按类型】命令组

All by Type【全部按类型】命令组主要包括如图 1.106 所示的 26 种类型。

3. Deselect All【取消选择全部】命令

Deselect All【取消选择全部】命令的快捷键为"Ctrl+D"，该命令没有选项面板。

4. Hierarchy【层次】命令

Hierarchy【层次】命令没有快捷键和选项面板。

5. Inverse【反转】命令

Inverse【反转】命令的快捷键为"Ctrl+Shift+I",没有选项面板。

6. Similar【类似】命令

Similar【类似】命令的图标为 ,Select Similar Options【选择类似对象选项】面板如图1.107所示。

图1.106 All by Type
【全部按类型】命令组面板

图1.107 Select Similar Options
【选择类似对象选项】面板

7. Grow【增长】命令

Grow【增长】命令的图标为 ,快捷键为">" ,没有选项面板。

8. Grow Along Loop【沿循环方向扩大】命令

Grow Along Loop【沿循环方向扩大】命令的快捷键为"Ctrl+>"。

9. Shrink【收缩】命令

Shrink【收缩】命令的图标为 ,快捷键为"<",没有选项面板。

10. Shrink Along loop【沿循环方向收缩】命令

Shrink Along loop【沿循环方向收缩】命令的快捷键为"Ctrl+<"。

11. Quick Select Sets【快速选择集】命令

Quick Select Sets【快速选择集】命令在默认情况下没有任何选择集。

————————Type【类型】————————

12. Object/Component【对象/组件】命令

Object/Component【对象/组件】命令的图标为▨，快捷键为▨，没有选项面板。

————————Polygons【多边形】————————

13. Components【组件】命令组

Components【组件】命令组主要包括如图1.108所示的7个命令。

14. Contiguous Edges【连续边】命令

Contiguous Edges【连续边】命令的图标为▨，Select Contiguous Edges Options【选择连续边选项】面板，如图1.109所示。

15. Shortest Edge Path Tool【最短边路径工具】命令

Shortest Edge Path Tool【最短边路径工具】命令的图标为▨，该命令没有选项面板。

16. Convert Selection【转化当前选择】命令组

Convert Selection【转化当前选择】命令组主要包括如图1.110所示的18个命令。

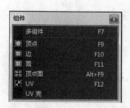
图1.108 Components【组件】命令组面板

图1.109 Select Contiguous Edges Options【选择连续边选项】面板

图1.110 Convert Selection【转化当前选择】命令组面板

17. Use Constraints…【使用约束…】命令

Use Constraints…【使用约束…】命令的图标为▨，Polygon Selection Constraints【多边形选择约束】面板如图1.111所示。

Propagation【传播】参数组主要包括 Off【禁用】、Shell【壳】、Border【边界】、Crease【折痕】、Angle【角度】、Edge Loop【循环边】和 Edge Ring【环形边】7个选项，默认 Off【禁用】选项。

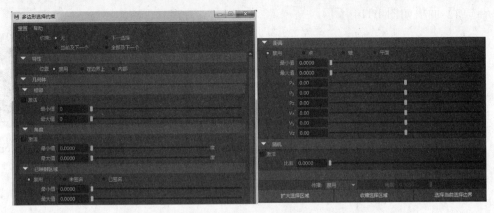

图 1.111　Polygon Selection Constraints【多边形选择约束】面板

────────────NURBS Curves【NURBS 曲线】────────────

18. Components【组件】命令组

Components【组件】命令组主要包括如图 1.112 所示的 4 个命令。

19. All CVs【所有 CV】命令

All CVs【所有 CV】命令的图标为 ，该命令没有选项面板。

20. First CV【第一个 CV】命令

First CV【第一个 CV】命令的图标为 ，该命令没有选项面板。

21. Last CV【最后一个 CV】命令

Last CV【最后一个 CV】命令的图标为 ，该命令没有选项面板。

22. Cluster Curve【簇曲线】命令

Cluster Curve【簇曲线】命令的图标为 ，该命令没有选项面板。

────────────UURBS Surfaces【NURBS 曲面】────────────

23. Components【组件】命令组

Components【组件】命令组主要包括如图 1.113 所示的 4 个命令。

24. CV Selection Boundary【CV 选择边界】命令

CV Selection Boundary【CV 选择边界】命令的图标为 ，该命令没有选项面板。

25. Surface Border【曲面边界】命令

Surface Border【曲面边界】命令的图标为 ，Select Surface Border Options【选择曲面

边界选项】面板如图 1.114 所示。

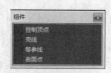

图 1.112　Components
【组件】命令组面板

图 1.113　Components
【组件】命令组面板

图 1.114　Select Surface Border Options
【选择曲面边界选项】面板

1.5　Modify【修改】菜单组

────────────── Transform【变换】──────────────

1. Transformation Tools【变换工具】命令组

Transformation Tools【变换工具】命令组主要包括如图 1.115 所示的 11 个命令。

图 1.115　Transformation Tools【变换工具】命令组面板

1）Move Tool【移动工具】命令

Move Tool【移动工具】命令的图标为，Tool Settings【工具设置】面板如图 1.116 所示。

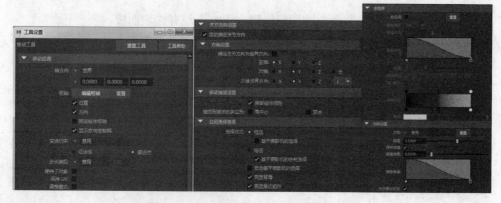

图 1.116　Tool Settings【工具设置】面板

第 1 章 Public【公共】模块菜单组

（1）Transform Constraint【变换约束】参数组：在该参数组中主要包括 Off【禁用】、Edge【边】和 Surface【曲面】3 个选项，默认 Off【禁用】选项。

（2）Step Snap【布长捕捉】参数组：在该参数组中主要包括 Off【禁用】、Relative【相对】和 Absolute【绝对】3 个选项，默认 Off【禁用】选项。

（3）Falloff mode【衰减模式】参数组：在该参数组中主要包括 Volume【体积】、Surface【表面】、Global【全局】和 Object【对象】4 个选项，默认 Volume【体积】选项。

（4）Interpolation【插值】参数组：在该参数组中主要包括 None【无】、Linear【线性】、Smooth【平滑】和 Spline【样条线】4 个选项，默认 None【无】选项。

（5）Symmetry【对称】参数组：在该参数组中主要包括 Off【禁用】、Object X【对象 X】、Object Y【对象 Y】、Object Z【对象 Z】、World X【世界 X】、World Y【世界 Y】、World Z【世界 Z】和 Topology【拓扑】8 个选项，默认 Off【禁用】选项。

2）Rotate Tool【旋转工具】命令

Rotate Tool【旋转工具】命令的图标为，Tool Settings【工具设置】面板如图 1.117 所示。

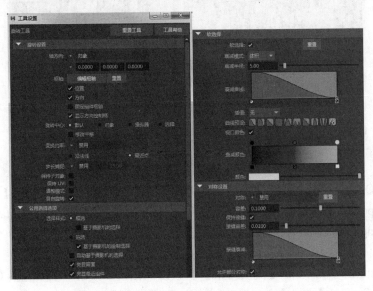

图 1.117　Tool Settings【工具设置】面板

（1）Axis Orientation【轴方向】参数组：在该参数组中主要包括 Object【对象】、World【世界】、Component【组件】、Gimbal【万向】和 Cusom【自定义】5 个选项，默认 Object【对象】选项。

（2）Transform Constraint【变换约束】参数组：在该参数组中主要包括 Off【禁用】、Edge【边】和 Surface【曲面】3 个选项，默认 Off【禁用】选项。

（3）Falloff mode【衰减模式】参数组：在该参数组中主要包括 Volume【体积】、Surface【表面】、Global【全局】和 Object【对象】4 个选项，默认 Volume【体积】选项。

（4）Interpolation【插值】参数组：在该参数组中主要包括 None【无】、Linear【线性】、Smooth【平滑】和 Spline【样条线】4 个选项，默认 None【无】选项。

（5）Symmetry【对称】参数组：在该参数组中主要包括 Off【禁用】、Object X【对象 X】、Object Y【对象 Y】、Object Z【对象 Z】、World X【世界 X】、World Y【世界 Y】、World Z【世界 Z】和 Topology【拓扑】8 个选项，默认 Off【禁用】选项。

3）Scale Tool【缩放工具】命令

Scale Tool【缩放工具】命令的图标为■，Tool Settings【工具设置】面板如图 1.118 所示。

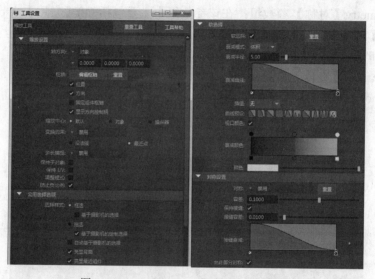

图 1.118　Tool Settings【工具设置】面板

（1）Axis Orientation【轴方向】参数组：在该参数组中主要提供了 Object【对象】、World【世界】、Component【组件】、Parent【父对象】、Normal【正常】、Along Rotation Axis【沿旋转轴】、Along Live Object Axis【沿激活对象的轴】和 Custom【自定义】8 个选项，默认【对象】选项。

（2）Transform Constraint【变换约束】参数组：在该参数组中主要包括 Off【禁用】、Edge【边】和 Surface【曲面】3 个选项，默认 Off【禁用】选项。

（3）Step Snap【布长捕捉】参数组：在该参数组中主要包括 Off【禁用】、Relative【相对】和 Absolute【绝对】3 个选项，默认 Off【禁用】选项。

（4）Falloff mode【衰减模式】参数组：在该参数组中主要包括 Volume【体积】、Surface【表面】、Global【全局】和 Object【对象】4 个选项，默认 Volume【体积】选项。

（5）Interpolation【插值】参数组：在该参数组中主要包括 None【无】、Linear【线性】、Smooth【平滑】和 Spline【样条线】4 个选项，默认 None【无】选项。

（6）Symmetry【对称】参数组：在该参数组中主要包括 Off【禁用】、Object X【对象 X】、Object Y【对象 Y】、Object Z【对象 Z】、World X【世界 X】、World Y【世界 Y】、World

第1章 Public【公共】模块菜单组

Z【世界 Z】和 Topology【拓扑】8 个选项，默认 Off【禁用】选项。

4）Type Manipulator【类型操纵器】命令

Type Manipulator【类型操纵器】命令的图标为 ，该命令没有选项面板。

5）Universal Manipulator【通用操纵器】命令

Universal Manipulator【通用操纵器】命令的图标为 ，快捷键为 Ctrl+T，Tool Settings【工具设置】面板如图 1.119 所示。

图 1.119　Tool Settings【工具设置】面板

6）Move Normal Tool【移动法线工具】命令

Move Normal Tool【移动法线工具】命令的图标为 ，Tool Settings【工具设置】面板如图 1.120 所示。

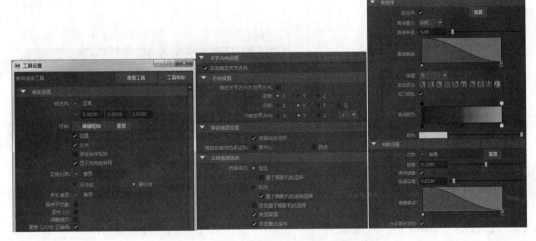

图 1.120　Tool Settings【工具设置】面板

（1）Axis Orientation【轴方向】参数组：在该参数组中主要提供了 Object【对象】、World【世界】、Component【组件】、Parent【父对象】、Normal【正常】、Along Rotation Axis【沿旋转轴】、Along Live Object Axis【沿激活对象的轴】和 Custom【自定义】8 个选项，默认【对象】选项。

（2）Transform Constraint【变换约束】参数组：在该参数组中主要包括 Off【禁用】、Edge【边】和 Surface【曲面】3 个选项，默认 Off【禁用】选项。

（3）Step Snap【布长捕捉】参数组：在该参数组中主要包括 Off【禁用】、Relative【相对】和 Absolute【绝对】3 个选项，默认 Off【禁用】选项。

（4）Falloff mode【衰减模式】参数组：在该参数组中主要包括 Volume【体积】、Surface【表面】、Global【全局】和 Object【对象】4 个选项，默认 Volume【体积】选项。

（5）Interpolation【插值】参数组：在该参数组中主要包括 None【无】、Linear【线性】、Smooth【平滑】和 Spline【样条线】4 个选项，默认 None【无】选项。

（6）Symmetry【对称】参数组：在该参数组中主要包括 Off【禁用】、Object X【对象 X】、Object Y【对象 Y】、Object Z【对象 Z】、World X【世界 X】、World Y【世界 Y】、World Z【世界 Z】和 Topology【拓扑】8 个选项，默认 Off【禁用】选项。

7）Move/Rotate/Scale Tool【移动/旋转/缩放工具】命令

Move/Rotate/Scale Tool【移动/旋转/缩放工具】命令的图标为■，没有选项面板。

8）Show Manipulator Tool【显示操纵器工具】命令

Show Manipulator Tool【显示操纵器工具】命令的图标为■，没有选项面板。

9）Default Object Manipulator【默认对象操纵器】命令组

Default Object Manipulator【默认对象操纵器】命令组主要包括如图 1.121 所示的 5 个命令。

10）Proportional Modification Tool【成比例修改工具】命令

Proportional Modification Tool【成比例修改工具】命令的图标为■，Tool Settings【工具设置】面板如图 1.122 所示。

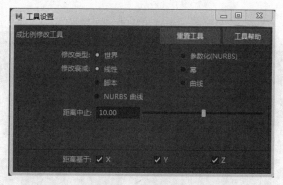

图 1.121　Default Object Manipulator
【默认对象操纵器】命令组面板

图 1.122　Tool Settings
【工具设置】面板

11）Soft Modification Tool【软修改工具】命令

Soft Modification Tool【软修改工具】命令的图标为■，Tool Settings【工具设置】面板如图 1.123 所示。

Falloff mode【衰减模式】参数组：在该参数组中主要包括 Volume【体积】和 Surface【表面】2 个选项，默认 Volume【体积】选项。

2．Reset Transformations【重置变换】命令

Reset Transformations【重置变换】命令的 Reset Transformations Options【重置变换选项】面板如图 1.124 所示。

第 1 章　Public【公共】模块菜单组

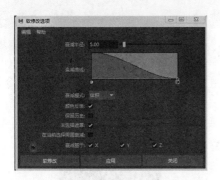

图 1.123　Tool Settings
【工具设置】面板

图 1.124　Reset Transformations Options
【重置变换选项】面板

3. Freeze Transformations【冻结变换】命令

Freeze Transformations【冻结变换】命令的 Freeze Transformations Options【冻结变换选项】面板如图 1.125 所示。

Lock normals【锁定法线】参数组：在该参数组中主要提供了 Never【从不】、Always【始终】和 Non-rigid Transformations Only【仅非刚性变换】选项，默认 Never【从不】选项。

4. Match Transformations【匹配变换】命令组（新增功能）

Match Transformations【匹配变换】命令组主要包括如图 1.126 所示的 5 个命令。

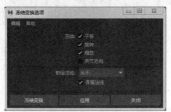

图 1.125　Freeze Transformations Options
【冻结变换选项】面板

图 1.126　Match Transformations
【匹配变换】命令组面板

———————————————— Pivot【枢轴】————————————————

5. Center Pivot【居中枢轴】命令

Center Pivot【居中枢轴】命令没有选项命令。

6. Bake Pivot【烘焙枢轴】命令

Bake Pivot【烘焙枢轴】命令 Bake Pivot Options【烘焙枢轴选项】面板，如图 1.127 所示。

―――――――――――――――Align【对齐】―――――――――――――――

7. Snap Align Objects【捕捉对齐对象】命令组

Snap Align Objects【捕捉对齐对象】命令组主要包括如图 1.128 所示的 5 个命令。

图 1.127　Bake Pivot Options
【烘焙枢轴选项】面板

图 1.128　Snap Align Objects
【捕捉对齐对象】命令组面板

1）Point to Point【点到点】命令

Point to Point【点到点】命令的图标为■，Snap Point to Point Options【点到点捕捉选项】面板如图 1.129 所示。

2）2Points to 2Points【2 点到 2 点】命令

2Points to 2Points【2 点到 2 点】命令的图标为■，Snap 2 Points to 2 Points Options【2 点到 2 点捕捉选项】面板如图 1.130 所示。

图 1.129　Snap Point to Point Options
【点到点捕捉选项】面板

图 1.130　Snap 2 Points to 2 Points Options
【2 点到 2 点捕捉选项】面板

3）3Points to 3points【3 点到 3 点】命令

3Points to 3points【3 点到 3 点】命令的图标为■，Snap 3 Points to 3 points Options【3 点到 3 点捕捉选项】面板如图 1.131 所示。

4）Align Objects【对齐对象】命令

Align Objects【对齐对象】命令的图标为■，Align Objects Options【对齐对象选项】面板如图 1.132 所示。

图 1.131　Snap 3Points to 3points Options
【3 点到 3 点捕捉选项】面板

图 1.132　Align Objects Options
【对齐对象选项】面板

Align to【对齐到】参数组：在该参数组中主要包括 Selection average【选择平均】和 Last selected Object【上一个选定对象】2 个选项，默认 Selection average【选择平均】选项。

5）Position Along Curve【沿曲线放置】命令

Position Along Curve【沿曲线放置】命令的图标为 ，该命令没有选项面板。

8. Align Tool【对齐工具】命令

Align Tool【对齐工具】命令的图标为 ，该命令没有选项面板。

9. Snap Together Tool【捕捉到一起工具】命令

Snap Together Tool【捕捉到一起工具】命令的图标为 ，Tool Settings【工具设置】面板如图 1.133 所示。

──────Nodes【节点】──────

10. Evaluate Nodes【节点求值】命令组

Evaluate Nodes【节点求值】命令组主要包括如图 1.134 所示的 15 个命令。

──────Naming【命名】──────

11. Prefix Hierarchy Names…【添加层次名称前缀…】命令

单击 Prefix Hierarchy Names…【添加层次名称前缀…】命令，即可弹出 Prefix Hierarchy【前缀层次】面板，如图 1.135 所示。

图 1.133　Tool Settings　　图 1.134　Evaluate Nodes　　图 1.135　Prefix Hierarchy
　【工具设置】面板　　　　　【节点求值】命令组面板　　　　【前缀层次】面板

12. Search and Replace Names…【搜索和替换名称…】命令

单击 Search and Replace Names…【搜索和替换名称…】命令，弹出 Search Replace Options【搜索替换选项】面板，如图 1.136 所示。

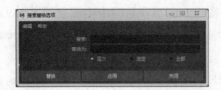

图 1.136 Search Replace Options【搜索替换选项】面板

―――――― Attributes【属性】――――――

13. Add Attribute…【添加属性…】命令

单击 Add Attribute…【添加属性…】命令，弹出 Add Attribute【添加属性】面板，如图 1.137 所示。

图 1.137 Add Attribute【添加属性】面板

提示：只有在选择对象的情况下单击 Add Attribute…【添加属性…】命令，才会弹出 Add Attribute【添加属性】面板。该面板主要有 3 个选项。

14. Edit Attribute…【编辑属性…】命令

Edit Attribute…【编辑属性…】命令的 Edit Attribute【编辑属性】面板，如图 1.138 所示。

15. Delete Attribute…【删除属性…】命令

Delete Attribute…【删除属性…】命令的 Delete Attribute【删除属性】面板，如图 1.139 所示。

提示：在 Delete Attribute【删除属性】面板中，只有给对象添加了属性之后，才会有删除属性的选项。

―――――――――――Objects【对象】―――――――――――

16. Make Live【激活】命令

Make Live【激活】命令的图标为■，该命令没有选项面板。

17. Replace Objects【替换对象】命令

Replace Objects【替换对象】命令的 Replace Objects Options【替换对象选项】面板，如图 1.140 所示。

图 1.138　Edit Attribute
【编辑属性】面板

图 1.139　Delete Attribute
【删除属性】面板

图 1.140　Replace Objects Options
【替换对象选项】面板

18. Convert【转化】命令组

Convert【转化】命令组主要包括如图 1.141 所示的 21 个命令。

1）NURBS to Polygons【NURBS 到多边形】命令

NURBS to Polygons【NURBS 到多边形】命令的图标为■，Convert NURBS to Polygons Options【将 NURBS 转化为多边形选项】面板，如图 1.142 所示。

2）NURBS to Subdiv【NURBS 到细分曲面】命令

NURBS to Subdiv【NURBS 到细分曲面】命令的图标为■，Convert NURBS/Polygons to Subdiv Options【将 NURBS/多边形转化为细分曲面选项】面板，如图 1.143 所示。

3）Polygons to Subdiv【多边形到细分曲面】命令

Polygons to Subdiv【多边形到细分曲面】命令的图标为■，Convert NURBS/Polygons to Subdiv Options【将 NURBS/多边形转化为细分曲面选项】面板，如图 1.144 所示。

图 1.141　Convert
【转化】命令组面板

图 1.142　Convert NURBS to Polygons Options
【将 NURBS 转化为多边形选项】面板

图 1.143　Convert NURBS/Polygons to Subdiv Options
【将 NURBS/多边形转化为细分曲面选项】面板

图 1.144　Convert NURBS/Polygons to Subdiv Options
【将 NURBS/多边形转化为细分曲面选项】面板

4）Smooth Mesh Preview to Polygons【平滑网络预览到多边形】命令

Smooth Mesh Preview to Polygons【平滑网络预览到多边形】命令的图标为 ，没有选项面板。

5）Polygon Edges to Curve【多边形边到曲线】命令

Polygon Edges to Curve【多边形边到曲线】命令的图标为 ，Poly to Curve Options【多边形到曲线选项】面板，如图 1.145 所示。

6）Type to Curves【曲线类型】命令

Type to Curves【曲线类型】命令没有选项面板。

7）Subdiv to Polygons【细分曲面到多边形】命令

Subdiv to Polygons【细分曲面到多边形】命令的图标为 ，Convert Subdiv to Polygons Options【将细分曲面转化为多边形选项】面板，如图 1.146 所示。

图 1.145　Poly To Curve Options
【多边形到曲线选项】面板

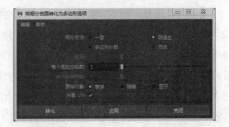

图 1.146　Convert Subdiv to Polygons Options
【将细分曲面转化为多边形选项】面板

8）Subdiv to NURBS【细分曲面到 UNRBS】命令

Subdiv to NURBS【细分曲面到 UNRBS】命令的图标为■，Convert Subdiv to NURBS Options【将细分曲面转化为 NURBS 选项】面板，如图 1.147 所示。

9）NURBS Curve to Bezier【NURBS 曲线到 Bezier】命令

NURBS Curve to Bezier【NURBS 曲线到 Bezier】命令没有选项面板。

10）Bezier Curve to NURBS【Bezier 曲线到 NURBS】命令

Bezier Curve to NURBS【Bezier 曲线到 NURBS】命令没有选项面板。

11）Paint Effects to Polygons【Paint Effects 到多边形】命令

Paint Effects to Polygons【Paint Effects 到多边形】命令的图标为■，Convert Paint Effects to Polygons Options【将 Paint Effects 转化为多边形选项】面板如图 1.148 所示。

图 1.147　Convert Subdiv to NURBS Options
【将细分曲面转化为 NURBS 选项】面板

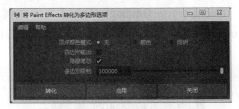
图 1.148　Convert Paint Effects to Polygons Options
【将 Paint Effects 转化为多边形选项】面板

12）Paint Effects to NURBS【Paint Effects 到 NURBS】命令

Paint Effects to NURBS【Paint Effects 到 NURBS】命令的图标为■，Convert Paint Effects to NURBS Options【将 Paint Effects 转化为 NURBS 选项】面板，如图 1.149 所示。

13）Paint Effects to Curves【Paint Effects 到曲线】命令

Paint Effects to Curves【Paint Effects 到曲线】命令的图标为■，Convert Paint Effects to Curve Options【将 Paint Effects 转化为曲线选项】面板，如图 1.150 所示。

图 1.149　Convert Paint Effects to NURBS Options
【将 Paint Effects 转化为 NURBS 选项】面板

图 1.150　Convert Paint Effects to Curve Options
【将 Paint Effects 转化为曲线选项】面板

14）Texture to Geometry【纹理到几何体】命令

Texture to Geometry【纹理到几何体】命令的图标为■，Texture to Geometry Options【纹理到几何体选项】面板，如图 1.151 所示。

【着色器模板】参数组：在该参数组中主要提供了 Default…【默认…】和【Lambert1】2 个选项，默认 Default…【默认…】选项。

15）Displacement to Polygons【置换到多边形】命令

Displacement to Polygons【置换到多边形】命令的图标为■，没有选项面板。

16）Displacement to Polygons with History【置换到多边形（带历史）】命令

Displacement to Polygons with History【置换到多边形（带历史）】命令的图标为■，没有选项面板。

17）Fluid to Polygons【流体到多边形】命令

Fluid to Polygons【流体到多边形】命令的图标为■，没有选项面板。

18）nParticle to Polygons【nParticle 到多边形】命令

nParticle to Polygons【nParticle 到多边形】命令的图标为■，没有选项面板。

19）Instance to Object【实例到对象】命令

Instance to Object【实例到对象】命令的图标为■，没有选项面板。

20）Geometry to Bounding Box【几何体到边界框】命令

Geometry to Bounding Box【几何体到边界框】命令的图标为■，Convert Geometry to Bounding Box Options【将几何体到边界框选项】面板，如图 1.152 所示。

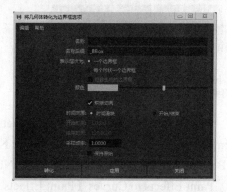

图 1.151　Texture to Geometry Options　　图 1.152　Convert Geometry to Bounding Box Options
【纹理到几何体选项】面板　　　　　　　　　【将几何体到边界框选项】面板

21）Convert XGen Primitives to Polygons【将 XGen 基本体转化为多边形】命令

Convert XGen Primitives to Polygons【将 XGen 基本体转化为多边形】命令的 Convert XGen Primitives to Polygons Options【将 XGen 基本体转化为多边形】面板，如图 1.153 所示。

图 1.153　Convert XGen Primitives to Polygons Options【将 XGen 基本体转化为多边形】面板

Paint Tool【绘制工具】

19. Paint Scripts Tool【绘制脚本工具】命令

Paint Scripts Tool【绘制脚本工具】命令的图标为，Tool Settings【工具设置】面板如图 1.154 所示。

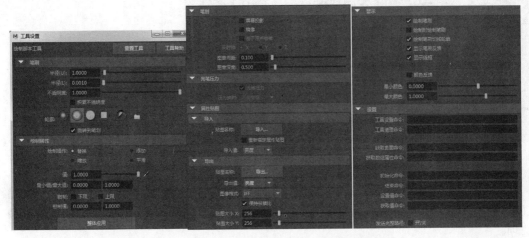

图 1.154　Tool Settings【工具设置】面板

（1）Import value【导入值】参数组：在该参数组中主要提供了 Luminance【亮度】、Alpha【Alpha】、Red【红】、Green【绿】和 Blue【蓝】5 个选项，默认 Luminance【亮度】选项。

（2）Import Value【导出值】参数组：在该参数组中主要提供了 Luminance【亮度】、【Alpha】、【RGB】和【RGBA】4 个选项，默认【亮度】选项。

（3）Image format【图像格式】参数组：在该参数组中主要提供了【GIF】、【SoftImage】、【RLA】、【TIFF】、【SGI】、Alias【锯齿】、【IFF】、【JPEG】和【EPS】9 个选项，默认【IFF】选项。

20. Paint Attributes Tool【绘制属性工具】命令

Paint Attributes Tool【绘制属性工具】命令的图标为，Tool Settings【工具设置】面板如图 1.155 所示。

（1）Vector index【向量索引】参数组：在该参数组中主要提供了【x/R】、【y/G】和【z/B】3 个选项，默认【x/R】选项。

（2）Import value【导入值】参数组：在该参数组中主要提供了 luminance【亮度】、【Alpha】、Red【红】、Green【绿】和 Blue【蓝】5 个选项，默认 Luminance【亮度】选项。

（3）Export format【导出值】参数组：在该参数组中主要提供了 luminance【亮度】、【Alpha】、【RGB】和【RGBA】4 个选项，默认 luminance【亮度】选项。

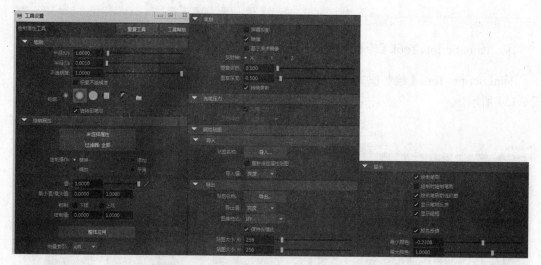

图 1.155　Tool Settings【工具设置】面板

（4）Image format【图像格式】参数组：在该参数组中主要提供了【GIF】、【SoftImage】、【RLA】、【TIFF】、【SGI】、Alias【锯齿】、【IFF】、【JPEG】和【EPS】9 个选项，默认【IFF】选项。

────────────── Assets【资源】──────────────

21. Asset【资源】命令组

Asset【资源】命令组主要包括如图 1.156 所示的 12 个命令和 1 个命令组。

1）Add to Asset【添加到资源】命令

Add to Asset【添加到资源】命令的 Add to Asset Options【添加到资源选项】面板，如图 1.157 所示。

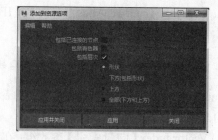

图 1.156　Asset【资源】命令组面板　　　图 1.157　Add to Asset Options【添加到资源选项】面板

2）Remove from Asset【从资源移除】命令

Remove from Asset【从资源移除】命令的 Remove from Asset Options【从资源移除选项】面板，如图 1.158 所示。

3）Export Proxy Asset…【导出代理资源…】命令

Export Proxy Asset…【导出代理资源…】命令的 Export Asset Proxy Options【导出资源

代理选项】面板，如图1.159所示。

图1.158 Remove from Asset Options
【从资源移除选项】面板

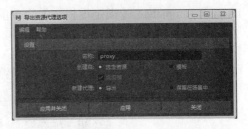

图1.159 Export Asset Proxy Options
【导出资源代理选项】面板

4）Publish Attributes【发布属性】命令

Publish Attributes【发布属性】命令的 Publish Attributes Options【发布属性选项】面板，如图1.160所示。

5）Unpublish Attributes【取消发布属性】命令

Unpublish Attributes【取消发布属性】命令没有选项面板。

6）Publish Connections【发布连接】命令

Publish Connections【发布连接】命令的 Publish Connections Options【发布连接选项】面板，如图1.161所示。

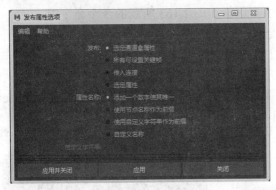

图1.160 Publish Attributes Options
【发布属性选项】面板

图1.161 Publish Connections Options
【发布连接选项】面板

7）Lock Unpublished Attributes【锁定未发布的属性】命令

Lock Unpublished Attributes【锁定未发布的属性】命令没有选项面板。

8）Unlock Unpublished Attributes【取消锁定未发布的属性】命令

Unlock Unpublished Attributes【取消锁定未发布的属性】命令没有选项面板。

9）Publish Node【发布节点】命令

Publish Node【发布节点】命令没有选项面板。

10）Unpublish Node【取消发布节点】命令

Unpublish Node【取消发布节点】命令没有选项面板。

11）Assign Template…【指定模板…】命令

Assign Template…【指定模板…】命令的 Assign Template Options【指定模板选项】面板，如图 1.162 所示。

图 1.162　Assign Template Options【指定模板选项】面板

12）Select Asset Contents【选定资源内容】命令

Select Asset Contents【选定资源内容】命令没有选项面板。

13）Advanced Assets【高级资源】命令组

Advanced Assets【高级资源】命令组主要包括如图 1.163 所示的 3 个命令组。

（1）Node Publishing【节点发布】命令组。

Node Publishing【节点发布】命令组主要包括如图 1.164 所示的 3 个命令。

① Publish as selection Transform【发布为选择变换】命令。

Publish as selection Transform【发布为选择变换】命令的 Publish Selection Transform Options【发布选择变换选项】面板，如图 1.165 所示。

图 1.163　Advanced Assets　　图 1.164　Node Publishing　　图 1.165　Publish Selection Transform Options
【高级资源】命令组面板　　【节点发布】命令组面板　　【发布选择变换选项】面板

② Publish Parent Anchor【发布父锚点】命令。

Publish Parent Anchor【发布父锚点】命令的 Publish Node Options【发布节点选项】面板，如图 1.166 所示。

③ Publish Child Anchor【发布子锚点】命令。

Publish Child Anchor【发布子锚点】命令的 Publish Node Options【发布节点选项】面板，如图 1.167 所示。

第1章 Public【公共】模块菜单组

图 1.166 Publish Node Options
【发布节点选项】面板

图 1.167 Publish Node Options
【发布节点选项】面板

（2）Node Unpublishing【节点取消发布】命令组。

Node Unpublishing【节点取消发布】命令组主要包括 Unpublish Selection Transform【取消发布选择变换】、Unpublish Parent Anchor【取消发布父锚点】和 Unpublish Child Anchor【取消发布子锚点】3 个命令，如图 1.168 所示。

（3）Set Current Asset【设定当前资源】命令组。

Set Current Asset【设定当前资源】命令组在默认情况下没有命令。

1.6 3Display【显示】菜单组

————————————Viewport【视口】————————————

1. Grid【栅格】命令

Grid【栅格】命令的 Grid Options【栅格选项】面板如图 1.169 所示。

2. Heads Up Display【题头显示】命令组

Heads Up Display【题头显示】命令组主要包括如图 1.170 所示的 23 个命令。

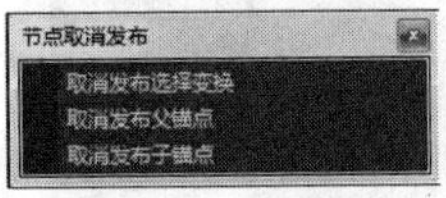

图 1.168 Node Unpublishing
【节点取消发布】命令组面板

图 1.169 Grid Options
【栅格选项】面板

图 1.170 Heads Up Display
【题头显示】命令组面板

Poly Count【多边形计数】命令的 Poly Count Options【多边形计数选项】面板，如图 1.171 所示。

Object【对象】

3. Hide【隐藏】命令组

Hide【隐藏】命令组主要包括如图 1.172 所示的 22 个命令。

1）Hide Geometry【隐藏几何体】命令组

Hide Geometry【隐藏几何体】命令组主要包括如图 1.173 所示的 8 个命令。

图 1.171　Poly Count Options
【多边形计数选项】面板

图 1.172　Hide
【隐藏】命令组面板

图 1.173　Hide Geometry
【隐藏几何体】命令组面板

2）Hide Kinematics【隐藏运动学】命令组

Hide Kinematics【隐藏运动学】命令组主要包括如图 1.174 所示的 3 个命令。

3）Hide Deformers【隐藏变形器】命令组

Hide Deformers【隐藏变形器】命令组主要包括如图 1.175 所示的 7 个命令。

4. Show【显示】命令组

Show【显示】命令组主要包括如图 1.176 所示的 23 个命令。

图 1.174　Hide Kinematics
【隐藏运动学】命令组面板

图 1.175　Hide Deformers
【隐藏变形器】命令组面板

图 1.176　Show
【显示】命令组面板

第 1 章 Public【公共】模块菜单组

1）Show Geometry【显示几何体】命令组

Show Geometry【显示几何体】命令组主要包括如图 1.177 所示的 8 个命令。

2）Show Kinematics【显示运动学】命令组

Show Kinematics【显示运动学】命令组主要包括如图 1.178 所示的 3 个命令。

3）Show Deformers【显示变形器】命令组

Show Deformers【显示变形器】命令组主要包括如图 1.179 所示的 7 个命令。

图 1.177　Show Geometry　　图 1.178　Show Kinematics　　图 1.179　Show Deformers
【显示几何体】命令组面板　　【显示运动学】命令组面板　　【显示变形器】命令组面板

5．Toggle Show/Hide【切换显示/隐藏】命令

Toggle Show/Hide【切换显示/隐藏】命令的 Toggle Visibility Options【切换可见性选项】面板，如图 1.180 所示。

6．Per Camera Visibility【根据摄影机可见性】命令组

Per Camera Visibility【根据摄影机可见性】命令组主要包括如图 1.181 所示的 6 个命令。

7．Wireframe Color…【线框颜色…】命令

单击 Wireframe Color…【线框颜色…】命令，弹出 Wireframe Color【线框颜】面板，如图 1.182 所示。

图 1.180　Toggle Visibility Options　　图 1.181　Per Camera Visibility　　图 1.182　Wireframe Color
【切换可见性选项】面板　　　　【根据摄影机可见性】命令组面板　　　【线框颜】面板

8．Object Display【对象显示】命令组

Object Display【对象显示】命令组主要包括如图 1.183 所示的 9 个命令。

9．Transform Display【变换显示】命令组

Transform Display【变换显示】命令组主要包括如图 1.184 所示的 4 个命令。

10. Polygons【多边形】命令组

Polygons【多边形】命令组主要包括如图 1.185 所示的 29 个命令。

图 1.183　Object Display
【对象显示】命令组面板　　图 1.184　Transform Display
【变换显示】命令组面板　　图 1.185　Polygons
【多边形】命令组面板

1）Culling Options【消隐选项】命令组

Culling Options【消隐选项】命令组主要包括如图 1.186 所示的 3 个命令。

2）Component IDs【组件 ID】命令组

Component IDs【组件 ID】命令组主要包括如图 1.187 所示的 4 个命令。

3）Metadata【元数据】命令组

Metadata【元数据】命令组主要包括如图 1.188 所示的 1 个命令。

图 1.186　Culling Options
【消隐选项】命令组面板　　图 1.187　Component IDs
【组件 ID】命令组面板　　图 1.188　Metadata
【元数据】命令组面板

11. NURBS【NURBS】命令组

NURBS【NURBS】命令组主要包括如图 1.189 所示的 12 个命令。

1）Custom【自定义】命令

Custom【自定义】命令的 Custom NURBS Display Options【自定义 NURBS 显示选项】面板如图 1.190 所示。

第 1 章 Public【公共】模块菜单组

图 1.189　NURBS
【NURBS】命令组面板

图 1.190　Custom NURBS Display Options
【自定义 NURBS 显示选项】面板

Scope【范围】参数组：主要包括 Active Objects【活动对象】、All Objects【所有对象】、New Curves【新曲线】和 New Surfaces【新曲面】4 个选项，默认 Active Objects【活动对象】选项。

2）Hull【壳线】命令

Hull【壳线】命令的 NURBS Smoothness（Hull）Options【NURBS 平滑（壳线）选项】面板，如图 1.191 所示。

3）Rough【粗糙】命令

Rough【粗糙】命令的 NURBS Smoothness（Rough）Options【NURBS 平滑度（粗糙）选项】面板，如图 1.192 所示。

图 1.191　NURBS Smoothness（Hull）Options
【NURBS 平滑（壳线）选项】面板

图 1.192　NURBS Smoothness（Rough）Options
【NURBS 平滑度（粗糙）选项】面板

4）Medium【中等】命令

Medium【中等】命令的 NURBS Smoothness（Medium）Options【NURBS 平滑度（中）选项】面板，如图 1.193 所示。

5）Fine【精细】命令

Fine【精细】命令的 NURBS Smoothness（Fine）Options【NURBS 平滑度（精细）选项】面板，如图 1.194 所示。

图 1.193　NURBS Smoothness（Medium）Options
【NURBS 平滑度（中）选项】面板

图 1.194　NURBS Smoothness（Fine）Options
【NURBS 平滑度（精细）选项】面板

6）Custom Smoothness【自定义平滑度】命令

Custom Smoothness【自定义平滑度】命令的 NURBS Smoothness（Custom）Options【NURBS 平滑度（自定义）选项】面板，如图 1.195 所示。

12. Animation【动画】命令组

Animation【动画】命令组主要包括如图 1.196 所示的 6 个命令。

13. Rendering【渲染】命令组

Rendering【渲染】命令组主要包括如图 1.197 所示的 3 个命令。

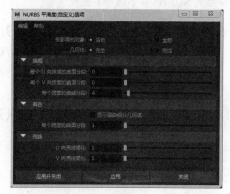

图 1.195　NURBS Smoothness（Custom）Options【NURBS 平滑度（自定义）选项】面板

图 1.196　Animation【动画】命令组面板

图 1.197　Rendering【渲染】命令组面板

1）Camera/Light Manipulator【摄影机/灯光操纵器】命令组

在 Camera/Light Manipulator【摄影机/灯光操纵器】命令组中只包括 Select Camera or Light【选择摄影机或灯光】命令。

2）Stroke Display Quality【笔划显示质量】命令组

Stroke Display Quality【笔划显示质量】命令组主要包括如图 1.198 所示的 6 个选项。

14. Frame All in All Views【在所有视图中框显示所有内容】命令

Frame All in All Views【在所有视图中框显示所有内容】命令的快捷键为 Shift+A，该命令没有选项面板。

15. Frame Selection in All Views【在所有视图中框显当前选择】命令（新增功能）

Frame Selection in All Views【在所有视图中框显当前选择】命令（新增功能）的快捷键为"Shift+F"，该命令没有选项面板。

16. Frame Selection with Children in Views【在所有视图中框显当前选择(包含子对象)】命令（新增功能）

Frame Selection with Children in Views【在所有视图中框显当前选择（包含子对象）】命令（新增功能）的快捷键为"Ctrl+Shift+F"，该命令没有选项面板。

1.7 Window【窗口】菜单组

1. Workspaces【工作区】命令组

Workspaces【工作区】命令组主要包括如图 1.199 所示的 20 个命令。

1）Maya Classic【Maya 经典】命令

Maya Classic【Maya 经典】命令的 Workspace layout options:Maya Classic【工作布局选项 Maya 经典】面板，如图 1.200 所示。

图 1.198　Stroke Display Quality【笔划显示质量】命令组面板

图 1.199　Workspaces【工作区】命令组面板

图 1.200　Workspace layout options:Maya Classic【工作布局选项 Maya 经典】面板

（1）Choose a Menu Set【选择菜单集】参数组：主要包括 Common Menu Set【常用菜单设置】、modeling Menu Set【建模菜单设置】、Rigging Menu Set【操作菜单设置】、Animation Menu Set【动画菜单设置】、Dynamics Menu Set【动态菜单设置】和 Rendering Menu Set【渲染菜单设置】6 个选项，默认 Modeling Menu Set【建模菜单设置】。

（2）Choose a Hotkey Set【选择热键集】参数组：该参数只包括 Maya_Default【Maya 默认】一个选项。

（3）Choose a VP 2.0 Preset【选择 VP 2.0 预设】参数组：该参数组只包括 Default set【默认设置】一个选项。

提示：Workspaces【工作区】命令组中的命令的 Workspaces Options【工作布局选项】面板的参数完全相同，在此就不再详细介绍，请读者参考 Workspace layout options:Maya Classic【工作布局选项 Maya 经典】面板参数即可。

──────────────── Editors【编辑器】────────────────

2. General Editors【常规编辑器】命令组

General Editors【常规编辑器】命令组主要包括如图 1.201 所示的 19 个命令。

1）Hypergraph:Hierarchy【Hypergraph:层次】命令

Hypergraph:Hierarchy【Hypergraph:层次】命令的快捷键为，Hyper Graph Options（DAG）【Hypergraph 选项（DAG）】面板如图 1.202 所示。

2）Hypergraph:Connections【Hypergraph:链接】命令

Hypergraph:Connections【Hypergraph:链接】命令快捷键为■，Hyper Graph Options（DAG）【Hypergraph 选项（DAG）】面板如图 1.203 所示。

图 1.201　General Editors　　图 1.202　Hyper Graph Options（DAG）　图 1.203　Hyper Graph Options（DAG）
【常规编辑器】命令组面板　　【Hypergraph 选项（DAG）】面板　　【Hypergraph 选项（DAG）】面板

3. Modeling Editors【建模编辑器】命令组

Modeling Editors【建模编辑器】命令组主要包括如图 1.204 所示的 5 个命令。

4. Animation Editors【动画编辑器】命令组

Animation Editors【动画编辑器】命令组主要包括如图 1.205 所示的 10 个命令。

5. Rendering Editor【渲染编辑器】命令组

Rendering Editor【渲染编辑器】命令组主要包括如图 1.206 所示的 10 个命令。

图 1.204　Modeling Editors　　图 1.205　Animation Dditors　　图 1.206　Rendering Editor
【建模编辑器】命令组面板　　【动画编辑器】命令组面板　　【渲染编辑器】命令组面板

【mental ray】命令组主要包括如图 1.207 所示的 5 个命令。

第1章 Public【公共】模块菜单组

6. Relation Editors【关系编辑器】命令组

Relation Editors【关系编辑器】命令组主要包括如图 1.208 所示的 11 个命令。

1）Light Linking【灯光链接】命令组

Light Linking【灯光链接】命令组主要包括如图 1.209 所示的 2 个命令。

图 1.207　mental ray
【mental ray】命令组面板　　　图 1.208　Relation Editors
【关系编辑器】命令组面板　　　图 1.209　Light Linking
【灯光链接】命令组面板

2）UV Linking【UV 链接】命令组

UV Linking【UV 链接】命令组主要包括如图 1.210 所示的 4 个命令。

7. UI Elements【UI 元素】命令组

UI Elements【UI 元素】命令组主要包括如图 1.211 所示的 10 个命令。

8. Settings/Preferences【设置/首选项】命令组

Settings/Preferences【设置/首选项】命令组主要包括如图 1.212 所示的 8 个命令。

图 1.210　UV Linking
【UV 链接】命令组面板　　　图 1.211　UI Elements
【UI 元素】命令组面板　　　图 1.212　Settings/Preferences
【设置/首选项】命令组面板

9. Outliner【大纲视图】命令

Outliner【大纲视图】命令没有选项面板。

10. Node Editor【节点编辑器】命令

Node Editor【节点编辑器】命令没有选项面板。

11. Playblast【播放预览】命令

Playblast【播放预览】命令的图标为，Playblast Options【播放预览选项】面板，如

图 1.213 所示。

（1）Format【格式】参数组：在该参数组中主要包括【avi】和【image】2 个选项，默认【avi】选项。

（2）Encoding【编码】参数组：在该参数组中主要包括【MS-RLE】、【MS-CRAM】、【MS-YUV】、【IYUV 编码解码器】、【Toshiba YUV411】和【None】6 个选项，默认【None】选项。

（3）Display Size【显示大小】参数组：在该参数组中主要包括 From Window【来自窗口】、From Render Settings【来自渲染设置】和 Custom【自定义】3 个选项，默认 From Window【来自窗口】选项。

12．Minimize Application【最小化应用程序】命令

Minimize Application【最小化应用程序】命令没有选项面板。

13．Raise Main Window【提升主窗口】命令

Raise Main Window【提升主窗口】命令没有选项面板。

14．Raise Application Windows【提升应用程序窗口】命令

Raise Application Windows【提升应用程序窗口】命令没有选项面板。

1.8 Cache【缓存】菜单组

1．Alembic Cache【Alembic 缓存】命令组

Alembic Cache【Alembic 缓存】命令组主要包括如图 1.214 所示的 7 个命令。

图 1.213 Playblast Options
【播放预览选项】面板

图 1.214 Alembic Cache
【Alembic 缓存】命令组面板

1）Open Alembic…【打开 Alembic…】命令

Open Alembic…【打开 Alembic…】命令没有选项面板。

2）Reference Alembic…【引用 Alembic…】命令

Reference Alembic…【引用 Alembic…】命令的 Reference Options【引用选项】面板，如图 1.215 所示。

图 1.215　Reference Options【引用选项】面板

3）Import Alembic【导入 Alembic…】命令

Import Alembic【导入 Alembic…】命令 Alembic Import【Alembic 导入】面板，如图 1.216 所示。

图 1.216　Alembic Import【Alembic 导入】面板

4）Replace Alembic…【替换 Alembic…】命令

Alembic Import…【替换 Alembic…】命令没有选项面板。

5）Export All to Alembic…【将所有内容导出到 Alembic…】命令

Export All to Alembic…【将所有内容导出到 Alembic…】命令的 Alembic Export【Alembic 导出】面板，如图 1.217 所示。

6）Export Selection to Alembic…【将当前选择导出到 Alembic…】命令

Export Selection to Alembic…【将当前选择导出到 Alembic…】命令的 Alembic Export【Alembic 导出】面板的参数与图 1.217 所示完全相同。

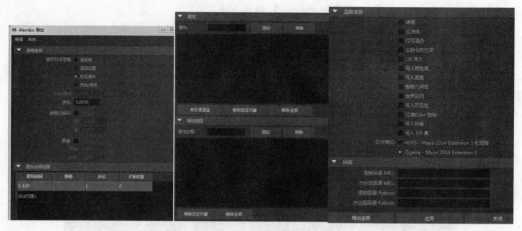

图 1.217　Alembic Export【Alembic 导出】面板

7）About Alembic…【关于 Alembic…】命令

About Alembic…【关于 Alembic…】命令没有选项面板。

2. Geometry【几何缓存】命令组

在 Geometry【几何缓存】命令组中，主要包括 13 个命令，如图 1.218 所示。

1）Create New Cache【创建新缓存】命令

Create New Cache【创建新缓存】命令的图标为 ，Create Geometry Cache Options【创建几何缓存选项】面板如图 1.219 所示。

图 1.218　Geometry
【几何缓存】命令组面板

图 1.219　Create Geometry Cache Options
【创建几何缓存选项】面板

2）Import Cache…【导入缓存…】命令

Import Cache…【导入缓存…】命令的图标为 ，该命令没有选项面板。

3）Export Cache…【导出缓存…】命令

Export Cache…【导出缓存…】命令的图标为 ，Export Geometry Cache Options【导出

几何缓存选项】面板，如图 1.220 所示。

4）Disable All Caches On Selected【禁用选定对象的所有缓存】命令

Disable All Caches On Selected【禁用选定对象的所有缓存】命令的图标为 ，该命令没有选项面板。

5）Enable All Caches On Selected【启用选定对象的所有缓存】命令

Enable All Caches On Selected【启用选定对象的所有缓存】命令的图标为 ，该命令没有选项面板。

6）replace Cache【替换缓存】命令

replace Cache【替换缓存】命令的图标为 ，Replace Geometry Cache Options【替换几何缓存选项】面板，如图 1.221 所示。

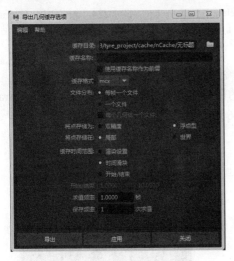

图 1.220　Export Geometry Cache Options
【导出几何缓存选项】面板

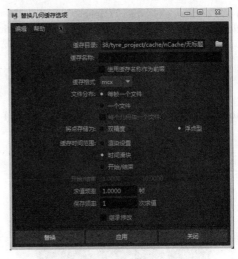

图 1.221　Replace Geometry Cache Options
【替换几何缓存选项】面板

7）Merge Cache【合并缓存】命令

Merge Cache【合并缓存】命令的图标为 ，Merge Geometry Cache Options【合并几何缓存选项】面板，如图 1.222 所示。

8）Delete Caches【删除缓存】命令

Delete Caches【删除缓存】命令的图标为 ，Delete Geometry Cache Options【删除几何缓存选项】面板，如图 1.223 所示。

9）Append to Cache【附加到缓存】命令

Append to Cache【附加到缓存】命令的图标为 ，Append to Geometry Cache Options【附加到几何缓存选项】面板，如图 1.224 所示。

10）Replace Cache Frame【替换缓存帧】命令

Replace Cache Frame【替换缓存帧】命令的图标为 ，Replace Geometry Cache Frames Options【替换几何缓存选项】面板，如图 1.225 所示。

图 1.222　Merge Geometry Cache Options【合并几何缓存选项】面板

图 1.223　Delete Geometry Cache Options【删除几何缓存选项】面板

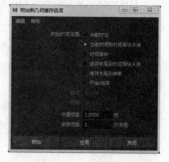

图 1.224　Append to Geometry Cache Options【附加到几何缓存选项】面板

11）Delete Cache Frame【删除缓存帧】命令

Delete Cache Frame【删除缓存帧】命令的图标为■，Delete Geometry Cache Frames Options【删除几何缓存选项】面板，如图 1.226 所示。

图 1.225　Replace Geometry Cache Frames Options【替换几何缓存选项】面板

图 1.226　Delete Geometry Cache Frames Options【删除几何缓存选项】面板

12）Delete History Ahead of Cache【删除缓存之前的历史】命令

Delete History Ahead of Cache【删除缓存之前的历史】命令的图标为■，该命令没有选项面板。

13）Paint Cache Weights Tool【绘制缓存权重工具】命令

Paint Cache Weights Tool【绘制缓存权重工具】命令的图标为■，Tool Settings【工具设置】面板如图 1.227 所示。

第 1 章 Public【公共】模块菜单组

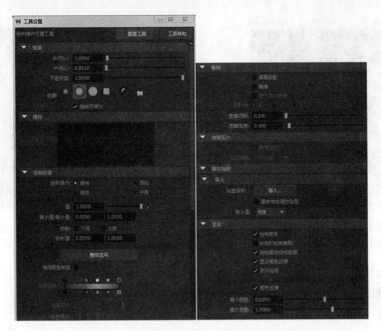

图 1.227　Tool Settings【工具设置】面板

Import value【导入值】参数组：主要包括 Luminance【亮度】、【Alpha】、Red【红】、Green【绿】和 Blue【蓝】5 个选项，默认 Luminance【亮度】选项。

3．GPU Cache【GPU 缓存】命令组

在 GPU Cache【GPU 缓存】命令组中，主要包括 4 个命令，如图 1.228 所示。

1）Import…【导入…】命令

Import…【导入…】命令的 GPU Cache Import Options【GPU 缓存导入选项】面板，如图 1.229 所示。

图 1.228　GPU Cache
【GPU 缓存】命令组面板

图 1.229　GPU Cache Import Options
【GPU 缓存导入选项】面板

2）Export All…【导出全部…】命令

Export All…【导出全部…】命令的 GPU Cache Export Options【GPU 缓存导出选项】面板，如图 1.230 所示。

3）Export Selection…【导出当前选择…】命令

Export Selection…【导出当前选择…】命令的 GPU Cache Export Options【GPU 缓存导出选项】面板，如图 1.231 所示。

图 1.230　GPU Cache Export Options
【GPU 缓存导出选项】面板

图 1.231　GPU Cache Export Options
【GPU 缓存导出选项】面板

4）Refresh All…【导出全部…】命令

Refresh All…【导出全部…】命令没有选项面板。

1.9　Help【帮助】菜单组

────────────── Find【查找】──────────────

1. Find Menu【查找菜单】命令

Find Menu【查找菜单】命令没有选项面板。

2. Autodesk Maya Help【Autodesk Maya 帮助】命令

Autodesk Maya Help【Autodesk Maya 帮助】命令的快捷键为 F1，该命令没有选项面板。

3. What's New【新特性】命令组

在 What's New【新特性】命令组中，主要包括【新特性列表】和【亮显新特性】2 个命令，如图 1.232 所示。

────────────── Learn【了解】──────────────

4. 1-Minute Startup Movies【1 分钟启动影片】命令

1-Minute Startup Movies【1 分钟启动影片】命令没有选项面板。

5. Maya Learning Channel【Maya 教程频道】命令

Maya Learning Channel【Maya 教程频道】命令没有选项面板。

6. Tutorials【教程】命令组

在 Tutorials【教程】命令组中，主要包括 2 个命令，如图 1.233 所示。

图 1.232　What's New【新特性】命令组面板 　　图 1.233　Tutorials【教程】命令组面板

──────────── Advance【前进】 ────────────

7. Learning Path【学习途径】命令

Learning Path【学习途径】命令没有选项面板。

8. Maya Scripting Reference【Maya 脚本参考】命令组

在 Maya Scripting Reference【Maya 脚本参考】命令组中，主要包括 3 个命令，如图 1.234 所示。

9. Maya Communities【Maya 社区】命令组

在 Maya Communities【Maya 社区】命令组中，主要包括 3 个命令，如图 1.235 所示。

10. Maya Resources and Tools【Maya 资源和工具】命令组

在 Maya Resources and Tools【Maya 资源和工具】命令组中，主要包括【浏览器安装助手】、【下载 Bonus Tools】和【下载植被】3 个命令，如图 1.236 所示。

图 1.234　Maya Scripting Reference【Maya 脚本参考】命令组面板 　　图 1.235　Maya Communities【Maya 社区】命令组面板 　　图 1.236　Maya Resources and Tools【Maya 资源和工具】命令组面板

──────────── Support【支持】 ────────────

11. Maya Services and Support【Maya 服务和支持】命令组

在 Maya Services and Support【Maya 服务和支持】命令组中，主要包括 3 个命令，如图 1.237 所示。

12. Expand【反馈】命令组

在 Expand【反馈】命令组中，主要包括 4 个命令，如图 1.238 所示。

──────────── Expand【展开】 ────────────

13. Creative Market【Creative Market】命令（新增功能）

Creative Market【Creative Market】命令的图标为，该命令没有选项面板。

14. Autodesk Exchange Apps 命令

Autodesk Exchange Apps 命令没有选项面板。

15. Try Other Autodesk Products【尝试使用其他 Autodesk 产品】命令组

在 Try Other Autodesk Products【尝试使用其他 Autodesk 产品】命令组中，主要包括 5 个命令，如图 1.239 所示。

图 1.237　Maya Services and Support【Maya 服务和支持】命令组面板 　　图 1.238　Expand【反馈】命令组面板　　图 1.239　Try Other Autodesk Products【尝试使用其他 Autodesk 产品】命令组面板

──────────── About【关于】────────────

16. About Maya【关于 Maya】命令

About Maya【关于 Maya】命令没有选项面板。

17. Maya Home Page【Maya 主页】命令

Maya Home Page【Maya 主页】命令没有选项面板。

第 2 章　Modeling【建模】模块菜单组

Modeling【建模】模块菜单组主要包括 Mesh【网络】、Edit Mesh【编辑网络】、Mesh Tools【网络工具】、Mesh Display【网络显示】、Curves【曲线】、Surfaces【曲面】、Deform【变形】、UV【UV】和 Generate【生成】9 个命令菜单组。

2.1　Mesh【网络】菜单命令组

———————————Combine【结合】———————————

1. Booleans【布尔】命令组

Booleans【布尔】命令组包括 Union【并集】、Difference【差集】和 Intersection【交集】3 个命令。

1）Union【并集】命令

Union【并集】命令的图标为▇。Union Operation Options【并集操作选项】面板，如图 2.1 所示。

2）Difference【差集】命令

Difference【差集】命令的图标为▇，Difference Operation Options【差集操作选项】面板，如图 2.2 所示。

图 2.1　Union Operation Options
【并集操作选项】面板

图 2.2　Difference Operation Options
【差集操作选项】面板

3）Intersection【交集】命令

Intersection【交集】命令的图标为▇，Intersection Operation Options【交集操作选项】面板，如图 2.3 所示。

2. Combine【结合】命令

Combine【结合】命令的图标为▣，Combine Options【结合选项】面板如图 2.4 所示。

图 2.3　Intersection Operation Options
　　　　【交集操作选项】面板

图 2.4　Combine Options
　　　　【结合选项】面板

3. Separate【分离】命令

Separate【分离】命令的图标为▣，该命令没有选项面板。

────────────Remesh【重新划分网络】────────────

4. Conform【一致】命令

Conform【一致】命令没有图标，Conform Options【一致选项】面板如图 2.5 所示。

5. Fill Hole【填充洞】命令

Fill Hole【填充洞】命令的图标为▣，该命令没有选项面板。

6. Reduce【减少】命令

Reduce【减少】命令的图标为▣，Reduce Options【减少选项】面板如图 2.6 所示。

图 2.5　Conform Options【一致选项】面板

图 2.6　Reduce Options【减少选项】面板

第 2 章　Modeling【建模】模块菜单组

Reduction method【减少方法】参数组：在该参数组中提供了 Percentage【百分比】、Vertex limit【顶点限制】和 Triangle limit【三角形限制】3 个选项，默认 Percentage【百分比】选项。

7. Smooth【平滑】命令

Smooth【平滑】命令的图标为▣，Smooth Options【平滑选项】面板如图 2.7 所示。

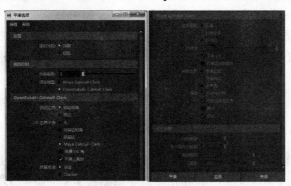

图 2.7　Smooth Options【平滑选项】面板

8. Triangulate【三角化】命令

Triangulate【三角化】命令的图标为▣，该命令没有选项面板。

9. Quadrangulate【四边形化】命令

Quadrangulate【四边形化】命令的快捷图标为▣，Quadrangulate Face Options【四边形化面选项】面板，如图 2.8 所示。

──────────── Mirror【镜像】────────────

10. Mirror【镜像】命令

Mirror【镜像】命令的图标为▣，Mirror Options【镜像选项】面板如图 2.9 所示。

图 2.8　Quadrangulate Face Options【四边形化面选项】面板　　图 2.9　Mirror Options【镜像选项】面板

（1）Mirror Axis Position【镜像轴位置】参数组：在该参数组中提供了 BounDing Box【边界框】、Object【对象】和 World【世界】3 个选项，默认 World【世界】选项。

（2）Mirror Direction【镜像方向】参数组：在该参数组中提供了【+】和【-】2 个选项，默认【-】选项。

（3）Border【边界】参数组：在该参数组中提供了 Merge Border Vertices【合并边界顶点】、Bridge Border Edges【桥接边界边】和 Do not Merge Border【不合并边界】3 个选项，默认 Merge Border Vertices【合并边界顶点】选项。

（4）Direction【方向】参数组：在该参数组中提供了 Local U【局部 U 向】、Local V【局部 V 向】、World U【世界 U 向】和 Word V【世界 V 向】4 个选项，默认 Local U【局部 U 向】选项。

―――――――――――― Transfer【传递】――――――――――――

11. Clipboard Actions【剪贴板操作】命令组

在 Clipboard Actions【剪贴板操作】命令组中包括 Copy Attributes【复制属性】、Paste Attributes【粘贴属性】和 Clear Clipboard【清空剪贴板】3 个命令。

1）Copy Attributes【复制属性】命令

Copy Attributes【复制属性】命令的图标为■，Copy Attributes Options【复制属性选项】面板，如图 2.10 所示。

2）Paste Attributes【粘贴属性】命令

Paste Attributes【粘贴属性】的图标为■，Paste Attributes Options【粘贴属性选项】面板，如图 2.11 所示。

3）Clear Clipboard【清空剪贴板】命令

Clear Clipboard【清空剪贴板】的图标为■，Clear Clipboard Options【清空剪贴板选项】面板，如图 2.12 所示。

图 2.10 Copy Attributes Options【复制属性选项】面板　　图 2.11 Paste Attributes Options【粘贴属性选项】面板　　图 2.12 Clear Clipboard Options【清空剪贴板选项】面板

12. Transfer Attributes【传递属性】命令

Transfer Attributes【传递属性】命令的图标为■，Transfer Attributes Options【传递属性选项】面板，如图 2.13 所示。

第 2 章　Modeling【建模】模块菜单组

13. Transfer Shading Sets【传递着色集】命令

Transfer Shading Sets【传递着色集】命令的图标为 ，Transfer Shading Sets Options【传递着色集选项】面板，如图 2.14 所示。

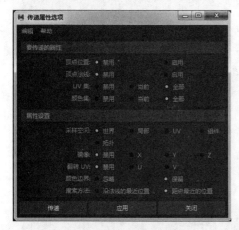

图 2.13　Transfer Attributes Options
【传递属性选项】面板

图 2.14　Transfer Shading Sets Options
【传递着色集选项】面板

14. Transfer Vertex Order【传递顶点属性】命令

Transfer Vertex Order【传递顶点属性】命令的图标为 ，该命令没有选项面板。

──────────── Optimize【优化】────────────

15. Clean up…【清理…】命令

Clean up…【清理…】命令的图标为 ，Cleanup Options【清理选项】面板如图 2.15 所示。

16. Smooth Proxy【平滑代理】命令组

在 Smooth Proxy【平滑代理】命令组包括 Subdiv Proxy【细分曲面代理】、Remove Subdiv Proxy Mirror【移除细分曲面代理镜像】、Crease Tool【折痕工具】、Toggle Proxy Display【切换代理显示】和 Both Proxy and Subdiv Display【代理和细分曲面同时显示】5 个命令。

1）Subdiv Proxy【细分曲面代理】命令

Subdiv Proxy【细分曲面代理】命令的图标为 ，快捷键为 Ctrl+~，Subdiv Proxy Options【细分曲面代理选项】面板，如图 2.16 所示。

2）Remove Subdiv Proxy Mirror【移除细分曲面代理镜像】命令

Remove Subdiv Proxy Mirror【移除细分曲面代理镜像】命令的图标为 ，Remove Subdiv Proxy Mirror Options【移除细分曲面代理镜像选项】面板，如图 2.17 所示。

Maya 2018 英汉速查手册

图 2.15　Cleanup Options
【清理选项】面板

图 2.16　Subdiv Proxy Options
【细分曲面代理选项】面板

3）Crease Tool【折痕工具】命令

Crease Tool【折痕工具】命令的图标为 ，Tool Settings【工具设置】面板如图 2.18 所示。

图 2.17　Remove Subdiv Proxy Mirror Options
【移除细分曲面代理镜像选项】面板

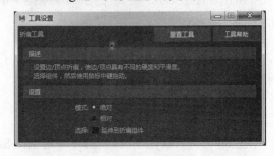

图 2.18　Tool Settings
【工具设置】面板

4）Toggle Proxy Display【切换代理显示】命令

Toggle Proxy Display【切换代理显示】命令的图标为 ，快捷键为 Ctrl+ ，该命令没有选项面板。

5）Both Proxy and Subdiv Display【代理和细分曲面同时显示】命令

Both Proxy and Subdiv Display【代理和细分曲面同时显示】命令的图标为 ，快捷键为 ，该命令没有选项面板。

2.2　Edit Mesh【编辑网络】菜单组

━━━━━━━━━━━━━━━Components【组件】━━━━━━━━━━━━━━━

1. Add Divisions【添加分段】命令

Add Divisions【添加分段】命令的图标为，Add Divisions to Edge Options【添加边的分段数选项】面板，如图 2.19 所示。

2. Bevel【倒角】命令

Bevel【倒角】命令的图标为，Bevel Options【倒角选项】面板如图 2.20 所示。

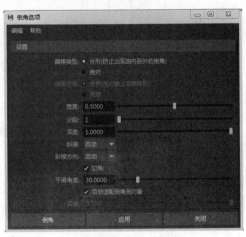

图 2.19　Add Divisions to Edge Options
【添加边的分段数选项】面板

图 2.20　Bevel Options
【倒角选项】面板

3. Bridge【桥接】命令

Bridge【桥接】命令的图标为，Bridge Options【桥接选项】面板如图 2.21 所示。

4. Circularize【圆形圆角】命令

Circularize【圆形圆角】命令的图标为，Polygon Circularize Options【多边形圆形圆角选项】面板，如图 2.22 所示。

Alignment【对齐】参数组主要包括 Automatic【自动】、Surface（Per-vertex）【曲面（逐顶点）】和 Surface（average）【曲面（平均）】3 个选项，默认 Automatic【自动】选项。

Maya 2018 英汉速查手册

图 2.21 Bridge Options
【桥接选项】面板

图 2.22 Polygon Circularize Options
【多边形圆形圆角选项】

5. Collapse【收拢】命令

Collapse【收拢】命令的图标为▦，该命令没有选项面板。

6. Connect【连接】命令

Connect【连接】命令的图标为▦，Connect Components Options【连接组件选项】面板，如图 2.23 所示。

7. Detach【分离】命令

Detach【分离】命令的图标▦，该命令没有选项面板。

8. Extrude【挤出】命令

Extrude【挤出】命令的图标为▦，快捷键为 Ctrl+E，Extrude Face Options【挤出面选项】面板，如图 2.24 所示。

图 2.23 Connect Components Options
【连接组件选项】面板

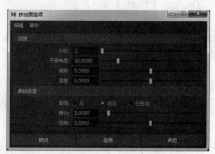

图 2.24 Extrude Face Options
【挤出面选项】面板

9. Merge【合并】命令

Merge【合并】命令的图标为▉，Merge Vertices Options【合并顶点选项】面板，如图2.25 所示。

10. Merge to Center【合并到中心】命令

Merge to Center【合并到中心】命令的图标为▉，该命令没有选项面板。

11. Transform【变换】命令

Transform【变换】命令的图标为▉，Transform Component-Vertex Options【变换组建-顶点选项】面板，如图2.26 所示。

图 2.25　Merge Vertices Options【合并顶点选项】面板

图 2.26　Transform Component-Vertex Options【变换组建-顶点选项】面板

12. Flip【翻转】命令（新增功能）

Flip【翻转】命令的图标为▉，该命令没有选项面板。

13. Symmetrize【对称】命令

Symmetrize【对称】命令的图标为▉，该命令没有选项面板。

──────── Vertex【顶点】 ────────

14. Average Vertices【平均化顶点】命令

Average Vertices【平均化顶点】命令的图标为▉，Average Vertices Options【平均化顶点选项】面板，如图2.27 所示。

15. Chamfer Vertices【切角顶点】命令

Chamfer Vertices【切角顶点】命令的图标▉，Chamfer Vertices Options【切角顶点选项】面板，如图2.28 所示。

16. Reorder Vertices【对顶点重新排序】命令（新增功能）

Reorder Vertices【对顶点重新排序】命令的图标为▉，该命令没有选项面板。

图 2.27　Average Vertices Options
【平均化顶点选项】面板

图 2.28　Chamfer Vertices Options
【切角顶点选项】面板

──────────────── Edge【边】────────────────

17. Delete Edge/Vertex【删除边/顶点】命令

Delete Edge/Vertex【删除边/顶点】命令图标为▇，快捷键为 Ctrl+Del 。

18. Edit Edge Flow【编辑边流】命令

Edit Edge Flow【编辑边流】命令的图标为▇，Edit Edge Flow Options【编辑边流选项】面板，如图 2.29 所示。

19. Flip Triangle Edge【翻转三角形边】命令

Flip Triangle Edge【翻转三角形边】命令图标为▇，该命令没有选项面板。

20. Spin Edge Backward【反向自旋边】命令

Spin Edge Backward【反向自旋边】命令的图标为▇，快捷键为 Ctrl+Alt+Left ，没有选项面板。

21. Spin Edge Forward【正向自旋边】命令

Spin Edge Forward【正向自旋边】命令的图标为▇，快捷键为 Ctrl+Alt+Right ，没有选项面板。

──────────────── Face【面】────────────────

22. Assign Invisible Faces【指定不可见面】命令

Assign Invisible Faces【指定不可见面】命令的图标为▇，Assign Invisible Faces Options【指定不可见面选项】面板，如图 2.30 所示。

图 2.29　Edit Edge Flow Options
【编辑边流选项】面板

图 2.30　Assign Invisible Faces Options
【指定不可见面选项】面板

第 2 章　Modeling【建模】模块菜单组

23. Duplicate【复制】命令

Duplicate【复制】命令的图标为，Duplicate Face Options【复制面选项】面板如图 2.31 所示。

24. Extract【提取】命令

Extract【提取】命令的图标为，Extract Options【提取选项】面板如图 2.32 所示。

图 2.31　Duplicate Face Options
【复制面选项】面板

图 2.32　Extract Options
【提取选项】面板

25. Poke【刺破】命令

Poke【刺破】命令的图标为，Poke Face Options【刺破面选项】面板如图 2.33 所示。

26. Wedge【楔形】命令（新增功能）

Wedge【楔形】命令的图标为，Wedge Face Options【楔形面选项】面板如图 2.34 所示。

图 2.33　Poke Face Options
【刺破面选项】面板

图 2.34　Wedge Face Options
【楔形面选项】面板

────────────── Curve【曲线】──────────────

27. Project Curve on Mesh【在网络上投影曲线】命令

Project Curve on Mesh【在网络上投影曲线】命令的图标为，Project Curve on Mesh Options【在网络上投影曲线选项】面板，如图 2.35 所示。

28. Split Mesh with Projected Curve【使用投影的曲线分割网络】命令

Split Mesh with Projected Curve【使用投影的曲线分割网络】命令图标为，Split Mesh with Projected Curve Options【使用投影的曲线分割网络选项】面板，如图 2.36 所示。

Maya 2018 英汉速查手册

图 2.35　Project Curve on Mesh Options
【在网络上投影曲线选项】面板

图 2.36　Split Mesh with Projected Curve Options
【使用投影的曲线分割网络选项】面板

2.3　Mesh Tools【网络工具】菜单组

1. Show Modeling Toolkit【显示建模工具包】命令

单击 Show Modeling Toolkit【显示建模工具包】命令在界面的右侧显示【建模工具】面板，如图 2.37 所示。此时，该命令变为 Hide Modeling Toolkit【隐藏建模工具包】命令。在此单击隐藏建模工具。

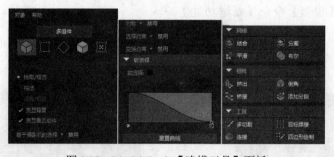

图 2.37　Model Tools【建模工具】面板

―――Tools【工具】―――

2. Append to Polygon【附加到多边形】命令

Append to Polygon【附加到多边形】命令图标为▇，Tool Settings【工具设置】面板如图 2.38 所示。

3. Connect【连接】命令

Connect【连接】命令的图标为▇，Tool Settings【工具设置】面板如图 2.39 所示。

4. Crease【折痕】命令

Crease【折痕】命令的图标为▇，Tool Settings【工具设置】面板如图 2.40 所示。

第 2 章　Modeling【建模】模块菜单组

图 2.38　Tool Settings【工具设置】面板

图 2.39　Tool Settings【工具设置】面板

5．Create Polygon【创建多边形】命令

Create Polygon【创建多边形】命令的图标为 ▩，Tool Settings【工具设置】面板如图 2.41 所示。

图 2.40　Tool Settings【工具设置】面板

图 2.41　Tool Settings【工具设置】面板

6．Insert Edge Loop【插入循环边】命令

Insert Edge Loop【插入循环边】命令的图标为 ▩，Tool Settings【工具设置】面板如图 2.42 所示。

7．Make Hole【生成洞】命令

Make Hole【生成洞】命令的图标为 ▩，Tool Settings【工具设置】面板如图 2.43 所示。

图 2.42　Tool Settings【工具设置】面板

图 2.43　Tool Settings【工具设置】面板

Merge mode【合并模式】参数组：在该参数组中提供了 First【第一个】、Middle【中间】、Second【第二个】、Project First【投影第一项】、Project Middle【投影中间项】、Project Second【投影第二项】和 None【无】7 个选项，默认 None【无】选项。

8. Multi-Cut【多切割】命令

Multi-Cut【多切割】命令的图标为，Tool Settings【工具设置】面板如图 2.44 所示。

9. Offset Edge Loop【偏移循环边】命令

Offset Edge Loop【偏移循环边】命令的图标，Tool Settings【工具设置】面板如图 2.45 所示。

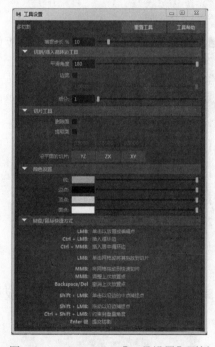

图 2.44　Tool Settings【工具设置】面板

图 2.45　Tool Settings【工具设置】面板

10. Paint Reduce Weights【绘制减少权重】命令

Paint Reduce Weights【绘制减少权重】命令图标为，Tool Settings【工具设置】面板如图 2.46 所示。

11. Paint Transfer Attributes【绘制传递属性】命令

Paint Transfer Attributes【绘制传递属性】命令图标为，Tool Settings【工具设置】面板如图 2.47 所示。

第 2 章　Modeling【建模】模块菜单组

图 2.46　Tool Settings【工具设置】面板

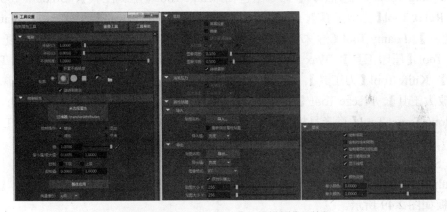

图 2.47　Tool Settings【工具设置】面板

（1）Vector index【向量索引】参数组：在该参数组中提供了【x/R】、【y/G】和【z/B】3 个选项，默认【x/R】选项。

（2）Import【导入值】参数组：在该参数组中提供了 Luminance【亮度】、Alpha【Alpha】、Red【红】、Green【绿】和 Blue【蓝】5 个选项，默认 Luminance【亮度】选项。

（3）Image format【图像格式】参数组：在该参数组中提供了【GIF】、【SoftImage】、【RlA】、【TIFF】、【SGI】、Alias【锯齿】、【IFF】、【JPEG】和【EPS】9 个选项，默认【IFF】选项。

12．Quad Draw【四边形绘制】命令

Quad Draw【四边形绘制】命令图标为 ，Tool Settings【工具设置】面板如图 2.48 所示。

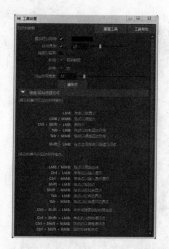

图 2.48　Tool Settings【工具设置】面板

13. Sculpting Tools【雕刻工具】命令组

Sculpting Tools【雕刻工具】命令组包括 Sculpt Tool【雕刻工具】、Smooth Tool【平滑工具】、Relax Tool【松弛工具】、Grab Tool【抓取工具】、Pinch Tool【收缩工具】、Flatten Tool【展平工具】、Foamy Tool【泡沫工具】、Spray Tool【喷射工具】、Repeat Tool【重复工具】、Imprint Tool【压印工具】、Wax Tool【上蜡工具】、Scrape Tool【刮擦工具】、Fill Tool【填充工具】、Knife Tool【刀工具】、Smear Tool【涂抹工具】、Bulge Tool【凸起工具】、Amplify Tool【放大工具】、Freeze Tool【冻结工具】、Smooth Target Tool【平滑目标工具】（新增功能）、Clone Target Tool【克隆目标工具】（新增功能）、Mask Target Tool【遮罩目标工具】（新增功能）和 Erase Target Tool【擦除目标工具】（新增功能）22 个命令。

1）Sculpt Tool【雕刻工具】命令

Sculpting Tools【雕刻工具】命令的图标为 ，快捷键为 Ctrl+1，Tool Settings【工具设置】面板如图 2.49 所示。

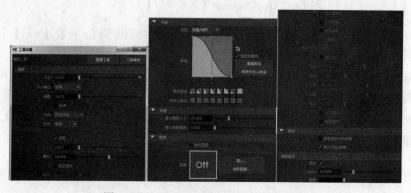

图 2.49　Tool Settings【工具设置】面板

第 2 章 Modeling【建模】模块菜单组

（1）Size Units【大小单位】参数组：在该参数组中提供了 World【世界】和 Screen Pixels【屏幕像素】2 个选项，默认 World【世界】选项。

（2）Direction【方向】参数组：在该参数组中提供了 Center Normal【中心法线】、Averaged Normal【平均化法线】、Vertex Normal【顶点法线】、Forward【前进】、Right【右侧】、【X】、【Y】、【Z】和 Camera【摄影机】9 个选项，默认 Vertex Normal【顶点法线】选项。

（3）Symmetry【对称】参数组：在该参数组中提供了 Off【禁用】、Object X【对象 X】、Object Y【对象 Y】、Object Z【对象 Z】、World X【世界 X】、World Y【世界 Y】和 World Z【世界 Z】7 个选项，默认 Off【禁用】选项。

（4）Type【类型】参数组：在该参数组中提供了 Surface/Volume【曲面/体积】、Surface【曲面】和 Volume【体积】3 个选项，默认 Surface/Volume【曲面/体积】选项。

（5）Draw Method【绘制方法】参数组：在该参数组中提供了 Continuous【连续】、Scale image from center【从中心缩放图像】和 Scale image from side【从一侧缩放图像】3 个选项，默认 Continuous【连续】选项。

2）Smooth Tool【平滑工具】命令

Smooth Tool【平滑工具】命令的图标为，快捷键为 Ctrl+2，Tool Settings【工具设置】面板如图 2.50 所示。

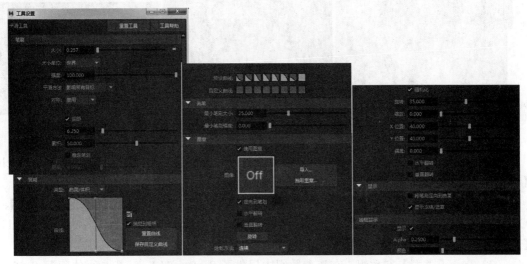

图 2.50　Tool Settings【工具设置】面板

（1）Size Units【大小单位】参数组：在该参数组中提供了 World【世界】和 Screen Pixels【屏幕像素】2 个选项，默认 World【世界】选项。

（2）Smooth Method【平滑方向】参数组：在该参数组中提供了 Affect all targets【影响所有目标】和 Affect Current target only【仅影响当前目标】2 个选项，默认 Affect all targets【影响所有目标】选项。

（3）Symmetry【对称】参数组：在该参数组中提供了 Off【禁用】、Object X【对象 X】、Object Y【对象 Y】、Object Z【对象 Z】、World X【世界 X】、World Y【世界 Y】和 World Z

【世界Z】7个选项，默认Off【禁用】选项。

（4）Type【类型】参数组：在该参数组中提供了Surface/Volume【曲面/体积】、Surface【曲面】和Volume【体积】3个选项，默认Surface/Volume【曲面/体积】选项。

（5）Draw Method【绘制方法】参数组：在该参数组中提供了Continuous【连续】、Scale image from center【从中心缩放图像】和Scale image from side【从一侧缩放图像】3个选项，默认Continuous【连续】选项。

3）Relax Tool【松弛工具】命令

Relax Tool【松弛工具】命令的图标为 ，快捷键为 Ctrl+3 ，Tool Settings【工具设置】面板如图2.51所示。

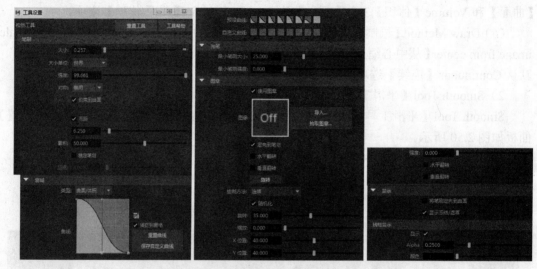

图2.51　Tool Settings【工具设置】面板

（1）Size Units【大小单位】参数组：在该参数组中提供了World【世界】和Screen Pixels【屏幕像素】2个选项，默认World【世界】选项。

（2）Symmetry【对称】参数组：在该参数组中提供了Off【禁用】、Object X【对象X】、Object Y【对象Y】、Object Z【对象Z】、World X【世界X】、World Y【世界Y】和World Z【世界Z】7个选项，默认Off【禁用】选项。

（3）Type【类型】参数组：在该参数组中提供了Surface/Volume【曲面/体积】、Surface【曲面】和Volume【体积】3个选项，默认Surface/Volume【曲面/体积】选项。

（4）Draw Method【绘制方法】参数组：在该参数组中提供了Continuous【连续】、Scale image from center【从中心缩放图像】和Scale image from side【从一侧缩放图像】3个选项，默认Continuous【连续】选项。

4）Grab Tool【抓取工具】命令

Grab Tool【抓取工具】命令的图标为 ，快捷键为 Ctrl+4 ，Tool Settings【工具设置】面板如图2.52所示。

第 2 章　Modeling【建模】模块菜单组

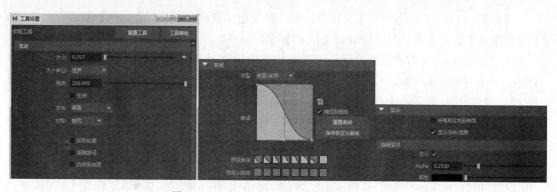

图 2.52　Tool Settings【工具设置】面板

（1）Size Units【大小单位】参数组：在该参数组中提供了 World【世界】和 Screen Pixels【屏幕像素】2 个选项，默认 World【世界】选项。

（2）Direction【方向】参数组：在该参数组中提供了 Screen【屏幕】、Center Normal【中心法线】、Averaged Normal【平均化法线】、【XY】、【XZ】和 Camera【YZ】6 个选项，默认 Screen【屏幕】选项。

（3）Symmetry【对称】参数组：在该参数组中提供了 Off【禁用】、Object X【对象 X】、Object Y【对象 Y】、Object Z【对象 Z】、World X【世界 X】、World Y【世界 Y】和 World Z【世界 Z】7 个选项，默认 Off【禁用】选项。

（4）Type【类型】参数组：在该参数组中提供了 Surface/Volume【曲面/体积】、Surface【曲面】和 Volume【体积】3 个选项，默认 Surface/Volume【曲面/体积】选项。

5）Pinch Tool【收缩工具】命令

Pinch Tool【收缩工具】命令的图标为 ，快捷键为 Ctrl+5，Tool Settings【工具设置】面板如图 2.53 所示。

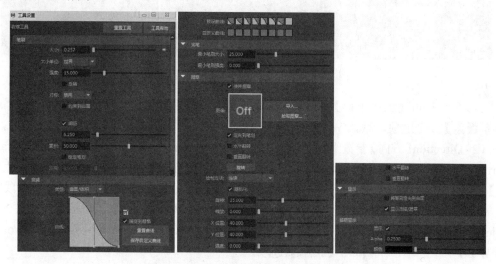

图 2.53　Tool Settings【工具设置】面板

（1）Size Units【大小单位】参数组：在该参数组中提供了 World【世界】和 Screen Pixels【屏幕像素】2 个选项，默认 World【世界】选项。

（2）Symmetry【对称】参数组：在该参数组中提供了 Off【禁用】、Object X【对象 X】、Object Y【对象 Y】、Object Z【对象 Z】、World X【世界 X】、World Y【世界 Y】和 World Z【世界 Z】7 个选项，默认 Off【禁用】选项。

（3）Type【类型】参数组：在该参数组中提供了 Surface/Volume【曲面/体积】、Surface【曲面】和 Volume【体积】3 个选项，默认 Surface/Volume【曲面/体积】选项。

（4）Draw Method【绘制方法】参数组：在该参数组中提供了 Continuous【连续】、Scale image from center【从中心缩放图像】和 Scale image from side【从一侧缩放图像】3 个选项，默认 Continuous【连续】选项。

6）Flatten Tool【展平工具】命令

Flatten Tool【展平工具】命令的图标为 ，快捷键为 Ctrl+6，Tool Settings【工具设置】面板如图 2.54 所示。

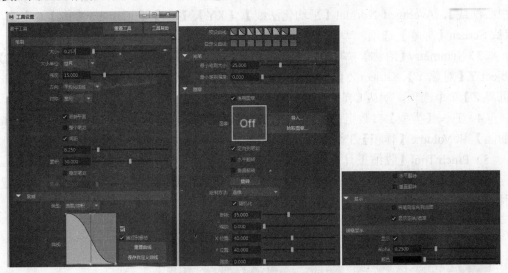

图 2.54　Tool Settings【工具设置】面板

（1）Size Units【大小单位】参数组：在该参数组中提供了 World【世界】和 Screen Pixels【屏幕像素】2 个选项，默认 World【世界】选项。

（2）Direction【方向】参数组：在该参数组中提供了 Center Normal【中心法线】、Averaged Normal【平均化法线】、Vertex Normal【顶点法线】、Forward【前进】、Right【右侧】、【X】、【Y】、【Z】和 Camera【摄影机】9 个选项，默认 Averaged Normal【平均化法线】选项。

（3）Symmetry【对称】参数组：在该参数组中提供了 Off【禁用】、Object X【对象 X】、Object Y【对象 Y】、Object Z【对象 Z】、World X【世界 X】、World Y【世界 Y】和 World Z【世界 Z】7 个选项，默认 Off【禁用】选项。

（4）Type【类型】参数组：在该参数组中提供了 Surface/Volume【曲面/体积】、Surface

第2章 Modeling【建模】模块菜单组

【曲面】和 Volume【体积】3 个选项，默认 Surface/Volume【曲面/体积】选项。

（5）Draw Method【绘制方法】参数组：在该参数组中提供了 Continuous【连续】、Scale image from center【从中心缩放图像】和 Scale image from side【从一侧缩放图像】3 个选项，默认 Continuous【连续】选项。

7）Foamy Tool【泡沫工具】命令

Foamy Tool【泡沫工具】命令图标为 ，快捷键为 Ctrl+7，Tool Settings【工具设置】面板如图 2.55 所示。

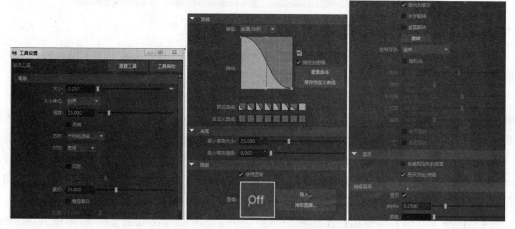

图 2.55　Tool Settings【工具设置】面板

（1）Size Units【大小单位】参数组：在该参数组中提供了 World【世界】和 Screen Pixels【屏幕像素】2 个选项，默认 World【世界】选项。

（2）Direction【方向】参数组：在该参数组中提供了 Center Normal【中心法线】、Averaged Normal【平均化法线】、Vertex Normal【顶点法线】、Forward【前进】、Right【右侧】、【X】、【Y】、【Z】和 Camera【摄影机】9 个选项，默认 Averaged Normal【平均化法线】选项。

（3）Symmetry【对称】参数组：在该参数组中提供了 Off【禁用】、Object X【对象 X】、Object Y【对象 Y】、Object Z【对象 Z】、World X【世界 X】、World Y【世界 Y】和 World Z【世界 Z】7 个选项，默认 Off【禁用】选项。

（4）Type【类型】参数组：在该参数组中提供了 Surface/Volume【曲面/体积】、Surface【曲面】和 Volume【体积】3 个选项，默认 Surface/Volume【曲面/体积】选项。

（5）Draw Method【绘制方法】参数组：在该参数组中提供了 Continuous【连续】、Scale image from center【从中心缩放图像】和 Scale image from side【从一侧缩放图像】3 个选项，默认 Continuous【连续】选项。

8）Spray Tool【喷射工具】命令

Spray Tool【喷射工具】命令的图标为 ，快捷键为 Ctrl+8，Tool Settings【工具设置】面板如图 2.56 所示。

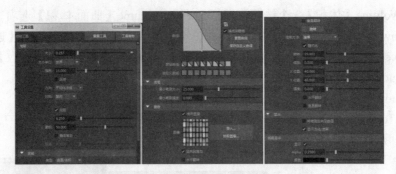

图2.56　Tool Settings【工具设置】面板

（1）Size Units【大小单位】参数组：在该参数组中提供了World【世界】和Screen Pixels【屏幕像素】2个选项，默认World【世界】选项。

（2）Direction【方向】参数组：在该参数组中提供了Center Normal【中心法线】、Averaged Normal【平均化法线】、Vertex Normal【顶点法线】、Forward【前进】、Right【右侧】、【X】、【Y】、【Z】和Camera【摄影机】9个选项，默认Averaged Normal【平均化法线】选项。

（3）Symmetry【对称】参数组：在该参数组中提供了Off【禁用】、Object X【对象X】、Object Y【对象Y】、Object Z【对象Z】、World X【世界X】、World Y【世界Y】和World Z【世界Z】7个选项，默认Off【禁用】选项。

（4）Type【类型】参数组：在该参数组中提供了Surface/Volume【曲面/体积】、Surface【曲面】和Volume【体积】3个选项，默认Surface/Volume【曲面/体积】选项。

（5）Draw Method【绘制方法】参数组：在该参数组中提供了Continuous【连续】、Scale image from center【从中心缩放图像】和Scale image from side【从一侧缩放图像】3个选项，默认Continuous【连续】选项。

9）Repeat Tool【重复工具】命令

Repeat Tool【重复工具】命令图标为▣，快捷键为 Ctrl+9 ，Tool Settings【工具设置】面板如图2.57所示。

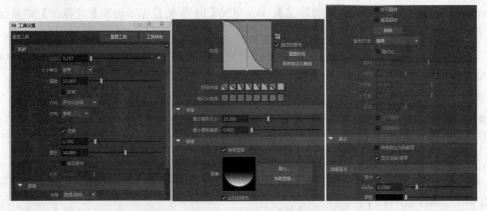

图2.57　Tool Settings【工具设置】面板

第 2 章　Modeling【建模】模块菜单组

（1）Size Units【大小单位】参数组：在该参数组中提供了 World【世界】和 Screen Pixels【屏幕像素】2 个选项，默认 World【世界】选项。

（2）Direction【方向】参数组：在该参数组中提供了 Center Normal【中心法线】、Averaged Normal【平均化法线】、Vertex Normal【顶点法线】、Forward【前进】、Right【右侧】、【X】、【Y】、【Z】和 Camera【摄影机】9 个选项，默认 Averaged Normal【平均化法线】选项。

（3）Symmetry【对称】参数组：在该参数组中提供了 Off【禁用】、Object X【对象 X】、Object Y【对象 Y】、Object Z【对象 Z】、World X【世界 X】、World Y【世界 Y】和 World Z【世界 Z】7 个选项，默认 Off【禁用】选项。

（4）Type【类型】参数组：在该参数组中提供了 Surface/Volume【曲面/体积】、Surface【曲面】和 Volume【体积】3 个选项，默认 Surface/Volume【曲面/体积】选项。

（5）Draw Method【绘制方法】参数组：在该参数组中提供了 Continuous【连续】、Scale image from center【从中心缩放图像】和 Scale image from side【从一侧缩放图像】3 个选项，默认 Continuous【连续】选项。

10）Imprint Tool【压印工具】命令

Imprint Tool【压印工具】命令的图标为，Tool Settings【工具设置】面板如图 2.58 所示。

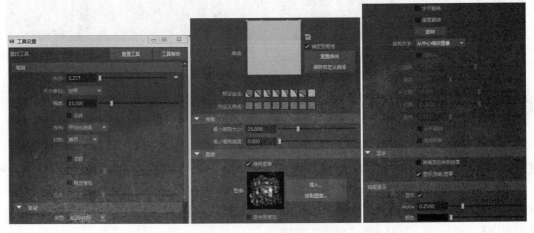

图 2.58　Tool Settings【工具设置】面板

（1）Size Units【大小单位】参数组：在该参数组中提供了 World【世界】和 Screen Pixels【屏幕像素】2 个选项，默认 World【世界】选项。

（2）Direction【方向】参数组：在该参数组中提供了 Center Normal【中心法线】、Averaged Normal【平均化法线】、Vertex Normal【顶点法线】、Forward【前进】、Right【右侧】、【X】、【Y】、【Z】和 Camera【摄影机】9 个选项，默认 Averaged Normal【平均化法线】选项。

（3）Symmetry【对称】参数组：在该参数组中提供了 Off【禁用】、Object X【对象 X】、Object Y【对象 Y】、Object Z【对象 Z】、World X【世界 X】、World Y【世界 Y】和 World Z【世界 Z】7 个选项，默认 Off【禁用】选项。

（4）Type【类型】参数组：在该参数组中提供了 Surface/Volume【曲面/体积】、Surface【曲面】和 Volume【体积】3 个选项，默认 Surface/Volume【曲面/体积】选项。

（5）Draw Method【绘制方法】参数组：在该参数组中提供了 Continuous【连续】、Scale image from center【从中心缩放图像】和 Scale image from side【从一侧缩放图像】3 个选项，默认 Continuous【连续】选项。

11）Wax Tool【上蜡工具】命令

Wax Tool【上蜡工具】命令的图标为 ，Tool Settings【工具设置】面板如图 2.59 所示。

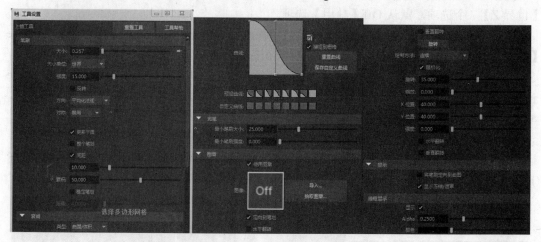

图 2.59　Tool Settings【工具设置】面板

（1）Size Units【大小单位】参数组：在该参数组中提供了 World【世界】和 Screen Pixels【屏幕像素】2 个选项，默认 World【世界】选项。

（2）Direction【方向】参数组：在该参数组中提供了 Center Normal【中心法线】、Averaged Normal【平均化法线】、Vertex Normal【顶点法线】、Forward【前进】、Right【右侧】、【X】、【Y】、【Z】和 Camera【摄影机】9 个选项，默认 Averaged Normal【平均化法线】选项。

（3）Symmetry【对称】参数组：在该参数组中提供了 Off【禁用】、Object X【对象 X】、Object Y【对象 Y】、Object Z【对象 Z】、World X【世界 X】、World Y【世界 Y】和 World Z【世界 Z】7 个选项，默认 Off【禁用】选项。

（4）Type【类型】参数组：在该参数组中提供了 Surface/Volume【曲面/体积】、Surface【曲面】和 Volume【体积】3 个选项，默认 Surface/Volume【曲面/体积】选项。

（5）Draw Method【绘制方法】参数组：在该参数组中提供了 Continuous【连续】、Scale image from center【从中心缩放图像】和 Scale image from side【从一侧缩放图像】3 个选项，默认 Continuous【连续】选项。

12）Scrape Tool【刮擦工具】命令

Scrape Tool【刮擦工具】命令的图标为 ，Tool Settings【工具设置】面板如图 2.60 所示。

第 2 章　Modeling【建模】模块菜单组

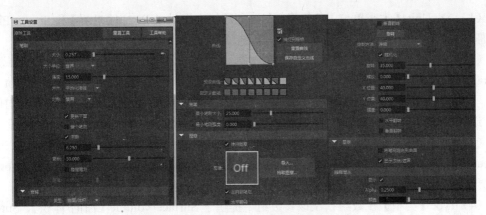

图 2.60　Tool Settings【工具设置】面板

（1）Size Units【大小单位】参数组：在该参数组中提供了 World【世界】和 Screen Pixels【屏幕像素】2 个选项，默认 World【世界】选项。

（2）Direction【方向】参数组：在该参数组中提供了 Center Normal【中心法线】、Averaged Normal【平均化法线】、Vertex Normal【顶点法线】、Forward【前进】、Right【右侧】、【X】、【Y】、【Z】和 Camera【摄影机】9 个选项，默认 Averaged Normal【平均化法线】选项。

（3）Symmetry【对称】参数组：在该参数组中提供了 Off【禁用】、Object X【对象 X】、Object Y【对象 Y】、Object Z【对象 Z】、World X【世界 X】、World Y【世界 Y】和 World Z【世界 Z】7 个选项，默认 Off【禁用】选项。

（4）Type【类型】参数组：在该参数组中提供了 Surface/Volume【曲面/体积】、Surface【曲面】和 Volume【体积】3 个选项，默认 Surface/Volume【曲面/体积】选项。

（5）Draw Method【绘制方法】参数组：在该参数组中提供了 Continuous【连续】、Scale image from center【从中心缩放图像】和 Scale image from side【从一侧缩放图像】3 个选项，默认 Continuous【连续】选项。

13）Fill Tool【填充工具】命令

Fill Tool【填充工具】命令的图标为■，Tool Settings【工具设置】面板如图 2.61 所示。

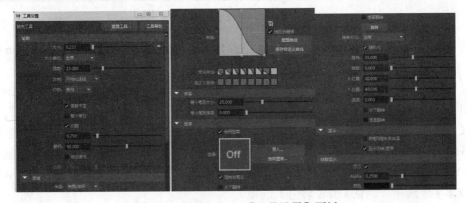

图 2.61　Tool Settings【工具设置】面板

（1）Size Units【大小单位】参数组：在该参数组中提供了 World【世界】和 Screen Pixels【屏幕像素】2 个选项，默认 World【世界】选项。

（2）Direction【方向】参数组：在该参数组中提供了 Center Normal【中心法线】、Averaged Normal【平均化法线】、Vertex Normal【顶点法线】、Forward【前进】、Right【右侧】、【X】、【Y】、【Z】和 Camera【摄影机】9 个选项，默认 Averaged Normal【平均化法线】选项。

（3）Symmetry【对称】参数组：在该参数组中提供了 Off【禁用】、Object X【对象 X】、Object Y【对象 Y】、Object Z【对象 Z】、World X【世界 X】、World Y【世界 Y】和 World Z【世界 Z】7 个选项，默认 Off【禁用】选项。

（4）Type【类型】参数组：在该参数组中提供了 Surface/Volume【曲面/体积】、Surface【曲面】和 Volume【体积】3 个选项，默认 Surface/Volume【曲面/体积】选项。

（5）Draw Method【绘制方法】参数组：在该参数组中提供了 Continuous【连续】、Scale image from center【从中心缩放图像】和 Scale image from side【从一侧缩放图像】3 个选项，默认 Continuous【连续】选项。

14）Knife Tool【刀工具】命令

Knife Tool【刀工具】命令的图标为，Tool Settings【工具设置】面板如图 2.62 所示。

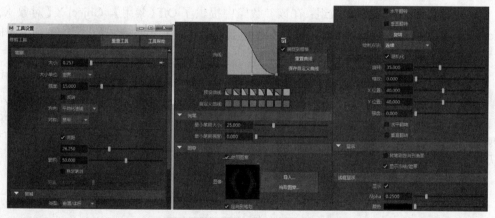

图 2.62　Tool Settings【工具设置】面板

（1）Size Units【大小单位】参数组：在该参数组中提供了 World【世界】和 Screen Pixels【屏幕像素】2 个选项，默认 World【世界】选项。

（2）Direction【方向】参数组：在该参数组中提供了 Center Normal【中心法线】、Averaged Normal【平均化法线】、Vertex Normal【顶点法线】、Forward【前进】、Right【右侧】、【X】、【Y】、【Z】和 Camera【摄影机】9 个选项，默认 Averaged Normal【平均化法线】选项。

（3）Symmetry【对称】参数组：在该参数组中提供了 Off【禁用】、Object X【对象 X】、Object Y【对象 Y】、Object Z【对象 Z】、World X【世界 X】、World Y【世界 Y】和 World Z【世界 Z】7 个选项，默认 Off【禁用】选项。

（4）Type【类型】参数组：在该参数组中提供了 Surface/Volume【曲面/体积】、Surface

第 2 章 Modeling【建模】模块菜单组

【曲面】和 Volume【体积】3 个选项,默认 Surface/Volume【曲面/体积】选项。

(5) Draw Method【绘制方法】参数组:在该参数组中提供了 Continuous【连续】、Scale image from center【从中心缩放图像】和 Scale image from side【从一侧缩放图像】3 个选项,默认 Continuous【连续】选项。

15) Smear Tool【涂抹工具】命令

Smear Tool【涂抹工具】命令的图标为,Tool Settings【工具设置】面板如图 2.63 所示。

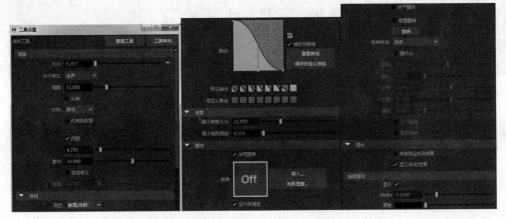

图 2.63 Tool Settings【工具设置】面板

(1) Size Units【大小单位】参数组:在该参数组中提供了 World【世界】和 Screen Pixels 【屏幕像素】2 个选项,默认 World【世界】选项。

(2) Symmetry【对称】参数组:在该参数组中提供了 Off【禁用】、Object X【对象 X】、Object Y【对象 Y】、Object Z【对象 Z】、World X【世界 X】、World Y【世界 Y】和 World Z 【世界 Z】7 个选项,默认 Off【禁用】选项。

(3) Type【类型】参数组:在该参数组中提供了 Surface/Volume【曲面/体积】、Surface 【曲面】和 Volume【体积】3 个选项,默认 Surface/Volume【曲面/体积】选项。

(4) Draw Method【绘制方法】参数组:在该参数组中提供了 Continuous【连续】、Scale image from center【从中心缩放图像】和 Scale image from side【从一侧缩放图像】3 个选项,默认 Continuous【连续】选项。

16) Bulge Tool【凸起工具】命令

Bulge Tool【凸起工具】命令的图标为 ,Tool Settings【工具设置】面板如图 2.64 所示。

(1) Size Units【大小单位】参数组:在该参数组中提供了 World【世界】和 Screen Pixels 【屏幕像素】2 个选项,默认 World【世界】选项。

(2) Symmetry【对称】参数组:在该参数组中提供了 Off【禁用】、Object X【对象 X】、Object Y【对象 Y】、Object Z【对象 Z】、World X【世界 X】、World Y【世界 Y】和 World Z 【世界 Z】7 个选项,默认 Off【禁用】选项。

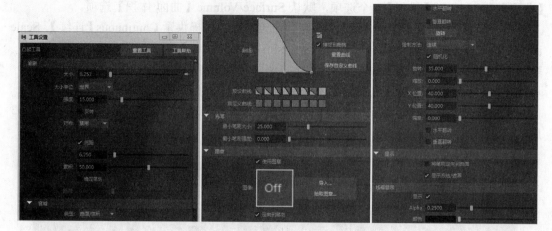

图 2.64　Tool Settings【工具设置】面板

（3）Type【类型】参数组：在该参数组中提供了 Surface/Volume【曲面/体积】、Surface【曲面】和 Volume【体积】3 个选项，默认 Surface/Volume【曲面/体积】选项。

（4）Draw Method【绘制方法】参数组：在该参数组中提供了 Continuous【连续】、Scale image from center【从中心缩放图像】和 Scale image from side【从一侧缩放图像】3 个选项，默认 Continuous【连续】选项。

17）Amplify Tool【放大工具】命令

Amplify Tool【放大工具】命令的图标为，Tool Settings【工具设置】面板如图 2.65 所示。

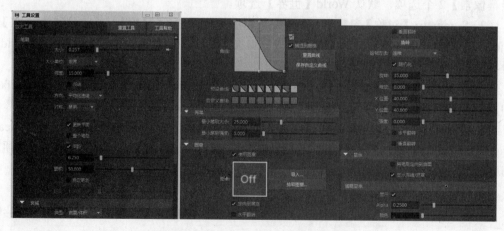

图 2.65　Tool Settings【工具设置】面板

（1）Size Units【大小单位】参数组：在该参数组中提供了 World【世界】和 Screen Pixels【屏幕像素】2 个选项，默认 World【世界】选项。

（2）Direction【方向】参数组：在该参数组中提供了 Center Normal【中心法线】、Averaged Normal【平均化法线】、Vertex Normal【顶点法线】、Forward【前进】、Right【右侧】、【X】、

第 2 章　Modeling【建模】模块菜单组

【Y】、【Z】和 Camera【摄影机】9 个选项，默认 Averaged Normal【平均化法线】选项。

（3）Symmetry【对称】参数组：在该参数组中提供了 Off【禁用】、Object X【对象 X】、Object Y【对象 Y】、Object Z【对象 Z】、World X【世界 X】、World Y【世界 Y】和 World Z【世界 Z】7 个选项，默认 Off【禁用】选项。

（4）Type【类型】参数组：在该参数组中提供了 Surface/Volume【曲面/体积】、Surface【曲面】和 Volume【体积】3 个选项，默认 Surface/Volume【曲面/体积】选项。

（5）Draw Method【绘制方法】参数组：在该参数组中提供了 Continuous【连续】、Scale image from center【从中心缩放图像】和 Scale image from side【从一侧缩放图像】3 个选项，默认 Continuous【连续】选项。

18）Freeze Tool【冻结工具】命令

Freeze Tool【冻结工具】命令的图标为█，快捷键为Ctrl+0，Tool Settings【工具设置】面板如图 2.66 所示。

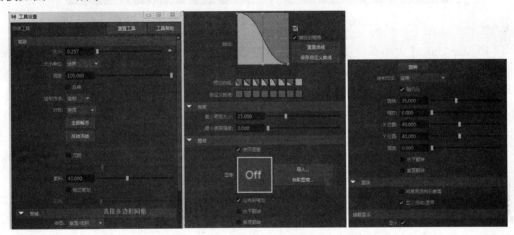

图 2.66　Tool Settings【工具设置】面板

（1）Size Units【大小单位】参数组：在该参数组中提供了 World【世界】和 Screen Pixels【屏幕像素】2 个选项，默认 World【世界】选项。

（2）Paint Method【绘制方法】参数组：在该参数组中提供了 Paint【绘制】和 Smooth【平滑】2 个选项，默认 Paint【绘制】选项。

（3）Symmetry【对称】参数组：在该参数组中提供了 Off【禁用】、Object X【对象 X】、Object Y【对象 Y】、Object Z【对象 Z】、World X【世界 X】、World Y【世界 Y】和 World Z【世界 Z】7 个选项，默认 Off【禁用】选项。

（4）Type【类型】参数组：在该参数组中提供了 Surface/Volume【曲面/体积】、Surface【曲面】和 Volume【体积】3 个选项，默认 Surface/Volume【曲面/体积】选项。

（5）Draw Method【绘制方法】参数组：在该参数组中提供了 Continuous【连续】、Scale image from center【从中心缩放图像】和 Scale image from side【从一侧缩放图像】3 个选项，默认 Continuous【连续】选项。

―――――――――――――― Shape Authoring【变形创作】――――――――――――――

19) Smooth Target Tool【平滑目标工具】命令（新增功能）

Smooth Target Tool【平滑目标工具】命令的图为，Tool Settings【工具设置】面板如图2.67 所示。

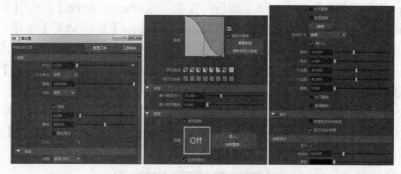

图 2.67　Tool Settings【工具设置】面板

（1）Size Units【大小单位】参数组：在该参数组中提供了 World【世界】和 Screen Pixels【屏幕像素】2 个选项，默认 World【世界】选项。

（2）Symmetry【对称】参数组：在该参数组中提供了 Off【禁用】、Object X【对象 X】、Object Y【对象 Y】、Object Z【对象 Z】、World X【世界 X】、World Y【世界 Y】和 World Z【世界 Z】7 个选项，默认 Off【禁用】选项。

（3）Type【类型】参数组：在该参数组中提供了 Surface/Volume【曲面/体积】、Surface【曲面】和 Volume【体积】3 个选项，默认 Surface/Volume【曲面/体积】选项。

（4）Draw Method【绘制方法】参数组：在该参数组中提供了 Continuous【连续】、Scale image from center【从中心缩放图像】和 Scale image from side【从一侧缩放图像】3 个选项，默认 Continuous【连续】选项。

20) Clone Target Tool【克隆目标工具】命令（新增功能）

Clone Target Tool【克隆目标工具】命令的图为，Tool Settings【工具设置】面板如图 2.68 所示。

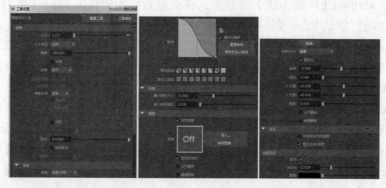

图 2.68　Tool Settings【工具设置】面板

第 2 章　Modeling【建模】模块菜单组

（1）Size Units【大小单位】参数组：在该参数组中提供了 World【世界】和 Screen Pixels【屏幕像素】2 个选项，默认 World【世界】选项。

（2）Symmetry【对称】参数组：在该参数组中提供了 Off【禁用】、Object X【对象 X】、Object Y【对象 Y】、Object Z【对象 Z】、World X【世界 X】、World Y【世界 Y】和 World Z【世界 Z】7 个选项，默认 Off【禁用】选项。

（3）Clone Method【克隆方法】参数组：在该参数组中提供了 Copy【复制】和 Add【相加】2 个选项，默认 Copy【复制】选项。

（4）Type【类型】参数组：在该参数组中提供了 Surface/Volume【曲面/体积】、Surface【曲面】和 Volume【体积】3 个选项，默认 Surface/Volume【曲面/体积】选项。

（5）Draw Method【绘制方法】参数组：在该参数组中提供了 Continuous【连续】、Scale image from center【从中心缩放图像】和 Scale image from side【从一侧缩放图像】3 个选项，默认 Continuous【连续】选项。

21）Mask Target Tool【遮罩目标工具】命令（新增功能）

Mask Target Tool【遮罩目标工具】命令的图标为，Tool Settings【工具设置】面板如图 2.69 所示。

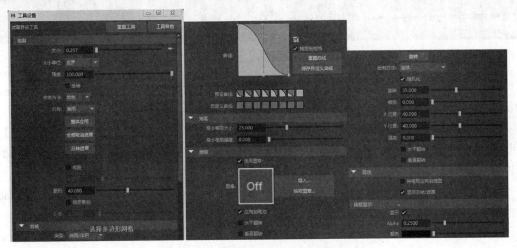

图 2.69　Tool Settings【工具设置】面板

（1）Size Units【大小单位】参数组：在该参数组中提供了 World【世界】和 Screen Pixels【屏幕像素】2 个选项，默认 World【世界】选项。

（2）Paint Method【绘制方法】参数组：在该参数组中提供了 Paint【绘制】和 Smooth【平滑】2 个选项，默认 Paint【绘制】选项。

（3）Symmetry【对称】参数组：在该参数组中提供了 Off【禁用】、Object X【对象 X】、Object Y【对象 Y】、Object Z【对象 Z】、World X【世界 X】、World Y【世界 Y】和 World Z【世界 Z】7 个选项，默认 Off【禁用】选项。

（4）Type【类型】参数组：在该参数组中提供了 Surface/Volume【曲面/体积】、Surface【曲面】和 Volume【体积】3 个选项，默认 Surface/Volume【曲面/体积】选项。

（5）Draw Method【绘制方法】参数组：在该参数组中提供了 Continuous【连续】、Scale image from center【从中心缩放图像】和 Scale image from side【从一侧缩放图像】3 个选项，默认 Continuous【连续】选项。

22）Erase Target Tool【擦除目标工具】命令（新增功能）

Erase Target Tool【擦除目标工具】命令图标为 ，Tool Settings【工具设置】面板如图 2.70 所示。

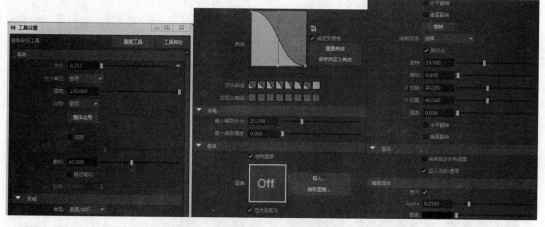

图 2.70　Tool Settings【工具设置】面板

（1）Size Units【大小单位】参数组：在该参数组中提供了 World【世界】和 Screen Pixels【屏幕像素】2 个选项，默认 World【世界】选项。

（2）Symmetry【对称】参数组：在该参数组中提供了 Off【禁用】、Object X【对象 X】、Object Y【对象 Y】、Object Z【对象 Z】、World X【世界 X】、World Y【世界 Y】和 World Z【世界 Z】7 个选项，默认 Off【禁用】选项。

（3）Type【类型】参数组：在该参数组中提供了 Surface/Volume【曲面/体积】、Surface【曲面】和 Volume【体积】3 个选项，默认 Surface/Volume【曲面/体积】选项。

（4）Draw Method【绘制方法】参数组：在该参数组中提供了 Continuous【连续】、Scale image from center【从中心缩放图像】和 Scale image from side【从一侧缩放图像】3 个选项，默认 Continuous【连续】选项。

14. Slide Edge【滑动边】命令

Slide Edge【滑动边】命令的图标为 ，Tool Settings【工具设置】面板如图 2.71 所示。

15. Target Weld【目标焊接】命令

Target Weld【目标焊接】命令的图标为 ，Tool Settings【工具设置】面板如图 2.72 所示。

第 2 章　Modeling【建模】模块菜单组

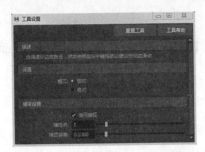

图 2.71　Tool Settings【工具设置】面板

图 2.72　Tool Settings【工具设置】面板

2.4　Mesh Display【网络显示】菜单组

――――Normals【法线】――――

1. Average【平均】命令

Average【平均】命令的图标为■，Average Normals Options【平均化法线选项】面板，如图 2.73 所示。

2. Conform【一致】命令

Conform【一致】命令的图标为■，该命令没有选项面板。

3. Reverse【反转】命令

Reverse【反转】命令的图标为■，Reverse Normals Options【反转法线选项】面板，如图 2.74 所示。

图 2.73　Average Normals Options
【平均化法线选项】面板

图 2.74　Reverse Normals Options
【反转法线选项】面板

4. Set Normal Angle…【设定法线角度…】命令

Set Normal Angle…【设定法线角度…】命令的图标为■，该命令没有选项面板。

5. Set to Face【设定为面】命令

Set to Face【设定为面】命令的图标为▇，Set To Face Normal Options【设定为面法线选项】面板，如图 2.75 所示。

6. Set Vertex Normal【设置顶点法线】命令

Set Vertex Normal【设置顶点法线】命令的图标为▇，Set Vertex Normal Options【设置顶点法线选项】面板，如图 2.76 所示。

图 2.75　Set To Face Normal Options　　图 2.76　Set Vertex Normal Options
【设定为面法线选项】面板　　　　　　　【设置顶点法线选项】面板

7. Harden Edge【硬化边】命令

Harden Edge【硬化边】命令的图标为▇，该命令没有选项面板。

8. Soften Edge【软化边】命令

Soften Edge【软化边】命令的图标为▇，该命令没有选项面板。

9. Soften/Harden Edges【软化/硬化边】命令

Soften/Harden Edges【软化/硬化边】命令图标为▇，Soften/Harden Edges Options【软化/硬化边选项】面板，如图 2.77 所示。

图 2.77　Soften/Harden Edges Options【软化/硬化边选项】面板

10. Lock Normals【锁定法线】命令

Lock Normals【锁定法线】命令的图标为▇，该命令没有选项面板。

11. Unlock Normals【解除锁定法线】命令

Unlock Normals【解除锁定法线】命令的图标为▇，该命令没有选项面板。

12. Vertex Normal Edit Tool【顶点法线编辑工具】命令

Vertex Normal Edit Tool【顶点法线编辑工具】命令的图标为■，Tool Settings【工具设置】面板，如图 2.78 所示。

———————————————Vertex Color【顶点颜色】———————————————

13. Apply Color【应用颜色】命令

Apply Color【应用颜色】命令的图标为■，Apply Color Options【应用颜色选项】面板，如图 2.79 所示。

图 2.78　Tool Settings【工具设置】面板　　图 2.79　Apply Color Options【应用颜色选项】面板

14. Paint Vertex Color Tool【绘制顶点颜色工具】命令

Paint Vertex Color Tool【绘制顶点颜色工具】命令的图标为■，Tool Settings【工具设置】面板，如图 2.80 所示。

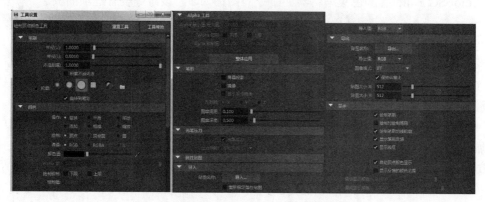

图 2.80　Tool Settings【工具设置】面板

（1）Export value【导出值】参数组：在该参数组中提供了 Luminance【亮度】、【Alpha】、【RGB】和【RGBA】4 个选项，默认【RGB】选项。

（2）Image format【图像格式】参数组：在该参数组中提供了【GIF】、【Soft Image】、【RLA】、【TIFF】、【SGI】、Alias【锯齿】、【IFF】、【JPEG】和【EPS】9 个选项，默认【IFF】选项。

―――――――――――――― Vertex Color Sets【顶点颜色集】――――――――――――――

15. Create Empty Set【创建空集】命令

Create Empty Set【创建空集】命令的图标为▦，Create Empty Color Set Options【创建空集颜色集选项】面板，如图 2.81 所示。

16. Delete Current Set【删除当前集】命令

Delete Current Set【删除当前集】命令的图标为▦，该命令没有选项面板。

17. Rename Current Set…【重命名当前集…】命令

Rename Current Set…【重命名当前集…】命令的图标为▦，该命令没有选项面板。

18. Modify Current Set【修改当前集】命令

Modify Current Set【修改当前集】命令的图标为▦，该命令没有选项面板。

19. Set Keyframe for Vertex Color【为顶点颜色设置关键帧】命令

Set Keyframe for Vertex Color【为顶点颜色设置关键帧】命令的图标为▦，该命令没有选项面板。

20. Color Set Editor【颜色集编辑器】命令

Color Set Editor【颜色集编辑器】命令的图标为▦，该命令没有选项面板。

―――――――――――――― Vertex Bake Sets【顶点烘焙集】――――――――――――――

21. Prelight（Maya）【预照明（Maya）】命令

Prelight（Maya）【预照明（Maya）】命令的图标为▦，Prelight Options【预照明选项】面板，如图 2.82 所示。

（1）Color blending【颜色混合】参数组：在该参数组中提供了 Overwrite【覆盖】、Add【相加】、Subtract【相减】、Multiply【相乘】、Divide【相除】、Average【平均】和 Don't overwrite【不覆盖】7 个选项，默认 Overwrite【覆盖】选项。

（2）Alpha blending【Alpha 混合】参数组：在该参数组中提供了 Overwrite【覆盖】、Add【相加】、Subtract【相减】、Multiply【相乘】、Divide【相除】、Average【平均】和 Don't overwrite【不覆盖】7 个选项，默认 Overwrite【覆盖】选项。

第 2 章　Modeling【建模】模块菜单组

图 2.81　Create Empty Color Set Options
【创建空集颜色集选项】面板

图 2.82　Prelight Options
【预照明选项】面板

22. Assign New Set【指定新集】命令

Assign New Set【指定新集】命令的图标为■，该命令没有选项面板。

23. Assign Existing Set【指定现有集】命令

Assign Existing Set【指定现有集】命令的图标为■，该命令在默认情况下没有二级命令，只有用户创建了命令集时，才会显示创建的命令。

24. Edit Assigned Set【编辑指定的集】命令

Edit Assigned Set【编辑指定的集】命令的图标为■，该命令没有选项面板。

──────────────Display Attributes【显示属性】──────────────

25. Toggle Display Colors Attribute【切换显示颜色属性】命令

Toggle Display Colors Attribute【切换显示颜色属性】命令的图标为■，该命令没有选项面板。

26. Color Material Channel【对材质通道上色】命令组

Color Material Channel【对材质通道上色】命令组包括■None【无】、■Ambient【环境光】、Ambient+Diffuse■【环境光+漫反射】、■Diffuse【漫反射】、■Specular【镜面反射】和■Emission【发射】6 个命令。这些命令都没有选项面板。

27. Material Blend Setting【材质混合设置】命令组

Material Blend Setting【材质混合设置】命令组包括■Overwrite【覆盖】、■Add【添加】、■Subtract【相减】、■Multiply【相乘】、■Divide【相除】、■Average【平均】和■Modulate

2x【相乘相除 2】7 个命令。这些命令都没有选项设置面板。

28. Per Instance Sharing【逐实例共享】命令组

Per Instance Sharing【逐实例共享】命令组包括■Select Shared Instances【选择共享实例】和■Share Instances【共享实例】2 个命令，这些命令都没有选项设置面板。

2.5 Curves【曲线】菜单组

————————————————Modify【修改】————————————————

1. Lock Length【锁定长度】命令

Lock Length【锁定长度】命令的图标为■，快捷键为■，该命令没有选项面板。

2. Unlock Length【解除锁定长度】命令

Unlock Length【解除锁定长度】命令的图标为■，快捷键为■，该命令没有选项面板。

3. Bend【弯曲】命令

Bend【弯曲】命令的图标为■，Bend Curves Options【弯曲曲线选项】面板，如图 2.83 所示。

4. Curl【卷曲】命令

Curl【卷曲】命令的图标为■，Curl Curves Options【卷曲曲线选项】面板，如图 2.84 所示。

图 2.83　Bend Curves Options
【弯曲曲线选项】面板

图 2.84　Curl Curves Options
【卷曲曲线选项】面板

5. Scale Curvature【缩放曲率】命令

Scale Curvature【缩放曲率】命令的图标为■，Scale Curvature Options【缩放曲率选项】面板，如图 2.85 所示。

6. Smooth【平滑】命令

Smooth【平滑】命令的图标为■，Smooth Curves Options【平滑曲线选项】面板，如图 2.86 所示。

第 2 章　Modeling【建模】模块菜单组

图 2.85　Scale Curvature Options
【缩放曲率选项】面板

图 2.86　Smooth Curves Options
【平滑曲线选项】面板

7. Straighten【拉直】命令

Straighten【拉直】命令的图标为，Straighten Curves Options【拉直曲线选项】面板，如图 2.87 所示。

―――――――――――Edit【编辑】―――――――――――

8. Duplicate Surface Curves【复制曲面曲线】命令

Duplicate Surface Curves【复制曲面曲线】命令的图标为，Duplicate Surface Curves Options【复制曲面曲线选项】面板，如图 2.88 所示。

图 2.87　Straighten Curves Options
【拉直曲线选项】面板

图 2.88　Duplicate Surface Curves Options
【复制曲面曲线选项】面板

9. Align【对齐】命令

Align【对齐】命令的图标为，Align Curves Options【对齐曲线选项】面板，如图 2.89 所示。

10. Add Points Tool【添加点工具】命令

Add Points Tool【添加点工具】命令的图标为，该命令没有选项面板。

11. Attach【附加】命令

Attach【附加】命令的图标为，Attach Curves Options【附加曲线选项】面板，如图 2.90 所示。

图 2.89　Align Curves Options
【对齐曲线选项】面板

图 2.90　Attach Curves Options
【附加曲线选项】面板

12. Detach【分离】命令

Detach【分离】命令的图标为▣，Detach Curves Options【分离曲线选项】面板，如图 2.91 所示。

13. Edit Curve Tool【编辑曲线工具】命令

Edit Curve Tool【编辑曲线工具】命令的图标为▣，该命令没有选项面板。

14. Move Seam【移动接缝】命令

Move Seam【移动接缝】命令的图标为▣，该命令没有选项面板。

15. Open/Close【开放/闭合】命令

Open/Close【开放/闭合】命令的图标▣，Open/Close Curve Options【开放/闭合曲线选项】面板，如图 2.92 所示。

图 2.91　Detach Curves Options
【分离曲线选项】面板

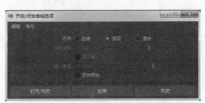
图 2.92　Open/Close Curve Options
【开放/闭合曲线选项】面板

16. Fillet【圆角】命令

Fillet【圆角】命令的图标为▣，Fillet Curve Options【圆角曲线选项】面板，如图 2.93 所示。

17. Cut【切割】命令

Cut【切割】命令的图标为▣，Cut Curve Options【切割曲线选项】面板，如图 2.94 所示。

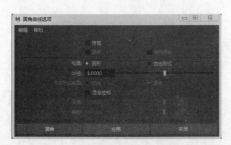

图 2.93　Fillet Curve Options【圆角曲线选项】面板　　图 2.94　Cut Curve Options【切割曲线选项】面板

第 2 章 Modeling【建模】模块菜单组

18. Intersect【相交】命令

Intersect【相交】命令的图标为■，Intersect Curves Options【曲线相交选项】面板，如图 2.95 所示。

19. Extend【延伸】命令组

Extend【延伸】命令组命令包括 Extend Curve【延伸曲线】和 Extend Curve On Surface【延伸曲面上的曲线】2 个命令。

1）Extend Curve【延伸曲线】命令

Extend Curve【延伸曲线】命令的图标■，Extend Curve Options【延伸曲线选项】面板，如图 2.96 所示。

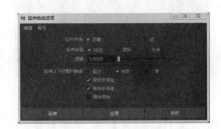

图 2.95　Intersect Curves Options
【曲线相交选项】面板

图 2.96　Extend Curve Options
【延伸曲线选项】面板

2）Extend Curve On Surface【延伸曲面上的曲线】命令

Extend Curve On Surface【延伸曲面上的曲线】命令的图标为■，Extend Curve On Surface Options【延伸曲面上的曲线选项】面板，如图 2.97 所示。

20. Insert Knot【插入结】命令

Insert Knot【插入结】命令的图标为■，Insert Knot Options【插入结选项】面板如图 2.98 所示。

图 2.97　Extend Curve On Surface Options
【延伸曲面上的曲线选项】面板

图 2.98　Insert Knot Options
【插入结选项】面板

21. Offset【偏移】命令组

Offset【偏移】命令组包括 Offset Curve【偏移曲线】和 Offset Curve On Surface【偏移

曲面上的曲线】2 个命令。

1）Offset Curve【偏移曲线】命令

Offset Curve【偏移曲线】命令的图标为 ，Offset Curve Options【偏移曲线选项】面板，如图 2.99 所示。

2）Offset Curve On Surface【偏移曲面上的曲线】命令

Offset Curve On Surface【偏移曲面上的曲线】命令的图标为 ，Offset Curve On Surface Options【偏移曲面上的曲线选项】面板，如图 2.100 所示。

图 2.99　Offset Curve Options
【偏移曲线选项】面板

图 2.100　Offset Curve On Surface Options
【偏移曲面上的曲线选项】面板

22. CV Hardness【CV 硬度】命令

CV Hardness【CV 硬度】命令的图标为 ，CV Hardness Options【CV 硬度选项】面板，如图 2.101 所示。

23. Fit B-Spline【拟合 B 样条线】命令

Fit B-Spline【拟合 B 样条线】命令的图标为 ，Fit B-Spline Optiona【拟合 B 样条线选项】面板，如图 2.102 所示。

图 2.101　CV Hardness Options
【CV 硬度选项】面板

图 2.102　Fit B-Spline Optiona
【拟合 B 样条线选项】面板

24. Project Tangent【投影切线】命令

Project Tangent【投影切线】命令的图标为 ，Project Tangent Options【投影切线选项】面板，如图 2.103 所示。

25. Smooth【平滑】命令

Smooth【平滑】命令的图标为 ，Smooth Curve Options【平滑曲线选项】面板，如图 2.104 所示。

图 2.103　Project Tangent Options
【投影切线选项】面板

图 2.104　Smooth Curve Options
【平滑曲线选项】面板

26. Bezier Curves【Bezier 曲线】命令组

Bezier Curves【Bezier 曲线】命令组包括 Anchor Presets【锚点预设】和 Tangent Options【切线选项】2 个命令组。

1）Anchor Presets【锚点预设】命令组

Anchor Presets【锚点预设】命令组包括【Bezier】、【Bezier 角点】和【角点】3 个命令，这 3 个命令都没有选项面板。

2）Tangent Options【切线选项】命令组

Tangent Options【切线选项】命令组包括 Smooth Anchor Tangents【光滑锚点切线】、Break Anchor Tangents【断开锚点切线】、Even Anchor Tangents【平坦锚点切线】和 Uneven Anchor Tangents【不平坦锚点切线】4 个命令，这 4 个命令都没有选项面板。

27. Rebuild【重建】命令

Rebuild【重建】命令的图标为 ，Rebuild Curve Options【重建曲线选项】面板，如图 2.105 所示。

28. Reverse Direction【反转方向】命令

Reverse Direction【反转方向】命令的图标为 ，Reverse Curve Options【反转曲线选项】面板，如图 2.106 所示。

图 2.105　Rebuild Curve Options
【重建曲线选项】面板

图 2.106　Reverse Curve Options
【反转曲线选项】面板

2.6 Surfaces【曲面】菜单组

―――――――――――――Create【创建】―――――――――――――

1. Loft【放样】命令

Loft【放样】命令的图标为▇，Loft Options【放样选项】面板，如图 2.107 所示。

2. Planar【平面】命令

Planar【平面】命令的图标为▇，Planar Trim Surface Options【平面修剪曲面选项】面板，如图 2.108 所示。

图 2.107　Loft Options
【放样选项】面板

图 2.108　Planar Trim Surface Options
【平面修剪曲面选项】面板

3. Revolve【旋转】命令

Revolve【旋转】命令的图标为▇，Revolve Options【旋转选项】面板，如图 2.109 所示。

4. Birail【双轨成形】命令组

Birai l【双轨成形】命令组包括 Birail1 Tool【双轨成形 1 工具】、Birail 2 Tool【双轨成形 2 工具】和 Birail 3+ Tool【双轨成形 3+工具】3 个命令。

1）Birail 1 Tool【双轨成形 1 工具】命令

Birail 1 Tool【双轨成形 1 工具】命令的图标为▇，Birail 1 Options【双轨成形 1 选项】面板，如图 2.110 所示。

2）Birail 2 Tool【双轨成形 2 工具】命令

Birail 2 Tool【双轨成形 2 工具】命令的图标为▇，Birail 2 Options【双轨成形 2 选项】面板，如图 2.111 所示。

3）Birail 3+Tool【双轨成形 3+工具】命令

Birail 3+Tool【双轨成形 3+工具】命令的图标为▇，Birail 3+Options【双轨成形 3+选项】面板，如图 2.112 所示。

第 2 章　Modeling【建模】模块菜单组

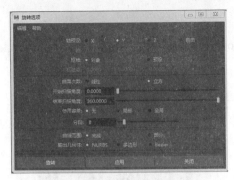

图 2.109　Revolve Options
【旋转选项】面板

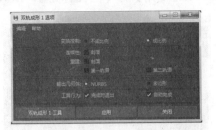

图 2.110　Birail 1 Options
【双轨成形 1 选项】面板

图 2.111　Birail 2 Options
【双轨成形 2 选项】面板

图 2.112　Birail 3+Options
【双轨成形 3+选项】面板

5. Extrude【挤出】命令

Extrude【挤出】命令的图标为 ，Extrude Options【挤出选项】面板如图 2.113 所示。

6. Boundary【边界】命令

Boundary【边界】命令的图标 ，Boundary Options【边界选项】面板如图 2.114 所示。

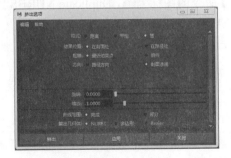

图 2.113　Extrude Options【挤出选项】面板

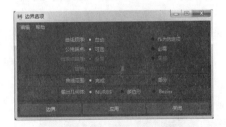

图 2.114　Boundary Options【边界选项】面板

7. Square【方形】命令

Square【方形】命令的图标为 ，Square Surface Options【方形曲面选项】面板，如图 2.115 所示。

8. Bevel【倒角】命令

Bevel【倒角】命令的图标为▇，Bevel Options【倒角选项】面板如图 2.116 所示。

图 2.115　Square Surface Options
【方形曲面选项】面板

图 2.116　Bevel Options
【倒角选项】面板

9. Bevel Plus【倒角+】命令

Bevel Plus【倒角+】命令的图标为▇，Bevel Plus Options【倒角+选项】选项如图 2.117 所示。

Outer Bevel style【外部倒角样式】和 Inner bevel style【内部倒角样式】都有 Straight Out【直出】、Straight In【直入】、Convex Out【凸出】、Convex In【凸入】、Concave Out【凹出】、Concave In【凹入】、Straight Side Edge【直侧边】、Straight Front Edge【直前边】、Straight Corner【直角点】、Convex Side Edge【凸侧边】、Convex Front Edge【凸前边】、Convex Corner【凸角点】、Concave Side Edge【凹侧边】、Concave Front Edge【凹前边】和 Convex Crease【凸折痕】16 种外部倒角样式。

────────── Edit NURBS Surfaces【编辑 NURBS 面片】──────────

10. Duplicate NURBS Patch【复制 NURBS 面片】命令

Duplicate NURBS Patch【复制 NURBS 面片】命令的图标为▇，Duplicate NURBS Patch Options【复制 NURBS 面片选项】面板，如图 2.118 所示。

图 2.117　Bevel Plus Options
【倒角+选项】面板

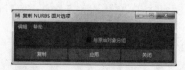

图 2.118　Duplicate NURBS Patch Options
【复制 NURBS 面片选项】面板

11. Align【对齐】命令

Align【对齐】命令的图标为█，Align Surfaces Options【对齐曲面选项】面板，如图 2.119 所示。

12. Attach【附加】命令

Attach【附加】命令的图标为█，Attach Surfaces Options【附加曲面选项】面板，如图 2.120 所示。

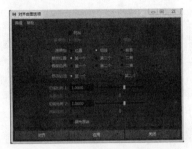

图 2.119　Align Surfaces Options
【对齐曲面选项】面板

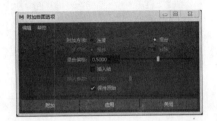

图 2.120　Attach Surfaces Options
【附加曲面选项】面板

13. Attach Without Moving【附加而不移动】命令

Attach Without Moving【附加而不移动】命令的图标为█，该命令没有选项面板。

14. Detach【分离】命令

Detach【分离】命令的图标为█，Detach Surfaces Options【分离曲面选项】面板如图 2.121 所示。

15. Move Seam【移动接缝】命令

Move Seam【移动接缝】命令的图标为█，该命令没有选项面板。

16. Open/Close【开放/闭合】命令

Open/Close【开放/闭合】命令的图标为█，Open/Close Surface Options【开放/闭合曲面选项】面板，如图 2.122 所示。

图 2.121　Detach Surfaces Options
【分离曲面选项】面板

图 2.122　Open/Close Surface Options
【开放/闭合曲面选项】面板

17. Intersect【相交】命令

Intersect【相交】命令的图标为▣，Intersect Surfaces Options【曲面相交选项】面板，如图2.123所示。

18. Project Curve on Surface【在曲面上投影曲线】命令

Project Curve on Surface【在曲面上投影曲线】命令的图标为▣，Project Curve on Surface Options【在曲面上投影曲线选项】面板，如图2.124所示。

图 2.123　Intersect Surfaces Options
【曲面相交选项】面板

图 2.124　Project Curve on Surface Options
【在曲面上投影曲线选项】面板

19. Trim Tool【修剪工具】命令

Trim Tool【修剪工具】命令的图标为▣，Tool Settings【工具设置】面板，如图2.125所示。

20. Untrim【取消修剪】命令

Untrim【取消修剪】命令的图标为▣，Untrim Options【取消修剪选项】面板，如图2.126所示。

图 2.125　Tool Settings【工具设置】面板

图 2.126　Untrim Options【取消修剪选项】面板

21. Extend【延伸】命令

Extend【延伸】命令的图标为▣，Extend Surface Options【延伸曲面选项】选项面板，如图2.127所示。

22. Insert Isoparms【插入等参线】命令

Insert Isoparms【插入等参线】命令的图标为▣，Insert Isoparms Options【插入等参线选项】面板，如图2.128所示。

第 2 章　Modeling【建模】模块菜单组

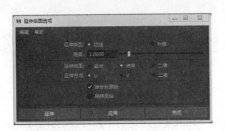

图 2.127　Extend Surface Options
【延伸曲面选项】面板

图 2.128　Insert Isoparms Options
【插入等参线选项】面板

23. Offset【偏移】命令

Offset【偏移】命令的图标为，Offset Surface Options【偏移曲面选项】面板如图 2.129 所示。

24. Round Tool【圆化工具】命令

Round Tool【圆化工具】命令的图标为，Tool Settings【工具设置】面板如图 2.130 所示。

图 2.129　Offset Surface Options【偏移曲面选项】面板

图 2.130　Tool Settings【工具设置】面板

25. Stitch【缝合】命令组

Stitch【缝合】命令组包括 Stitch Surface Points【缝合曲面点】、Stitch Edges Tool【缝合边工具】和 Global Stitch【全局缝合】3 个命令。

1）Stitch Surface Points【缝合曲面点】命令

Stitch Surface Points【缝合曲面点】命令图标为，Stitch Surface Points【缝合曲面点选项】，如图 2.131 所示。

2）Stitch Edges Tool【缝合边工具】命令

Stitch Edges Tool【缝合边工具】命令的图标为，Tool Settings【工具设置】面板如图 2.132 所示。

图 2.131　Stitch Surface Points【缝合曲面点选项】面板

图 2.132　Tool Settings【工具设置】面板

3) Global Stitch【全局缝合】命令

Global Stitch【全局缝合】命令的图标为 ，Global Stitch Options【全局缝合选项】面板，如图 2.133 所示。

26. Surface Fillet【曲面圆角】命令组

Surface Fillet【曲面圆角】命令组包括 Circular Fillet【圆形圆角】、Freeform Fillet【自由形式圆角】和 Fillet Blend Tool【圆角混合工具】3 个命令。

1) Circular Fillet【圆形圆角】命令

Circular Fillet【圆形圆角】命令的图标为 ，Circular Fillet Options【圆形圆角选项】面板，如图 2.134 所示。

图 2.133　Global Stitch Options
【全局缝合选项】面板

图 2.134　Circular Fillet Options
【圆形圆角选项】面板

2) Freeform Fillet【自由形式圆角】命令

Freeform Fillet【自由形式圆角】命令的图标为 ，Freeform Fillet Options【自由形式圆角选项】面板，如图 2.135 所示。

3) Fillet Blend Tool【圆角混合工具】命令

Fillet Blend Tool【圆角混合工具】命令的图标为 ，Fillet Blend Options【圆角混合选项】面板，如图 2.136 所示。

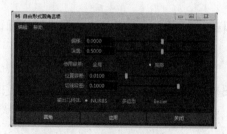
图 2.135　Freeform Fillet Options
【自由形式圆角选项】面板

图 2.136　Fillet Blend Options
【圆角混合选项】面板

27. Sculpt Geometry Tool【雕刻几何体工具】命令

Sculpt Geometry Tool【雕刻几何体工具】命令的图标为 ，Tool Settings【工具设置】面板，如图 2.137 所示。

第 2 章　Modeling【建模】模块菜单组

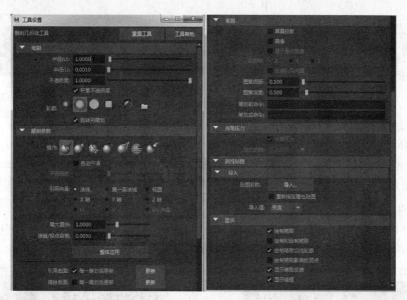

图 2.137　Tool Settings【工具设置】面板

Import Value【导入值】参数组：在该参数组中提供了 Luminance【亮度】、【Alpha】、Red【红】、Green【绿】和 Blue【蓝】5 个选项，默认 Luminance【亮度】选项。

28. Surface Editing【曲面编辑】命令组

Surface Editing【曲面编辑】命令组包括 Surface Editing Tool【曲面编辑工具】、Break Tangent【断开切线】和 Smooth Tangent【平滑切线】3 个命令。

1）Surface Editing Tool【曲面编辑工具】命令

Surface Editing Tool【曲面编辑工具】命令的图标为 ，Tool Settings【工具设置】面板，如图 2.138 所示。

2）Break Tangent【断开切线】命令

Break Tangent【断开切线】命令的图标为 ，该命令没有选项面板。

3）Smooth Tangent【平滑切线】命令

Smooth Tangent【平滑切线】命令的图标为 ，该命令没有选项面板。

29. Booleans【布尔】命令组

Booleans【布尔】命令组包括 Union Tool【并集工具】、Difference Tool【差集工具】和 Intersection Tool【交集工具】3 个命令。

1）Union Tool【并集工具】命令

Union Tool【并集工具】命令的图标为 ，NURBS Boolean Union Options【NURBS 布尔并集选项】面板，如图 2.139 所示。

图2.138 Tool Settings
【工具设置】面板

图2.139 NURBS Boolean Union Options
【NURBS布尔并集选项】面板

2）Difference Tool【差集工具】命令

Difference Tool【差集工具】命令的图标为■，NURBS Boolean Difference Options【NURBS布尔差集选项】面板，如图2.140所示。

3）Intersection Tool【交集工具】命令

Intersection Tool【交集工具】命令的图标为■，NURBS Boolean Intersection Options【NURBS布尔交集选项】面板，如图2.141所示。

图2.140 NURBS Boolean Difference Options
【NURBS布尔差集选项】面板

图2.141 NURBS Boolean Intersection Options
【NURBS布尔交集选项】面板

30. Rebuild【重建】命令

Rebuild【重建】命令的图标为■，Rebuild Surface Options【重建曲面选项】面板，如图2.142所示。

31. Reverse Direction【反转方向】命令

Reverse Direction【反转方向】命令的图标为■，Reverse Surface Direction Options【反转曲面方向选项】面板，如图2.143所示。

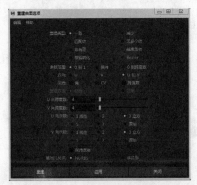

图2.142 Rebuild Surface Options
【重建曲面选项】面板

图2.143 Reverse Surface Direction Options
【反转曲面方向选项】面板

第 2 章　Modeling【建模】模块菜单组

2.7　Deform【变形】菜单组

———— Create【创建】————

1. Blend Shape【融合变形】命令

Blend Shape【融合变形】命令的图标为 ，Blend Shape Options【融合变形选项】面板，如图 2.144 所示。

图 2.144　Blend Shape Options【融合变形选项】面板

Deformation order【变形顺序】参数组：在该参数组中提供了 Automatic【自动】、Pre-deformation【变形前】、Post-deformation【变形后】、After【之后】、Split【分割】和 Parallel【平行】6 个选项，默认 Automatic【自动】选项。

2. Cluster【簇】命令

Cluster【簇】命令的图标为 ，Cluster Options【簇选项】面板，如图 2.145 所示。

图 2.145　Cluster Options【簇选项】面板

Deformation order【变形顺序】参数组：在该参数组中提供了 Default【默认】、Before【之前】、After【之后】、Split【分割】和 Parallel【平行】5 个选项，默认 Default【默认】选项。

3. Curve Warp【曲线扭曲】命令（新增功能）

Curve Warp【曲线扭曲】命令没有图标和选项面板。

4. Delta Mush【Delta Mush】命令

Delta Mush【Delta Mush】命令的 Create Delta Mush Deformer Options【创建 Delta Mush 变形器选项】面板，如图 2.146 所示。

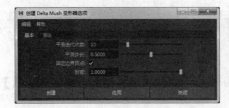

图 2.146　Create Delta Mush Deformer Options【创建 Delta Mush 变形器选项】面板

Deformation Order【变形顺序】参数组：在该参数组中提供了 Default【默认】、Before【之前】、After【之后】、Split【分割】和 Parallel【平行】5 个选项，默认 Default【默认】选项。

5. Lattice【晶格】命令

Lattice【晶格】命令的图标为 ，Lattice Options【晶格选项】面板，如图 2.147 所示。

图 2.147　Lattice Options【晶格选项】面板

Deformation order【变形顺序】参数组：在该参数组中提供了 Default【默认】、Before【之前】、After【之后】、Split【分割】和 Parallel【平行】5 个选项，默认 Default【默认】选项。

6. Wrap【包裹】命令

Wrap【包裹】命令的图标为 ，Wrap Options【包裹选项】面板，如图 2.148 所示。

Fall off mode【衰减模式】参数组：在该参数组中提供了 Volume【体积】和 Surface【表面】2 个选项，默认 Volume【体积】选项。

7. Shrink Wrap【收缩包裹】命令

Shrink Wrap【收缩包裹】命令的图标为 ，ShrinkWrap Options【收缩包裹选项】面板，如图 2.149 所示。

图 2.148 Wrap Options【包裹选项】面板

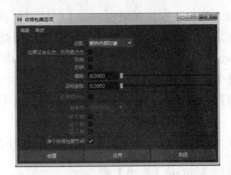

图 2.149 Shrink Wrap Options【收缩包裹选项】面板

（1）Projection【投影】参数组：在该参数组中提供了 Toward Inner Object【朝向内部对象】、Toward Center【朝向中心】、Parallel To Axes【平行于轴】、Vertex Normals【顶点法线】和 Closest【最近】5 个选项，默认 Parallel To Axes【平行于轴】选项。

（2）Axis Reference【轴参照】参数组：在该参数组中提供了 Target Local【目标局部】、Deformed Local【变形局部】和 Global【全局】3 个选项，默认 Target Local【目标局部】选项。

8. Wire【线】命令

Wire【线】命令的图标为 ，Tool Settings【工具设置】面板如图 2.150 所示。

Deformation order【变形顺序】参数组：在该参数组中提供了 Default【默认】、Before【之前】、After【之后】、Split【分割】和 Parallel【平行】5 个选项，默认 Default【默认】选项。

9. Wrinkle【褶皱】命令

Wrinkle【褶皱】命令的图标为 ，Tool Settings【工具设置】面板如图 2.151 所示。

图 2.150 Tool Settings【工具设置】面板

图 2.151 Tool Settings【工具设置】面板

10. Pose Space Deformation【姿势空间变形】命令组（新增功能）

Pose Space Deformation【姿势空间变形】命令组中包括了 Create Pose Interpolator【创建姿势插值器】和 Pose Editor【姿势编辑器】2 个命令。

1）Create Pose Interpolator【创建姿势插值器】命令

Create Pose Interpolator【创建姿势插值器】命令的图标为▥，Create Pose Interpolator Options【创建姿势插值器选项】面板，如图2.152所示。

【驱动着扭曲轴】参数组：在该参数组中提供了 X Axis【X 轴】、Y Axis【Y 轴】和 Z Axis【Z 轴】3个选项，默认 X Axis【X 轴】选项。

2）Pose Editor【姿势编辑器】命令

Pose Editor【姿势编辑器】命令的图标为▥，该命令没有选项面板。

11. Muscle【肌肉】命令组

在 Muscle【肌肉】命令组中，包括 Muscles/Bones【肌肉/骨骼】命令组、Simple Muscles【简单肌肉】命令组、Skin Setup【蒙皮设置】命令组、Muscle Setup【肌肉对象】命令组、▥Paint Muscle Weights…【绘制肌肉权重…】命令、Weighting【权重】命令组、Direction【方向】命令组、Displace【置换】命令组、Smart Collision【智能碰撞】命令组、Self/Multi Collision【自碰撞/多对象碰撞】命令组、Caching【缓存】命令组、Selection【选择】命令组和 Bonus Rigging【附加装备】命令组。

1）Muscles/Bones【肌肉/骨骼】命令组

在 Muscles/Bones【肌肉/骨骼】命令组中包括9个命令，如图2.153所示。

图2.152　Create Pose Interpolator Options
【创建姿势插值器选项】面板

图2.153　Muscles/Bones
【肌肉/骨骼】命令组面板

2）Simple Muscles【简单肌肉】命令组

在 Simple Muscles【简单肌肉】命令组中包括6个命令，如图2.154所示。

3）Skin Setup【蒙皮设置】命令组

在 Skin Setup【蒙皮设置】命令组中包括14个命令，如图2.155所示。

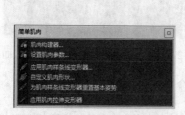

图2.154　Simple Muscles 简单肌肉】命令组面板

图2.155　Skin Setup【蒙皮设置】命令组面板

4）Muscle Setup【肌肉对象】命令组

在 Muscle Setup【肌肉对象】命令组中包括 7 个命令，如图 2.156 所示。

5）Weighting【权重】命令组

在 Weighting【权重】命令组中包括 5 个命令，如图 2.157 所示。

6）Direction【方向】命令组

在 Direction【方向】命令组中包括 3 个命令，如图 2.158 所示。

图 2.156 Muscle Setup
【肌肉对象】命令组面板

图 2.157 Weighting
【权重】命令组面板

图 2.158 Direction
【方向】命令组面板

7）Displace【置换】命令组

在 Displace【置换】命令组中包括 6 个命令，如图 2.159 所示。

8）Smart Collision【智能碰撞】命令组

在 Smart Collision【智能碰撞】命令组中包括 3 个命令，如图 2.160 所示。

9）Self/Multi Collision【自碰撞/多对象碰撞】命令组

在 Self/Multi Collision【自碰撞/多对象碰撞】命令组中包括 5 个命令，如图 2.161 所示。

图 2.159 Displace
置换】命令组面板

图 2.160 Smart Collision
【智能碰撞】面板

图 2.161 Self/Multi Collision
【自碰撞/多对象碰撞】命令组面板

10）Caching【缓存】命令组

在 Caching【缓存】命令组中包括 4 个命令，如图 2.162 所示。

11）Selection【选择】命令组

在 Selection【选择】命令组中包括 4 个命令，如图 2.163 所示。

图 2.162 Caching【缓存】命令组面板

图 2.163 Selection【选择】命令组面板

12）Bonus Rigging【附加装备】命令组

在 Bonus Rigging【附加装备】命令组中包括 3 个命令，如图 2.164 所示。

图 2.164　Bonus Rigging【附加装备】命令组面板

12. Nonlinear【非线性】命令组

在 Nonlinear【非线性】命令组中，包括 Bend【弯曲】、Flare【扩张】、Sine【正弦】、Squash【挤压】、Twist【扭曲】和 Wave【波浪】6 个命令。

1）Bend【弯曲】命令

Bend【弯曲】命令的图标为 ，Create Bend Deformer Options【创建弯曲变形器选项】面板，如图 2.165 所示。

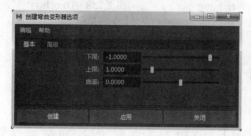 

图 2.165　Create Bend Deformer Options【创建弯曲变形器选项】面板

Deformation order【变形顺序】参数组：在该参数组中提供了 Default【默认】、Before【之前】、After【之后】、Split【分割】和 Parallel【平行】5 个选项，默认 Default【默认】选项。

2）Flare【扩张】命令

Flare【扩张】命令的图标为 ，Create Flare Deformer Options【创建扩张变形器选项】面板，如图 2.166 所示。

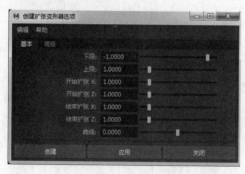 

图 2.166　Create Flare Deformer Options【创建扩张变形器选项】面板

Deformation order【变形顺序】参数组：在该参数组中提供了 Default【默认】、Before【之前】、After【之后】、Split【分割】和 Parallel【平行】5 个选项，默认 Default【默认】

第 2 章 Modeling【建模】模块菜单组

选项。

3）Sine【正弦】命令

Sine【正弦】命令的图标为 ，Create Sine Deformer Options【创建正弦变形器选项】面板，如图 2.167 所示。

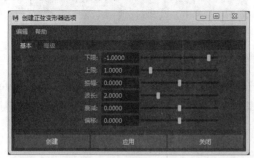 

图 2.167　Create Sine Deformer Options【创建正弦变形器选项】面板

Deformation order【变形顺序】参数组：在该参数组中提供了 Default【默认】、Before【之前】、After【之后】、Split【分割】和 Parallel【平行】5 个选项，默认 Default【默认】选项。

4）Squash【挤压】命令

Squash【挤压】命令的图标为 ，Create Squash Deformer Options【创建挤压变形器选项】面板，如图 1.168 所示。

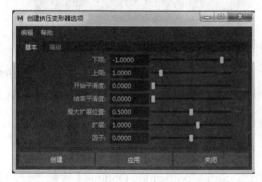 

图 2.168　Create Squash Deformer Options【创建挤压变形器选项】面板

Deformation order【变形顺序】参数组：在该参数组中提供了 Default【默认】、Before【之前】、After【之后】、Split【分割】和 Parallel【平行】5 个选项，默认 Default【默认】选项。

5）Twist【扭曲】命令

Twist【扭曲】命令的图标为 ，Create Twist Deformer Options【创建扭曲变形器选项】面板，如图 2.169 所示。

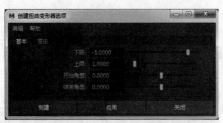

图 2.169　Create Twist Deformer Options【创建扭曲变形器选项】面板

Deformation order【变形顺序】参数组：在该参数组中提供了 Default【默认】、Before【之前】、After【之后】、Split【分割】和 Parallel【平行】5 个选项，默认 Default【默认】选项。

6）Wave【波浪】命令

Wave【波浪】命令的图标为　，Create Wave Deformer Options【创建波浪变形器选项】面板，如图 2.170 所示。

图 2.170　Create Wave Deformer Options【创建波浪变形器选项】面板

Deformation order【变形顺序】参数组：在该参数组中提供了 Default【默认】、Before【之前】、After【之后】、Split【分割】和 Parallel【平行】5 个选项，默认 Default【默认】选项。

13. Soft Modification【软修改】命令

Soft Modification【软修改】命令的图标为　，Soft Modification Options【软修改选项】面板，如图 2.171 所示。

图 2.171　Soft Modification Options【软修改选项】面板

Fall off mode【衰减模式】参数组：在该参数组中提供了 Volume【体积】和 Surface【表面】2 个选项，默认 Volume【体积】选项。

14. Sculpt【雕刻】命令

Sculpt【雕刻】命令的图标为，Sculpt Options【雕刻选项】面板如图 2.172 所示。

图 2.172　Sculpt Options【雕刻选项】面板

Deformation order【变形顺序】参数组：在该参数组中提供了 Default【默认】、Before【之前】、After【之后】、Split【分割】和 Parallel【平行】5 个选项，默认 Default【默认】选项。

15. Texture【纹理】命令

Texture【纹理】命令的图标为，Texture Options【纹理选项】面板如图 2.173 所示。

图 2.173　Texture Options【纹理选项】面板

（1）Point Space【点空间】参数组：在该参数组中提供了 World【世界】、Local【局部】和【UV】3 个选项，默认【UV】选项。

（2）Direction【方向】参数组：在该参数组中提供了 Normal【法线】、Handle【控制柄】和 Vector【向量】3 个选项，默认 Handle【控制柄】选项。

（3）Vector Space【向量空间】参数组：在该参数组中提供了 Object【对象】、World【世界】和 Tangent【切线】3 个选项，默认 Object【对象】选项。

（4）Deformation order【变形顺序】参数组：在该参数组中提供了 Default【默认】、Before【之前】、After【之后】、Split【分割】和 Parallel【平行】5 个选项，默认 Default【默认】选项。

16. Jiggle【抖动】命令组

在 Jiggle【抖动】命令组中，包括 Jiggle Deformer【抖动变形器】、Jiggle Disk Cache【抖动磁盘缓存】和 Jiggle Disk Cache Attributes【抖动磁盘缓存属性】3 个命令。

1）Jiggle Deformer【抖动变形器】命令

Jiggle Deformer【抖动变形器】命令的图标为 ，Jiggle Deformer Options【抖动变形器选项】面板，如图 2.174 所示。

图 2.174　Jiggle Deformer Options【抖动变形器选项】面板

2）Jiggle Disk Cache【抖动磁盘缓存】命令

Jiggle Disk Cache【抖动磁盘缓存】命令的图标为 ，Disk Cache Options【磁盘缓存选项】面板，如图 2.175 所示。

3）Jiggle Disk Cache Attributes【抖动磁盘缓存属性】命令

Jiggle Disk Cache Attributes【抖动磁盘缓存属性】命令的图标为 ，该命令没有选项面板。

17. Point On Curve【曲线上的点】命令

Point On Curve【曲线上的点】命令的图标为 ，Point On Curve Options【曲线上的点选项】面板，如图 2.176 所示。

图 2.175　Disk Cache Options
【磁盘缓存选项】面板

图 2.176　Point On Curve Options
【曲线上的点选项】面板

Edit【编辑】
---

18. Blend Shape【融合变形】命令组

在 Blend Shape【融合变形】命令组中，包括 Add【添加】、Remove【移除】、Swap【交换】、Bake Topology To Targets【将拓扑烘焙到目标】和 Edit Normalization Groups【编辑规

格化组】5个命令。

1）Add【添加】命令

Add【添加】命令的图标为![icon]，Add Blend Shape Target Options【添加融合变形目标选项】面板，如图2.177所示。

Post-Deformation【变形后】参数组：在该参数组中，提供了Tangent Space【切线空间】和Transform Space【变换空间】2个选项，默认Transform Space【切线空间】选项。

2）Remove【移除】命令

Remove【移除】命令的图标为![icon]，Remove Blend Shape Target Options【移除融合变形目标选项】面板，如图2.178所示。

图2.177　Add Blend Shape Target Options
【添加融合变形目标选项】面板

图2.178　Remove Blend Shape Target Options
【移除融合变形目标选项】面板

3）Bake Topology To Targets【将拓扑烘焙到目标】命令

Bake Topology To Targets【将拓扑烘焙到目标】命令的图标为![icon]，该命令没有选项面板。

19．Lattice【晶格】命令组

在Lattice【晶格】命令组中，包括如图2.179所示的2个命令。

20．Wrap【包裹】命令组

在Wrap【包裹】命令组中，包括如图2.180所示的2个命令。

21．Shrink Wrap【收缩包裹】命令组

在Shrink Wrap【收缩包裹】命令组中，包括如图2.181所示的6个命令。

图2.179　Lattice
【晶格】命令组面板

图2.180　Wrap
【包裹】命令组面板

图2.181　Shrink Wrap
【收缩包裹】命令组面板

22. Wire【线】命令组

在 Wire【线】命令组中，包括如图 2.182 所示的 7 个命令。

1）Add【添加】命令

Add【添加】命令的图标为 ，Add Wire Options【添加线选项】面板如图 2.183 所示。

图 2.182　Wire【线】命令组面板

图 2.183　Add Wire Options【添加线选项】面板

2）Add Holder【添加限制曲线】命令

Add Holder【添加限制曲线】命令的图标为 ，Add Wire Holder Options【添加线框限制曲线选项】面板，如图 2.184 所示。

3）Parent Base Wire【父基础线】命令

Parent Base Wire【父基础线】命令的图标为 ，Parent Base Wire Options【父基础线选项】面板，如图 2.185 所示。

图 2.184　Add Wire Holder Options
【添加线框限制曲线选项】面板

图 2.185　Parent Base Wire Options
【父基础线选项】面板

4）Remove【移除】命令

Remove【移除】命令的图标为 ，Remove Wire Options【移除线选项】面板，如图 2.186 所示。

5）Reset【重置】命令

Reset【重置】命令的图标为 ，Reset Wire Options【重置线选项】面板，如图 2.187 所示。

图 2.186　Remove Wire Options
【移除线选项】面板

图 2.187　Reset Wire Options
【重置线选项】面板

6）Show Base Wire【显示基础线】命令

Show Base Wire【显示基础线】命令的图标为■，该命令没有选项面板。

7）Wire Dropoff Locator【线衰减定位器】命令

Wire Dropoff Locator【线衰减定位器】命令的图标为■，Wire Dropoff Locator Options【线衰减定位器】面板，如图 2.188 所示。

23．Edit Membership Tool【编辑成员身份】命令

Edit Membership Tool【编辑成员身份】命令的图标为■，该命令没有选项面板。

24．Prune Membership【删除成员身份】命令组

在 Prune Membership【删除成员身份】命令组中，包括如图 2.189 所示的 4 个命令。

25．Mirror Deformer Weights【镜像变形器权重】命令

Mirror Deformer Weights【镜像变形器权重】命令的图标为■，Mirror Deformer Weights Options【镜像变形器权重选项】面板，如图 2.190 所示。

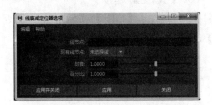
图 2.188　Wire Dropoff Locator Options【线衰减定位器】面板

图 2.189　Prune Membership【删除成员身份】命令组面板

图 2.190　Mirror Deformer Weights Options【镜像变形器权重选项】面板

26．Display Intermediate Objects【显示中间对象】命令

Display Intermediate Objects【显示中间对象】命令的图标为■，该命令没有选项面板。

27．Hide Intermediate Objects【隐藏中间对象】命令

Hide Intermediate Objects【隐藏中间对象】命令的图标为■，该命令没有选项面板。

──────────────── Paint Weights【绘制权重】────────────────

28．Blend Shape【融合变形】命令

Blend Shape【融合变形】命令的图标为■，Tool Settings【工具设置】面板，如图 2.191 所示。

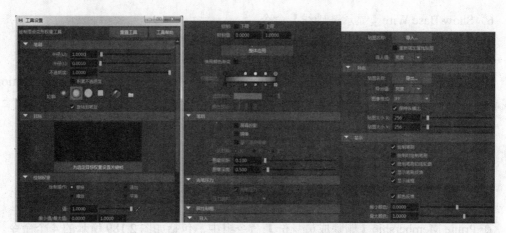

图 2.191　Tool Settings【工具设置】面板

（1）Import value【导入值】参数组：在该参数组中，提供了 Luminance【亮度】、Alpha【Alpha】、Red【红】、Green【绿】和 Blue【蓝】5 个选项，默认 Luminance【亮度】选项。

（2）Export value【导出值】参数组：在该参数组中，提供了 Luminance【亮度】、Alpha【Alpha】、RGB【RGB】和 RGBA【RGBA】4 个选项。默认【亮度】选项。

（3）Image format【图像格式】参数组：在该参数组中，提供了【GIF】、【SoftImage】、【RLA】、【TIFF】、【SGI】、Alias【锯齿】、【IFF】、【JPEG】和【EPS】9 个选项。默认【IFF】选项。

29. Cluster【簇】命令

Cluster【簇】命令的图标为 ，Tool Settings【工具设置】面板，如图 2.192 所示。

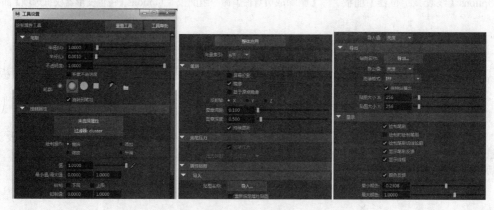

图 2.192　面板

（1）Vector index【向量索引】参数组：在该参数组中提供了【x/R】、【y/Y】和【z/B】3 个选项。默认【x/R】选项。

（2）Import value【导入值】参数组：在该参数组中，提供了 Luminance【亮度】、【Alpha】、

Red【红】、Green【绿】和 Blue【蓝】5 个选项，默认 Luminance【亮度】选项。

（3）Export【导出值】参数组：在该参数组中，提供了 Luminance【亮度】、【Alpha】、【RGB】和【RGBA】4 个选项。默认 Luminance【亮度】选项。

（4）【图像格式】参数组：在该参数组中，提供了【GIF】、【SoftImage】、【RLA】、【TIFF】、【SGI】、Alias【锯齿】、【IFF】、【JPEG】和【EPS】9 个选项。默认【IFF】选项。

30. Delta Mush【Delta Mush】命令

Delta Mush【Delta Mush】命令的 Tool Settings【工具设置】面板，如图 2.193 所示。

图 2.193　Tool Settings【工具设置】面板

（1）Vector index【向量索引】参数组：在该参数组中提供了【x/R】、【y/Y】和【z/B】3 个选项，默认【x/R】选项。

（2）Import value【导入值】参数组：在该参数组中，提供了 Luminance【亮度】、【Alpha】、Red【红】、Green【绿】和 Blue【蓝】5 个选项，默认 Luminance【亮度】选项。

（3）Export【导出值】参数组：在该参数组中，提供了 Luminance【亮度】、【Alpha】、【RGB】和【RGBA】4 个选项。默认 Luminance【亮度】选项。

（4）【图像格式】参数组：在该参数组中，提供了【GIF】、【SoftImage】、【RLA】、【TIFF】、【SGI】、Alias【锯齿】、【IFF】、【JPEG】和【EPS】9 个选项，默认【IFF】选项。

31. Lattice【晶格】命令

Lattice【晶格】命令的 Tool Settings【工具设置】面板，如图 2.194 所示。

（1）Vector Index【向量索引】参数组：在该参数组中，提供了【x/R】、【y/G】和【z/B】3 个选项，默认【x/R】选项。

（2）Import value【导入值】参数组：在该参数组中，提供了 Luminance【亮度】、【Alpha】、Red【红】、Green【绿】和 Blue【蓝】5 个选项，默认 Luminance【亮度】选项。

（3）Export value【导出值】参数组：在该参数组中，提供了 Luminance【亮度】、Alpha【Alpha】、RGB【RGB】和 RGBA【RGBA】4 个选项，默认 Luminance【亮度】选项。

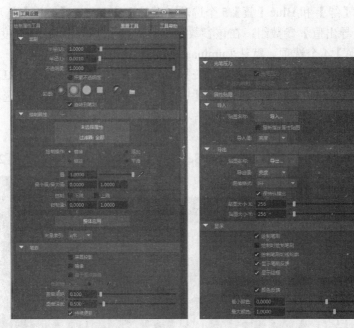

图 2.194 Tool Settings【工具设置】面板

（4）【图像格式】参数组：在该参数组中，提供了【GIF】、【SoftImage】、【RLA】、【TIFF】、【SGI】、Alias【锯齿】、【IFF】、【JPEG】和【EPS】9个选项，默认【IFF】选项。

32. Shrink Wrap【收缩包裹】命令

Shrink Wrap【收缩包裹】命令的图标为 ，Tool Settings【工具设置】面板，如图 2.195 所示。

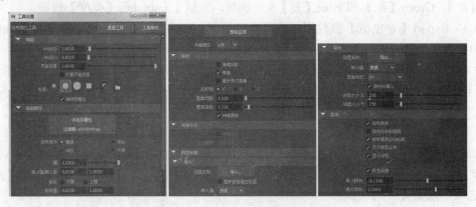

图 2.195 Tool Settings【工具设置】面板

（1）Vector Index【向量索引】参数组：在该参数组中提供了【x/R】、【y/Y】和【z/B】3个选项，默认【x/R】选项。

（2）Import value【导入值】参数组：在该参数组中，提供了 Luminance【亮度】、【Alpha】、

第 2 章 Modeling【建模】模块菜单组

Red【红】、Green【绿】和 Blue【蓝】5 个选项，默认 Luminance【亮度】选项。

（3）Export【导出值】参数组：在该参数组中，提供了 Luminance【亮度】、【Alpha】、【RGB】和【RGBA】4 个选项。默认 Luminance【亮度】选项。

（4）【图像格式】参数组：在该参数组中，提供了【GIF】、【SoftImage】、【RLA】、【TIFF】、【SGI】、Alias【锯齿】、【IFF】、【JPEG】和【EPS】9 个选项，默认【IFF】选项。

33. Wire【线】命令

Wire【线】命令的图标为，Tool Settings【工具设置】面板，如图 2.196 所示。

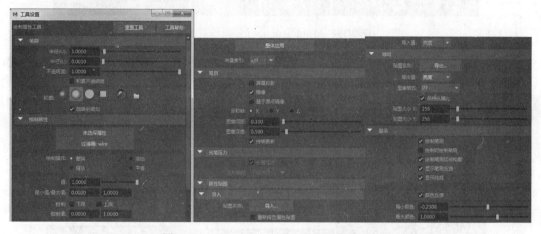

图 2.196 Tool Settings【工具设置】面板

（1）Vector index【向量索引】参数组：在该参数组中提供了【x/R】、【y/Y】和【z/B】3 个选项，默认【x/R】选项。

（2）Import value【导入值】参数组：在该参数组中，提供了 Luminance【亮度】、【Alpha】、Red【红】、Green【绿】和 Blue【蓝】5 个选项，默认 Luminance【亮度】选项。

（3）Export【导出值】参数组：在该参数组中，提供了 Luminance【亮度】、【Alpha】、【RGB】和【RGBA】4 个选项，默认 Luminance【亮度】选项。

（4）【图像格式】参数组：在该参数组中，提供了【GIF】、【SoftImage】、【RLA】、【TIFF】、【SGI】、Alias【锯齿】、【IFF】、【JPEG】和【EPS】9 个选项，默认【IFF】选项。

34. Nonlinear【非线性】命令

Nonlinear【非线性】命令的 Tool Settings【工具设置】面板，如图 2.197 所示。

（1）Vector index【向量索引】参数组：在该参数组中提供了【x/R】、【y/Y】和【z/B】3 个选项，默认【x/R】选项。

（2）Import value【导入值】参数组：在该参数组中，提供了 Luminance【亮度】、【Alpha】、Red【红】、Green【绿】和 Blue【蓝】5 个选项，默认 Luminance【亮度】选项。

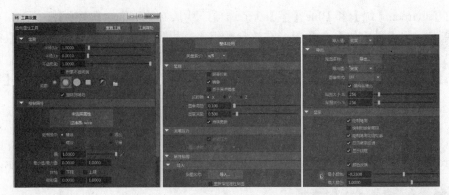

图 2.197　Tool Settings【工具设置】面板

（3）Export【导出值】参数组：在该参数组中，提供了 Luminance【亮度】、【Alpha】、【RGB】和【RGBA】4 个选项，默认 Luminance【亮度】选项。

（4）【图像格式】参数组：在该参数组中，提供了【GIF】、【SoftImage】、【RLA】、【TIFF】、【SGI】、Alias【锯齿】、【IFF】、【JPEG】和【EPS】9 个选项，默认【IFF】选项。

35．Jiggle【抖动】命令

Jiggle【抖动】命令的图标为 ，Tool Settings【工具设置】面板，如图 2.198 所示。

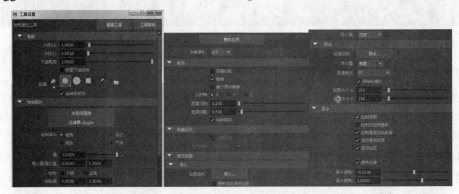

图 2.198　Tool Settings【工具设置】面板

（1）Vector index【向量索引】参数组：在该参数组中提供了【x/R】、【y/Y】和【z/B】3 个选项，默认【x/R】选项。

（2）Import value【导入值】参数组：在该参数组中，提供了 Luminance【亮度】、【Alpha】、Red【红】、Green【绿】和 Blue【蓝】5 个选项，默认 Luminance【亮度】选项。

（3）Export【导出值】参数组：在该参数组中，提供了 Luminance【亮度】、【Alpha】、【RGB】和【RGBA】4 个选项，默认 Luminance【亮度】选项。

（4）【图像格式】参数组：在该参数组中，提供了【GIF】、【SoftImage】、【RLA】、【TIFF】、【SGI】、Alias【锯齿】、【IFF】、【JPEG】和【EPS】9 个选项，默认【IFF】选项。

第 2 章 Modeling【建模】模块菜单组

36. Texture Deformer【纹理变形器】命令

Texture Deformer【纹理变形器】命令的图标为，Tool Settings【工具设置】面板，如图 2.199 所示。

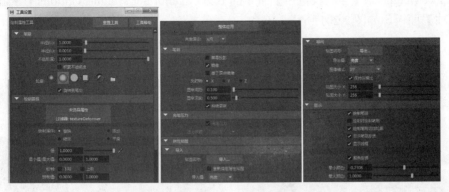

图 2.199 Tool Settings【工具设置】面板

（1）Vector index【向量索引】参数组：在该参数组中提供了【x/R】、【y/Y】和【z/B】3 个选项。默认【x/R】选项。

（2）Import value【导入值】参数组：在该参数组中，提供了 Luminance【亮度】、【Alpha】、Red【红】、Green【绿】和 Blue【蓝】5 个选项，默认 Luminance【亮度】选项。

（3）Export【导出值】参数组：在该参数组中，提供了 Luminance【亮度】、【Alpha】、【RGB】和【RGBA】4 个选项，默认 Luminance【亮度】选项。

（4）【图像格式】参数组：在该参数组中，提供了【GIF】、【SoftImage】、【RLA】、【TIFF】、【SGI】、Alias【锯齿】、【IFF】、【JPEG】和【EPS】9 个选项，默认【IFF】选项。

37. Set Membership【集成员身份】命令

Set Membership【集成员身份】命令的图标为，Tool Settings【工具设置】面板，如图 2.200 所示。

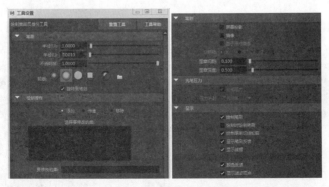

图 2.200 Tool Settings【工具设置】面板

―――――――――――――Weights【权重】―――――――――――――

38. Export Weights…【导出权重…】命令

Export Weights…【导出权重…】命令的 Export Deformer Weights Options【导出变形器权重选项】面板，如图 2.201 所示。

39. Import Weights…【导入权重…】命令

Import Weights…【导入权重…】命令的 Import Deformer Weights Options【导入变形器权重选项】面板，如图 2.202 所示。

图 2.201　Export Deformer Weights Options
【导出变形器权重选项】面板

图 2.202　Import Deformer Weights Options
【导入变形器权重选项】面板

Mapping method【映射方法】参数组：在该参数组中提供了 Index【索引】、Over【过】和 Nearest【最近】3 个选项，默认【索引】选项。

2.8　【UV】菜单组

―――――――――――――Editors【编辑器】―――――――――――――

1. UV Editor【UV 编辑器】命令

UV Editor【UV 编辑器】命令的图标为▣，UV Editor【UV 编辑器】面板，如图 2.203 所示。

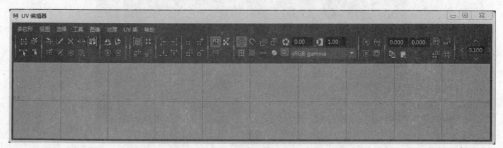

图 2.203　UV Editor【UV 编辑器】面板

2. UV Set Editor【UV集编辑器】命令

UV Set Editor【UV集编辑器】命令的图标为■，UV Set Editor【UV集编辑器】面板，如图2.204所示。

────────────Create【创建】────────────

3. Assign Shader【指定着色器】命令

Assign Shader【指定着色器】命令没有选项面板。

4. Automatic【自动】命令

Automatic【自动】命令的图标为■，Polygon Automatic Mapping Options【多边形自动映射选项】面板，如图2.205所示。

图2.204　UV Set Editor　　　　图2.205　Polygon Automatic Mapping Options
　　　【UV集编辑器】面板　　　　　　　　　【多边形自动映射选项】面板

（1）Planes【平面】参数组：在该参数组中提供了【3】、【4】、【5】、【6】、【8】和【12】6个选项，默认【6】选项。

（2）Shell layout【壳布局】参数组：在该参数组中提供了Overlap【重叠】、Along U【沿U方向】、Into square【置于方形】和Tile【平铺】6个选项，默认Into square【置于方形】选项。

（3）Spacing Presets【间距预设】参数组：在该参数组中提供了Custom【自定义】、2014 Map【1024贴图】、512 Map【512贴图】、256 Map【256贴图】、128 Map【128贴图】、64 Map【64贴图】和32 Map【32贴图】6个选项，默认512 Map【512贴图】选项。

5. Best Plane【最佳平面】命令

Best Plane【最佳平面】命令的图标为■，该命令没有选项面板。

6. Best Plane Texturing Tool【最佳平面纹理工具】命令

Best Plane Texturing Tool【最佳平面纹理工具】命令的图标为■，该命令没有选项面板。

7. Camera-Based【基于摄影机】命令

Camera-Based【基于摄影机】命令的图标为■，Create UVs Based On Camera Options【基于摄影机创建 UV 选项】面板，如图 2.206 所示。

8. Cylindrical【圆柱形】命令

Cylindrical【圆柱形】命令的图标为■，Cylindrical Mapping Options【圆柱形映射选项】面板，如图 2.207 所示。

图 2.206　Create UV Based On Camera Options
　　　　【基于摄影机创建 UV 选项】面板

图 2.207　Cylindrical Mapping Options
　　　　【圆柱形映射选项】面板

9. Planar【平面】命令

Planar【平面】命令的图标为■，Planar【平面映射选项】面板，如图 2.208 所示。

10. Spherical【球形】命令

Spherical【球形】命令的图标为■，Spherical【球形映射选项】面板，如图 2.209 所示。

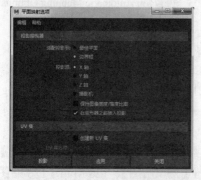

图 2.208　Planar【平面映射选项】面板

图 2.209　Spherical【球形映射选项】面板

第 2 章　Modeling【建模】模块菜单组

11. Contour Stretch【轮廓拉伸】命令

Contour Stretch【轮廓拉伸】命令的图标为，Contour Stretch Mapping Options【轮廓拉伸映射选项】面板，如图 2.210 所示。

12. Normal-Based【基于法线】命令

Normal-Based【基于法线】命令的图标为，Normal Based Projection Options【基于法线的投影选项】面板，如图 2.211 所示。

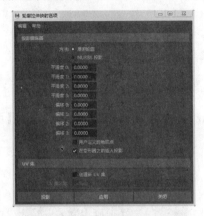

图 2.210　Contour Stretch Mapping Options
【轮廓拉伸映射选项】面板

图 2.211　Normal Based Projection Options
【基于法线的投影选项】面板

---------- Modify【修改】 ----------

13. Align【对齐】命令

Align【对齐】命令的图标为，Align UV Options【对齐 UV 选项】面板，如图 2.212 所示。

14. Flip【翻转】命令

Flip【翻转】命令的图标为，Flip UV Options【翻转 UV 选项】面板，如图 2.213 所示。

图 2.212　Align UV Options
【对齐 UV 选项】面板

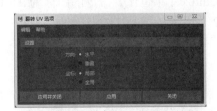

图 2.213　Flip UV Options
【翻转 UV 选项】面板

15. Grid【栅格】命令

Grid【栅格】命令的图标为▦，Grid UV Options【栅格 UV 选项】面板，如图 2.214 所示。

Map size presets【贴图大小预设】参数组：在该参数组中提供了 Custom/Non Square【自定义/非方形】、1024 Map【1024 贴图】、512 Map【512 贴图】、256 Map【256 贴图】、128 Map【128 贴图】、64 Map【64 贴图】、32 Map【32 贴图】和 16 Map【16 贴图】8 个选项，默认 Custom/Non Square【自定义/非方形】选项。

16. Layout Rectangle【排布矩形】命令

Layout Rectangle【排布矩形】命令的图标为▦，该命令没有选项面板。

17. Normalize【规格化】命令

Normalize【规格化】命令的图标为▦，Normalize UV Options【规格化 UV 选项】面板，如图 2.215 所示。

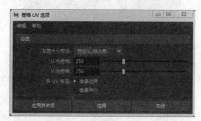
图 2.214　Grid UV Options
【栅格 UV 选项】面板

图 2.215　Normalize UV Options
【规格化 UV 选项】面板

18. Rotate【旋转】命令

Rotate【旋转】命令的图标为▦，Rotate UV Options【旋转 UV 选项】面板，如图 2.216 所示。

19. Unitize【单位化】命令

Unitize【单位化】命令的图标为▦，Unitize UV Options【单位化 UV 选项】面板，如图 2.217 所示。

图 2.216　Rotate UV Options
【旋转 UV 选项】面板

图 2.217　Unitize UV Options
【单位化 UV 选项】面板

第 2 章 Modeling【建模】模块菜单组

20. Map UV Border【映射 UV 边界】命令

Map UV Border【映射 UV 边界】命令的图标为▦，Map UV Border Options【映射 UV 边界选项】面板，如图 2.218 所示。

21. Straighten UV Border【拉直 UV 边界】命令

Straighten UV Border【拉直 UV 边界】命令的图标为▦，Straighten UV Border Options【拉直 UV 边界选项】面板，如图 2.219 所示。

图 2.218　Map UV Border Options
【映射 UV 边界选项】面板

图 2.219　Straighten UV Border Options
【拉直 UV 边界选项】面板

22. Optimize【优化】命令

Optimize【优化】命令的图标为▦，Optimize UV Options【优化 UV 选项】面板，如图 2.220 所示。

Map Size（Pixels）【贴图大小（像素）】参数组：在该参数组中提供了【4096】、【2048】、【1024】、【512】、【256】、【128】、【64】和【32】8 个选项，默认【1024】选项。

23. Unfold【展开】命令

Unfold【展开】命令的图标为▦，Unfold UVs Options【展开 UV 选项】面板，如图 2.221 所示。

Map Size（Pixels）【贴图大小（像素）】参数组：在该参数组中提供了【4096】、【2048】、【1024】、【512】、【256】、【128】、【64】和【32】8 个选项，默认【1024】选项。

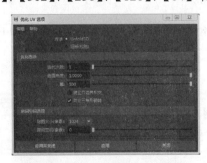

图 2.220　Optimize UV Options
【优化 UV 选项】面板

图 2.221　Unfold UV Options
【展开 UV 选项】面板

24. Layout【排布】命令（新增功能）

Layout【排布】命令的图标为▣，Layout UV Options【排布 UV 选项】面板，如图 2.222 所示。

（1）Shell Pre-Rotation【壳前期旋转】参数组：在该参数组中提供了 Off【禁用】、Horizontal【水平】、Vertical【垂直】、Align X to V【沿 X 对齐到 V】、Align Y to V【沿 Y 对齐到 V】和 Align Z to V【沿 Z 对齐到 V】6 个选项，默认 Off【禁用】选项。

（2）Shell Pre-Scaling【壳前缩放】参数组：在该参数组中提供了 Off【禁用】、Preserve 3D Ratios【保留三维比】和 Preserve UV Ratios【保留 UV 比】3 个选项，默认 Off【禁用】选项。

（3）Texture Map Size【纹理贴图大小】参数组：在该参数组中提供了【4096】、【2048】、【1024】、【512】、【256】、【128】、【64】、【32】和 Custom【自定义】9 个选项，默认【1024】选项。

（4）Shell Distribution【壳分布】参数组：在该参数组中提供了 Distribute【分布】和 Shell Centers【壳中心】2 个选项。默认 Distribute【分布】选项。

（5）Packing Region【紧缩区域】参数组：在该参数组中提供了 Full square【完整方形】、Top half【上半部】、Top Left【左上】、Top right【右上】、Bottom half【下半部】、Bottom left【左下】、Bottom right【右下】、Left half【左半部】、right half【右半部】和 custom【自定义】10 个选项。默认 Full square【完整方形】选项。

（6）Scale Mode【缩放模式】参数组：在该参数组中提供了 Off【禁用】、Uniform【一致】和 Non-Uniform【非均匀】3 个选项，默认 Uniform【一致】选项。

25. Warp Image…【扭曲图像…】命令

Warp Image…【扭曲图像…】命令的图标为▣，Warp Image Options【扭曲图像选项】面板，如图 2.223 所示。

———————————— Edit【编辑】————————————

26. Auto Seams【自动接缝】命令（新增功能）

Auto Seams【自动接缝】命令的图标为▣，Auto Seams Options【自动接缝选项】面板，如图 2.224 所示。

27. Cut UV Edges【切割 UV 边】命令

Cut UV Edges【切割 UV 边】命令的图标为▣，该命令没有选项面板。

28. Delete UVs【删除 UV】命令

Delete UVs【删除 UV】命令的图标为▣，该命令没有选项面板。

第 2 章 Modeling【建模】模块菜单组

图 2.222　Layout UV Options
【排布 UV 选项】面板

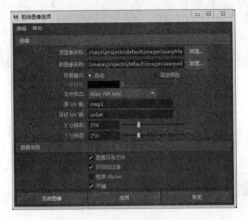

图 2.223　Warp Image Options
【扭曲图像选项】面板

29. Merge UV【合并 UV】命令

Merge UV【合并 UV】命令的图标为▥，Merge UV Options【合并 UV 选项】面板，如图 2.225 所示。

图 2.224　Auto Seams Options
【自动接缝选项】面板

图 2.225　Merge UV Options
【合并 UV 选项】面板

30. 3D Cut and Sew UV Tool【3D 切割和缝合 UV 工具】命令

3D Cut and Sew UV Tool【3D 切割和缝合 UV 工具】命令的图标为▥，Tool Settings【工具设置】面板，如图 2.226 所示。

31. Move and Sew UV Edges【移动并缝合 UV 边】命令

Move and Sew UV Edges【移动并缝合 UV 边】命令的图标为▥，Move and Sew UV Options【移动并缝合 UV 选项】面板，如图 2.227 所示。

图 2.226　Tool Settings
【工具设置】面板

图 2.227　Move and Sew UV Options
【移动并缝合 UV 选项】面板

32. Sew UV Edges【缝合 UV 边】命令

Sew UV Edges【缝合 UV 边】命令的图标为 ，该命令没有选项面板。

33. Split UV【分割 UV】命令

Split UV【分割 UV】命令的图标为 ，该命令没有选项面板。

34. 3D UV Grab Tool【3D UV 抓取工具】命令

3D UV Grab Tool【3D UV 抓取工具】命令的图标为 ，Tool Settings【工具设置】面板，如图 2.228 所示。

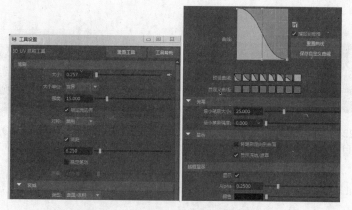

图 2.228　Tool Settings【工具设置】面板

（1）Size Units【大小单位】参数组：在该参数组中提供了 World【世界】和 Screen Pixels【屏幕像素】2 个选项，默认 World【世界】选项。

（2）Symmetry【对称】参数组：在该参数组中提供了 Off【禁用】、Object X【对象 X】、Object Y【对象 Y】、Object Z【对象 Z】、World X【世界 X】、World Y【世界 Y】和 World Z【世界 Z】7 个选项，默认 Off【禁用】选项。

第 2 章 Modeling【建模】模块菜单组

（3）Type【类型】参数组：在该参数组中提供了 Surface/Volume【曲面/体积】、Surface【曲面】和 Volume【体积】3 个选项，默认 Surface/Volume【曲面/体积】选项。

―――――――――――――――― Sets【集】――――――――――――――――

35. Copy UV to UV Set【将 UV 复制到 UV 集】命令组

在 Copy UV to UV Set【将 UV 复制到 UV 集】命令组中，默认情况下只有一个【复制到新 UV 集】的命令，命令图标为 ，Copy Current UV Set Options【复制当前 UV 集选项】面板，如图 2.229 所示。

36. Create Empty UV Set【创建空 UV 集】命令

Create Empty UV Set【创建空 UV 集】命令的图标为 ，Create UV Set Options【创建 UV 集选项】面板，如图 2.230 所示。

图 2.229 Copy Current UV Set Options
【复制当前 UV 集选项】面板

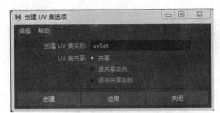

图 2.230 Create UV Set Options
【创建 UV 集选项】面板

37. Delete Current UV Set【删除当前 UV 集】命令

Delete Current UV Set【删除当前 UV 集】命令的图标为 ，该命令没有面板选项。

38. Rename Current UV Set…【重命名当前 UV 集…】命令

Rename Current UV Set…【重命名当前 UV 集…】命令的图标为 ，Rename Current UV Set Options【重命名当前 UV 集选项】面板，如图 2.231 所示。

39. Set Current UV Set…【设定当前 UV 集…】命令

Set Current UV Set…【设定当前 UV 集…】命令的图标为 ，Set Current UV Set Options【设定当前 UV 集选项】面板，如图 2.232 所示。

图 2.231 Rename Current UV Set Options
【重命名当前 UV 集选项】面板

图 2.232 Set Current UV Set Options
【设定当前 UV 集选项】面板

40. Per Instance Sharing【逐实共享】命令组

在 Per Instance Sharing【逐实共享】命令组中主要包括▣【选择共享实例】、▣【共享实例】和▣【将选定实例作为当前实例】3 个命令。这 3 个命令都没有选项面板。

2.9　Generate【生成】菜单组

1. XGen Editor【XGen 编辑器】命令

XGen Editor【XGen 编辑器】命令的图标为▣，单击该命令后在界面右侧弹出 XGen Editor【XGen 编辑器】面板，如图 2.233 所示。

2. XGen Library【XGen 库】命令

XGen Library【XGen 库】命令的图标为▣，XGen Library Window【XGen 库窗口】面板，如图 2.234 所示。

图 2.233　XGen Editor【XGen 编辑器】面板　　图 2.234　XGen Library Window【XGen 库窗口】面板

3. Create Description…【创建描述…】命令

Create Description…【创建描述…】命令的图标为▣，该命令没有选项面板。

4. Import Collections or Descriptions…【导入集合或描述…】命令

Import Collections or Descriptions…【导入集合或描述…】命令的图标为▣，Import Collections or Descriptions【导入集合或描述】面板，如图 2.235 所示。

图 2.235　Import Collections or Descriptions【导入集合或描述】面板

第 2 章 Modeling【建模】模块菜单组

5. Export Collections or Descriptions…【导出集合或描述…】命令

Export Collections or Descriptions…【导出集合或描述…】命令没有选项面板。

6. Export Selection as Archive（s）…【将当前选择项导出为归档…】命令

Export Selection as Archive（s）…【将当前选择项导出为归档…】命令的 Export Selection as Archive【将当前选择项导出为归档】面板，如图 2.236 所示。

图 2.236　Export Selection as Archive【将当前选择项导出为归档】面板

Destination Folder【细节级别】参数组：在该参数组中提供了 Disabled【已禁用】、Automatic（Poly Reduce%）【自动（多边形减少%）】和 Manual（LOD suffix）【手动（LOD 后缀）】3 个选项，默认 Disabled【已禁用】选项。

7. Batch Convert Scenes to Archives…【场景批量转化为归档…】命令

Batch Convert Scenes to Archives…【场景批量转化为归档…】命令的 Batch Convert Scenes to XGen Archives【将场景批量转化为 XGen 归档文件】面板，如图 2.237 所示。

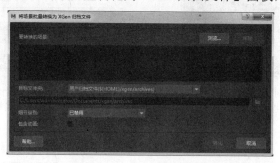

图 2.237　Batch Convert Scenes to XGen Archives【将场景批量转化为 XGen 归档文件】面板

（1）Destination Folder【目标文件夹】参数组：在该参数组中提供了 User Archives（\${HOME}/xgen/archives）【用户归档文件（\${HOME}/xgen/archives）】、Local Archives（\${PROJECT}/xgen/archives）【本地归档文件（\${PROJECT}/xgen/archives）】、Global Archives（\${XGEN_LOCATION}/presets/archives）【全局归档文件（\${XGEN_LOCATION}/presets/archives）】和 Custom【自定义】4 个选项，默认 User Archives（\${HOME}/xgen/archives）【用户归档文件（\${HOME}/xgen/archives）】选项。

（2）Level of Detail【细节级别】参数组：在该参数组中提供了 Disabled【已禁用】、

Automatic（PolyReduce%）【自动（多边形减少%）】和 Manual（LOD Suffix）【手动（LOD 后缀）】3 个选项，默认 Disabled【已禁用】选项。

8. Convert XGen Primitives to Polygons…【将 XGen 基本体转化为多边形…】命令（新增功能）

Convert XGen Primitives to Polygons…【将 XGen 基本体转化为多边形…】命令的图标为 , Convert XGen Primitives to Polygons Options【将 XGen 基本体转化为多边形的选项】面板，如图 2.238 所示。

UV Layout【UV 布局类型】参数组：在该参数组中提供了【2×2】、【3×3】和【4×4】3 个选项，默认【2×2】选项。

9. Convert to Interactive Groom…【转化为交互式修饰…】命令（新增功能）

Convert to Interactive Groom…【转化为交互式修饰…】命令的图标为 , Convert to Interactive Groom Options【转化为交互式修饰选项】面板，如图 2.239 所示。

图 2.238 Convert XGen Primitives to Polygons Options
【将 XGen 基本体转化为多边形的选项】面板

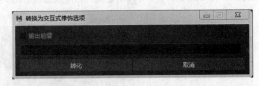

图 2.239 Convert to Interactive Groom Options
【转化为交互式修饰选项】面板

10. Import Preset…【导入预设…】命令

Import Preset…【导入预设…】命令的 Import Preset【导入预设】面板，如图 2.240 所示。

11. Export as Preset…【导出为预设…】命令

Export as Preset…【导出为预设…】命令，在默认情况下没有选项面板。

12. Create Interactive Groom Splines【创建交互式修饰样条线】命令（新增功能）

Create Interactive Groom Splines【创建交互式修饰样条线】命令的图标为 , Create Interactive Groom Splines Options【创建交互式修饰样条线选项】面板，如图 2.241 所示。

第 2 章　Modeling【建模】模块菜单组

图 2.240　Import Preset　　　　　　图 2.241　Create Interactive Groom Splines Options
　　　【导入预设】面板　　　　　　　　　　【创建交互式修饰样条线选项】面板

13. Interactive Groom Editor【交互式修饰编辑器】命令

Interactive Groom Editor【交互式修饰编辑器】命令的图标为，Interactive Groom Editor【交互式修饰编辑器】面板，如图 2.242 所示。

14. Interactive Groom Tools【交互式修饰工具】命令组（新增功能）

Interactive Groom Tools【交互式修饰工具】命令组包括如图 2.243 所示的 13 个命令。

图 2.242　Interactive Groom Editor　　　图 2.243　Interactive Groom Tools
　　【交互式修饰编辑器】面板　　　　　　　　　【交互式修饰工具】面板

1）Density Brush【密度笔刷】命令

Density Brush【密度笔刷】命令的图标为 ，快捷键为 Ctrl+1，Tool Settings【工具设置】面板，如图 2.244 所示。

（1）Paint Operation【绘制操作】参数组：在该参数组中提供了 Set【集】、Increase【增加】、Decrease【减少】和 Smooth【平滑】4 个选项，默认 Increase【增加】选项。

（2）Falloff Type【衰减类型】参数组：在该参数组中只提供了 Surface【曲面】选项。

（3）Symmetry【对称】参数组：在该参数组中提供了 Off【禁用】、Object X【对象 X】、Object Y【对象 Y】、Object Z【对象 Z】、World X【世界 X】、World Y【世界 Y】和 World Z【世界 Z】7 个选项，默认 Off【禁用】选项。

2）Place Brush【放置笔刷】命令

Place Brush【放置笔刷】命令的图标为 ，Tool Settings【工具设置】面板，如图 2.245 所示。

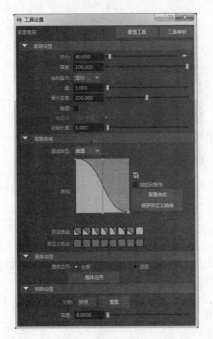

图 2.244 Tool Settings【工具设置】面板

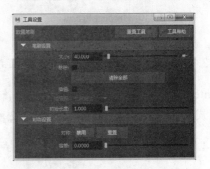

图 2.245 Tool Settings【工具设置】面板

Symmetry【对称】参数组：在该参数组中提供了 Off【禁用】、Object X【对象 X】、Object Y【对象 Y】、Object Z【对象 Z】、World X【世界 X】、World Y【世界 Y】和 World Z【世界 Z】7 个选项，默认 Off【禁用】选项。

3）Length Brush【长度笔刷】命令

Length Brush【长度笔刷】命令的图标为 ，Tool Settings【工具设置】面板，如图 2.246 所示。

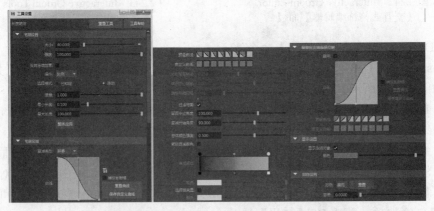

图 2.246 Tool Settings【工具设置】面板

（1）Operation【操作】参数组：在该参数组中提供了 Scale【比例】和 Smooth【平滑】2 个选项，默认 Scale【比例】选项。

（2）Falloff Type【衰减类型】参数组：在该参数组中提供了 Screen【屏幕】、Volume【体积】和 Surface【曲面】选项，默认 Screen【屏幕】选项。

（3）Symmetry【对称】参数组：在该参数组中提供了 Off【禁用】、Object X【对象 X】、Object Y【对象 Y】、Object Z【对象 Z】、World X【世界 X】、World Y【世界 Y】和 World Z【世界 Z】7 个选项，默认 Off【禁用】选项。

4）Cut Brush【修剪笔刷】命令

Cut Brush【修剪笔刷】命令的图标为 ，Tool Settings【工具设置】面板，如图 2.247 所示。

图 2.247　Tool Settings【工具设置】面板

（1）Cut Mode【修剪模式】参数组：在该参数组中提供了 Screen【屏幕】和 Volume【体积】2 个选项，默认 Screen【屏幕】选项。

（2）Symmetry【对称】参数组：在该参数组中提供了 Off【禁用】、Object X【对象 X】、Object Y【对象 Y】、Object Z【对象 Z】、World X【世界 X】、World Y【世界 Y】和 World Z【世界 Z】7 个选项，默认 Off【禁用】选项。

5）Width Brush【宽度笔刷】命令

Width Brush【宽度笔刷】命令的图标为 ，Tool Settings【工具设置】面板，如图 2.248 所示。

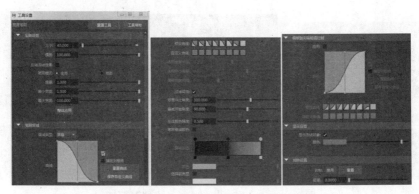

图 2.248　Tool Settings【工具设置】面板

（1）Falloff Type【衰减类型】参数组：在该参数组中提供了 Screen【屏幕】、Volume【体积】和 Surface【曲面】3 个选项，默认 Screen【屏幕】选项。

（2）Symmetry【对称】参数组：在该参数组中提供了 Off【禁用】、Object X【对象 X】、Object Y【对象 Y】、Object Z【对象 Z】、World X【世界 X】、World Y【世界 Y】和 World Z【世界 Z】7 个选项，默认 Off【禁用】选项。

6）Comb Brush【梳理笔刷】命令

Comb Brush【梳理笔刷】命令的图标为 ，快捷键为 Ctrl+2，Tool Settings【工具设置】面板，如图 2.249 所示。

图 2.249　Tool Settings【工具设置】面板

（1）Falloff Type【衰减类型】参数组：在该参数组中提供了 Screen【屏幕】和 Volume【体积】2 个选项，默认 Screen【屏幕】选项。

（2）Symmetry【对称】参数组：在该参数组中提供了 Off【禁用】、Object X【对象 X】、Object Y【对象 Y】、Object Z【对象 Z】、World X【世界 X】、World Y【世界 Y】和 World Z【世界 Z】7 个选项，默认 Off【禁用】选项。

7）Grab Brush【抓取笔刷】命令

Grab Brush【抓取笔刷】命令的图标为 ，快捷键为 Ctrl+3，Tool Settings【工具设置】面板，如图 2.250 所示。

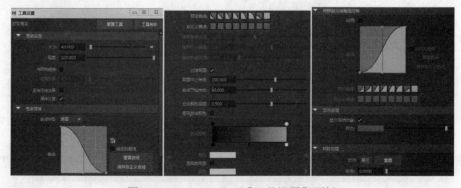

图 2.250　Tool Settings【工具设置】面板

第 2 章　Modeling【建模】模块菜单组

（1）Falloff Type【衰减类型】参数组：在该参数组中提供了 Screen【屏幕】和 Volume【体积】2 个选项，默认 Screen【屏幕】选项。

（2）Symmetry【对称】参数组：在该参数组中提供了 Off【禁用】、Object X【对象 X】、Object Y【对象 Y】、Object Z【对象 Z】、World X【世界 X】、World Y【世界 Y】和 World Z【世界 Z】7 个选项，默认 Off【禁用】选项。

8）Smooth Brush【平滑笔刷】命令

Smooth Brush【平滑笔刷】命令的图标为 ，快捷键为 Ctrl+6，Tool Settings【工具设置】面板，如图 2.251 所示。

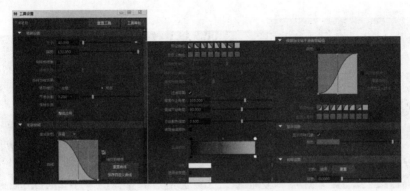

图 2.251　Tool Settings【工具设置】面板

（1）Falloff Type【衰减类型】参数组：在该参数组中提供了 Screen【屏幕】、Volume【体积】和 Surface【曲面】3 个选项，默认 Screen【屏幕】选项。

（2）Symmetry【对称】参数组：在该参数组中提供了 Off【禁用】、Object X【对象 X】、Object Y【对象 Y】、Object Z【对象 Z】、World X【世界 X】、World Y【世界 Y】和 World Z【世界 Z】7 个选项，默认 Off【禁用】选项。

9）Noise Brush【噪波笔刷】命令

Noise Brush【噪波笔刷】命令的图标为 ，快捷键为 Ctrl+7，Tool Settings【工具设置】面板，如图 2.252 所示。

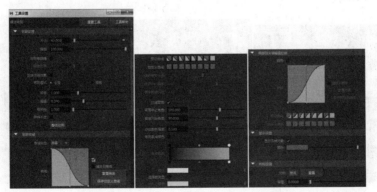

图 2.252　Tool Settings【工具设置】面板

（1）Falloff Type【衰减类型】参数组：在该参数组中提供了 Screen【屏幕】、Volume【体积】和 Surface【曲面】3 个选项，默认 Screen【屏幕】选项。

（2）Symmetry【对称】参数组：在该参数组中提供了 Off【禁用】、Object X【对象 X】、Object Y【对象 Y】、Object Z【对象 Z】、World X【世界 X】、World Y【世界 Y】和 World Z【世界 Z】7 个选项，默认 Off【禁用】选项。

10）Clump Brush【成束笔刷】命令

Clump Brush【成束笔刷】命令的图标为，快捷键为 Ctrl+8，Tool Settings【工具设置】面板，如图 2.253 所示。

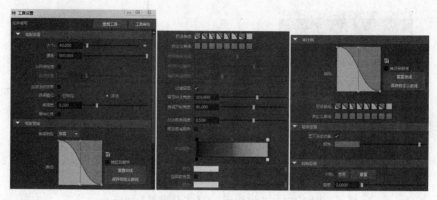

图 2.253　Tool Settings【工具设置】面板

（1）Falloff Type【衰减类型】参数组：在该参数组中提供了 Screen【屏幕】、Volume【体积】和 Surface【曲面】3 个选项，默认 Screen【屏幕】选项。

（2）Symmetry【对称】参数组：在该参数组中提供了 Off【禁用】、Object X【对象 X】、Object Y【对象 Y】、Object Z【对象 Z】、World X【世界 X】、World Y【世界 Y】和 World Z【世界 Z】7 个选项，默认 Off【禁用】选项。

11）Part Brush【分隔笔刷】命令

Part Brush【分隔笔刷】命令的图标为，快捷键为 Ctrl+9，Tool Settings【工具设置】面板，如图 2.254 所示。

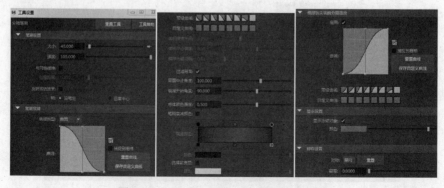

图 2.254　Tool Settings【工具设置】面板

（1）Falloff Type【衰减类型】参数组：在该参数组中只提供了 Surface【曲面】选项。

（2）Symmetry【对称】参数组：在该参数组中提供了 Off【禁用】、Object X【对象 X】、Object Y【对象 Y】、Object Z【对象 Z】、World X【世界 X】、World Y【世界 Y】和 World Z【世界 Z】7 个选项，默认 Off【禁用】选项。

12）Freeze Brush【冻结笔刷】命令

Freeze Brush【冻结笔刷】命令的图标为 ，快捷键为 Ctrl+0，Tool Settings【工具设置】面板，如图 2.255 所示。

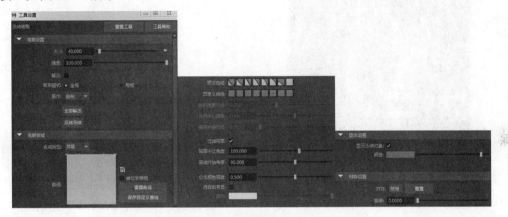

图 2.255　Tool Settings【工具设置】面板

（1）Operation【操作】参数组：在该参数组中提供了 Paint【绘制】和 Smooth【平滑】2 个选项，默认 Paint【绘制】选项。

（2）Falloff Type【衰减类型】参数组：在该参数组中提供了 Screen【屏幕】、Volume【体积】和 Surface【曲面】3 个选项，默认 Screen【屏幕】选项。

（3）Symmetry【对称】参数组：在该参数组中提供了 Off【禁用】、Object X【对象 X】、Object Y【对象 Y】、Object Z【对象 Z】、World X【世界 X】、World Y【世界 Y】和 World Z【世界 Z】7 个选项，默认 Off【禁用】选项。

13）Select Brush【选择笔刷】命令

Select Brush【选择笔刷】命令的图标为 ，Tool Settings【工具设置】面板如图 2.256 所示。

（1）Selection Type【选择类型】参数组：在该参数组中提供了 Screen【屏幕】、Volume【体积】和 Surface【曲面】3 个选项，默认 Screen【屏幕】选项。

（2）Symmetry【对称】参数组：在该参数组中提供了 Off【禁用】、Object X【对象 X】、Object Y【对象 Y】、Object Z【对象 Z】、World X【世界 X】、World Y【世界 Y】和 World Z【世界 Z】7 个选项，默认 Off【禁用】选项。

15．Cache【缓存】命令组

Cache【缓存】命令组包括如图 2.257 所示的 8 个命令。

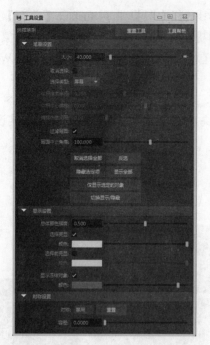

图 2.256　Tool Settings【工具设置】面板

图 2.257　Cache【缓存】命令组面板

1）Create New Cache【创建新缓存】命令

Create New Cache【创建新缓存】命令的图标为■，Create Cache Options【创建缓存选项】面板，如图 2.258 所示。

2）Import Cache【导入缓存】命令

Import Cache【导入缓存】命令的图标为■，Import Cache Options【导入缓存选项】面板，如图 2.259 所示。

图 2.258　Create Cache Options
【创建缓存选项】面板

图 2.259　Import Cache Options
【导入缓存选项】面板

3）Export Cache【导出缓存】命令

Export Cache【导出缓存】命令的图标为 ，Export Cache Options【导出缓存选项】面板，如图 2.260 所示。

4）Disable All Caches On Selected【禁用选定对象的所有缓存】命令

Disable All Caches On Selected【禁用选定对象的所有缓存】命令的图标为 ，该命令没有选项面板。

5）Enable All Caches On Selected【启用选定对象的所有缓存】命令

Enable All Caches On Selected【启用选定对象的所有缓存】命令的图标为 ，该命令没有选项面板。

6）Replace Cache【替换缓存】命令

Replace Cache【替换缓存】命令的图标为 ，Replace Cache Options【替换缓存选项】面板，如图 2.261 所示。

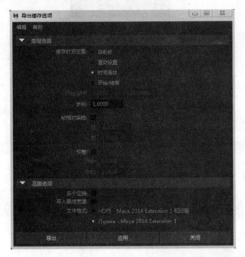
图 2.260　Export Cache Options
【导出缓存选项】面板

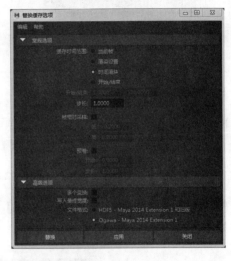
图 2.261　Replace Cache Options
【替换缓存选项】面板

7）Delete Cache【删除缓存】命令

Delete Cache【删除缓存】命令的图标为 ，Delete Cache Options【删除缓存选项】面板，如图 2.262 所示。

8）Delete Nodes Ahead of Cache【删除缓存之前的节点】命令

Delete Nodes Ahead of Cache【删除缓存之前的节点】命令的图标为 ，该命令没有选项面板。

16. Display HUD【显示 HUD】命令（新增功能）

该命令没有图标和选项面板，只有一个复选框。若勾选复选框，则在视图的左上角显示有关 HUD 信息。

―――――――――――――――Cloud Services【云服务】―――――――――――――――

17. Character Generator【角色生成器】命令

Character Generator【角色生成器】命令的 Character Generator【角色生成器】面板，如图 2.263 所示。

图 2.262　Delete Cache Options
【删除缓存选项】面板

图 2.263 Character Generator
【角色生成器】面板

18. ReCap【ReCap】命令

ReCap【ReCap】命令没有图标和选项面板。

―――――――――――――――Paint Effects【Paint 特效】―――――――――――――――

19. Paint Effects Tool【Paint Effects 工具】命令

Paint Effects Tool【Paint Effects 工具】命令的图标为 ，Tool Settings【工具设置】面板，如图 2.264 所示。

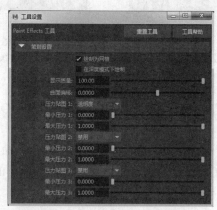

图 2.264　Tool Settings【工具设置】面板

在 Pressure map1【压力贴图 1】、Pressure map2【压力贴图 2】和 Pressure map3【压力贴图 3】参数组，提供了 Off【禁用】、Scale【缩放】、Width【宽度】、Softness【柔和度】、

Color【颜色】、Transparency【透明度】、Tube Width【管宽度】、Tube Length【管长度】、Incandescence【白炽度】、Glow spread【辉光扩散】、Tubes per Step【每步长管数】、Elevation【仰角】、Azimuth【方位角】、Path Follow【路径跟随】、Path Attract【路径吸引】、Random【随机】、Wiggle【抖动】、Curl【卷曲】、Noise【噪波】、Turbulence【湍流】、Num Twigs【细枝数】、Num Leaves【叶子数】、Num Petals【花瓣数】和 Surface Offset【曲面偏移】24 个选项。Pressure Map1【压力贴图 1】参数组默认 Transparency【透明度】选项。其他两个参数组默认 Off【禁用】选项。

20. Make Paintable【使可绘制】命令

Make Paintable【使可绘制】命令的图标为，该命令没有选项面板。

21. Get Brush…【获取笔刷…】命令

Get Brush…【获取笔刷…】命令的图标为，Content Browser【内容浏览器】面板，如图 2.265 所示。

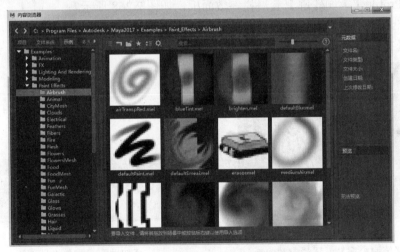

图 2.265　Content Browser【内容浏览器】面板

22. Template Brush Settings…【模板笔刷设置…】命令

Template Brush Settings…【模板笔刷设置…】的图标为，快捷键为 Ctrl+B，Paint Effects Brush Settings【Paint Effects 笔刷设置】面板，如图 2.266 所示。

Brush Type【笔刷类型】参数组：在该参数组中提供了 Paint【绘制】、Smear【涂抹】、Blur【模糊】、Erase【擦除】、Thin line【细线】和 Mesh【网络】6 个选项，默认 Paint【绘制】选项。

在 Paint Effects Brush Settings【Paint Effects 笔刷设置】面板中包括了 16 个卷展栏参数。

1）Channels【通道】卷展栏参数

Channels【通道】卷展栏参数面板，如图 2.267 所示。

图 2.266 Paint Effects Brush Settings
【Paint Effects 笔刷设置】面板

图 2.267 Channels
【通道】卷展栏参数面板

2）Brush Profile【笔刷轮廓】卷展栏参数

Brush Profile【笔刷轮廓】卷展栏参数面板，如图 2.268 所示。

3）Twist【扭曲】卷展栏参数

Twist【扭曲】卷展栏参数面板，如图 2.269 所示。

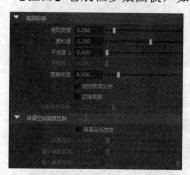

图 2.268 Brush Profile
【笔刷轮廓】卷展栏参数面板

图 2.269 Twist
【扭曲】卷展栏参数面板

4）Thin Line Multi Streaks【细线多条纹】卷展栏参数

Thin Line Multi Streaks【细线多条纹】卷展栏参数面板，如图 2.270 所示。

5）Mesh【网络】卷展栏参数

Mesh【网络】卷展栏参数面板如图 2.271 所示。

Interpolation【插值】参数组：在该参数组中提供了 None【无】、Linear【线性】、Smooth【平滑】和 Spline【样条线】4 个选项。在 Environment【环境】卷展栏中默认 Smooth【平滑】选项，在 Reflection Roll off【反射滚转】卷展栏中默认 Linear【线性】选项。

6）Shading【着色】卷展栏参数

Shading【着色】卷展栏参数面板如图 2.272 所示。

第 2 章　Modeling【建模】模块菜单组

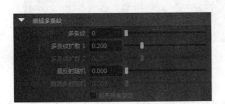

图 2.270　Thin Line Multi Streaks
【细线多条纹】卷展栏参数面板

图 2.271　Mesh
【网络】卷展栏参数面板

7）Texturing【纹理】卷展栏参数

Texturing【纹理】卷展栏参数面板，如图 2.273 所示。

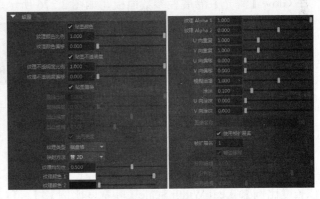

图 2.272　Shading
【着色】卷展栏参数面板

图 2.273　Texturing
【纹理】卷展栏参数面板

（1）Texture Type【纹理类型】参数组：在该参数组中提供了 Checker【棋盘格】、U Ramp【U 向渐变】、V Ramp【V 向渐变】、Fractal【分形】和 File【文件】5 个选项，默认 Checker【棋盘格】选项。

（2）Map Method【映射方法】参数组：在该参数组中提供了 Full View【完整视图】、

Brush start【笔刷开始】、Tube 2D【管 2D】和 Tube 3D【管 3D】4 个选项，默认 Tube 2D【管 2D】选项。

8）Illumination【照明】卷展栏参数

Illumination【照明】卷展栏参数面板，如图 2.274 所示。

9）Shadow Effects【阴影效果】卷展栏参数

Shadow Effects【阴影效果】卷展栏参数面板，如图 2.275 所示。

图 2.274 Illumination
【照明】卷展栏参数面板

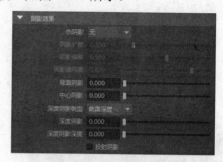

图 2.275 Shadow Effects
【阴影效果】卷展栏参数面板

（1）Fake Shadow【伪阴影】参数组：在该参数组中提供了 None【无】、2D Offset【2D 偏移】和 3D Cast【3D 投射】3 个选项，默认 None【无】选项。

（2）Depth Shadow Type【深度阴影】参数组：在该参数组中提供了 Surface Depth【曲面深度】和 Path Dist【路径距离】2 个选项，默认 Surface Depth【曲面深度】选项。

10）Glow【辉光】卷展栏参数

Glow【辉光】卷展栏参数面板，如图 2.276 所示。

图 2.276 Glow【辉光】卷展栏参数面板

11）Tubes【管】卷展栏参数

Tubes【管】卷展栏参数面板，如图 2.277 所示。

（1）Interpolation【插值】参数组：在 Width Scale【宽度比例】、Twig Length Scale【细枝长度比例】、Leaf Width Scale【叶宽度比例】、Leaf Curl【叶卷曲】、Petal Width Scale【叶瓣宽度比例】和 Petal Curl【花瓣卷曲】6 卷展栏中的 Interpolation【插值】参数组提供了相同选项，即 None【无】、linear【线性】、Smooth【平滑】和 Spline【样条线】4 个选项，默认 Linear【线性】选项。

（2）Tube Direction【管方向】参数组：在该参数组中提供了 Along Normal【沿法线】和 Along Path【沿路经】2 个选项，默认 Along Normal【沿法线】选项。

第 2 章　Modeling【建模】模块菜单组

图 2.277　Tubes【管】卷展栏参数面板

（3）Simplify Method【简化方法】参数组：在该参数组中提供了 Tubes and Segments【每步长管数】、Segments【分段】和 Tubes and Segments【管和分段】3 个选项，默认 Tubes and Segments【管和分段】选项。

（4）在 Leaves【叶位置】和 flowers【花位置】2 个参数组中，提供了相同的参数选项，即 On All【全部分支上的叶】、On Secondary Branches Only【仅次分支上的叶】和 On Twigs Only【仅细枝上的叶】3 个选项，默认 On ALL【全部分支上的叶】选项。

（5）Outside【碰撞方法】参数组：在该参数组中提供了 Outside【外部】、Inside【内部】和 Both Sides【两侧】3 个选项，默认 Outside【外部】选项。

（6）Turbulence Type【湍流类型】参数组：在该参数组中提供了 Off【禁用】、Local Force【局部力】、World Force【世界力】、Local Displacement【局部置换】、World Displacement【世界置换】、Grass Wind【草风】和 Tree Wind【树风】7 个选项，默认 Off【禁用】选项。

（7）在 Color Length Map【颜色长度贴图】、Transp Length Map【透明长度贴图】、Incand

Length Map【白炽度长度贴图】、Width Length Map【宽度长度贴图】和 Split Length Map【分割长度贴图】5 个参数组中提供了相同的参数选项,即 length【长度】和 max Length【最大长度】2 个选项,默认 length【长度】选项。

12) Gaps【间隙】卷展栏参数

Gaps【间隙】卷展栏参数面板,如图 2.278 所示。

13) Flow Animation【流动画】卷展栏参数

Flow Animation【流动画】卷展栏参数面板,如图 2.279 所示。

图 2.278　Gaps【间隙】卷展栏参数面板　　图 2.279　Flow Animation【流动画】卷展栏参数面板

14) Node Behavior【节点行为】卷展栏参数

Node Behavior【节点行为】卷展栏参数面板,如图 2.280 所示。

在 Node Behavior【节点状态】参数组中,提供了 Normal【正常】、Has No Effect【无效果】、Blocking【阻塞】、Waiting-Normal【等待-正常】、Waiting-Has No Effect【等待-无效果】和 Waiting-Blocking【等待-阻塞】6 个选项。默认 Normal【正常】选项。

15)【UUID】卷展栏参数

【UUID】卷展栏参数面板,如图 2.281 所示。

图 2.280　Node Behavior【节点行为】卷展栏参数面板　　图 2.281　【UUID】卷展栏参数面板

16) Extra Attributes【附加属性】卷展栏参数

Extra Attributes【附加属性】卷展栏参数面板,在默认情况下没有任何参数,如图 2.282 所示。

23. Reset Template Brush【重置模板笔刷】命令

Reset Template Brush【重置模板笔刷】命令的图标为，该命令没有选项面板。

24. Flip Tube Direction【翻转管方向】命令

Flip Tube Direction【翻转管方向】命令的图标为，该命令没有选项面板。

25. Make Collide【使碰撞】命令

Make Collide【使碰撞】命令的图标为，该命令没有选项面板。

第 2 章　Modeling【建模】模块菜单组

26. Paint on Paintable Objects【在可绘制对象上绘制】命令

该命令没有图标和选项面板。

27. Paint on View Plane【在视图平面上绘制】命令

该命令没有图标和选项面板。

28. Apply Settings to Last Stroke【将设置应用于上一笔划】命令

该命令没有图标和选项面板。

29. Get Settings from Selected Stroke【从选定笔划获取设置】命令

该命令没有图标和选项面板。

30. Apply Settings to Selected Strokes【将设置应用于选定笔划】命令

该命令没有图标和选项面板。

31. Share One Brush【共享一个笔刷】命令

该命令没有图标和选项面板。

32. Remove Brush Sharing【移除笔刷共享】命令

该命令没有图标和选项面板。

33. Select Brush/Stroke Names Containing…【选择包含指定字符的笔刷/笔划名称…】命令

Select Brush/Stroke Names Containing…【选择包含指定字符的笔刷/笔划名称…】面板，如图 2.283 所示。

图 2.282　Extra Attributes 【附加属性】卷展栏参数面板

图 2.283　Select Brush/Stroke Names Containing… 【选择包含指定字符的笔刷/笔划名称…】面板

34. Create Modifier【创建修改器】命令

该命令没有图标和选项面板。

35. Set Modifier Fill Object【设置修改器填充对象】命令

该命令没有图标和选项面板。

36. Brush Animation【笔刷动画】命令组

Brush Animation【笔刷动画】命令组，如图 2.284 所示。
1）Loop Brush Animation【循环笔刷动画】命令
Loop Brush Animation【循环笔刷动画】命令的 Brush Animation Looping Options【笔刷动画循环选项】面板，如图 2.285 所示。

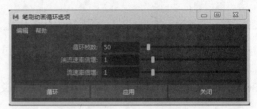

图 2.284　Brush Animation　　　　图 2.285　Brush Animation Looping Options
　　【笔刷动画】命令组面板　　　　　　　　【笔刷动画循环选项】面板

2）Make Brush Spring【生成笔刷弹簧】命令
Make Brush Spring【生成笔刷弹簧】命令的 Make Brush Spring Options【生成笔刷弹簧选项】面板，如图 2.286 所示。

3）Bake Spring Animation【烘焙弹簧动画】命令
Bake Spring Animation【烘焙弹簧动画】命令的 Bake Spring Animations【烘焙笔刷弹簧动画】面板，如图 2.287 所示。

图 2.286　Make Brush Spring Options　　图 2.287　Bake Spring Animations
　　【生成笔刷弹簧选项】面板　　　　　　　【烘焙笔刷弹簧动画】面板

37. Cure Utilities【曲线工具】命令组

Cure Utilities【曲线工具】命令组面板，如图 2.288 所示。
Make Pressure Curve【生成压力曲线】命令的 Make Pressure Curve Options【生成压力曲线选项】面板如图 2.289 所示。

第 2 章 Modeling【建模】模块菜单组

图 2.288 Cure Utilities
【曲线工具】命令组面板

图 2.289 Make Pressure Curve Options
【生成压力曲线选项】面板

38. Auto Paint【自动绘制】命令组

Auto Paint【自动绘制】命令组面板，如图 2.290 所示。

1）Paint Grid【栅格绘制】命令

Paint Grid【栅格绘制】命令的 Auto Paint Grid【自动绘制栅格】面板，如图 2.291 所示。

2）Paint Random【随机绘制】命令

Paint Random【随机绘制】命的 Auto Paint Random【自动随机绘制】面板，如图 2.292 所示。

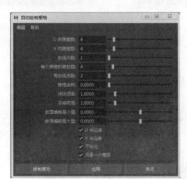

图 2.290 Auto Paint
【自动绘制】命令组
面板

图 2.291 Auto Paint Grid
【自动绘制栅格】面板

图 2.292 Auto Paint Random
【自动随机绘制】面板

39. Paint Effects Globats【Paint Effects 全局参数】命令

Paint Effects Globats【Paint Effects 全局参数】命令的 Paint Effects Globats【Paint Effects 全局参数】面板，如图 2.293 所示。

在 Node Behavior【节点状态】参数组中，提供了 Normal【正常】、Has No Effect【无效果】、Blocking【阻塞】、Waiting-Normal【等待-正常】、Waiting-Has No Effect【等待-无效果】和 Waiting-Blocking【等待-阻塞】6 个选项，默认 Normal【正常】选项。

40. Mesh Quality Attributes【网络质量属性】命令

Mesh Quality Attributes【网络质量属性】命令的选项面板，会根据选择的多边形网络的笔划或 Paint Effect 不同而有所不同。

41. Preset Blending【预设混合】命令

Preset Blending【预设混合】命令的 Brush Preset Blend【笔刷预设混合】面板，如图 2.294 所示。

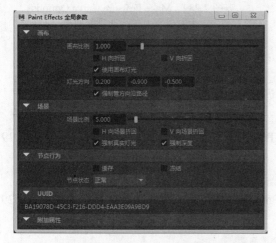

图 2.293　Paint Effects Globats
【Paint Effects 全局参数】面板

图 2.294　Brush Preset Blend
【笔刷预设混合】面板

42. Save Brush Preset…【保存笔刷预设…】命令

Save Brush Preset…【保存笔刷预设…】命令的 Save Brush Preset【保存笔刷预设】面板，如图 2.295 所示。

图 2.295　Save Brush Preset【保存笔刷预设】面板

第 3 章　Rigging【装备】模块菜单组

装备模块菜单组主要包括 Skeleton【骨架】、Skin【蒙皮】、Deform【变形】、Constrain【约束】和 Control【控制】5 个菜单组。建模模块菜单组、装备模块菜单组和动画模块菜单组中都包含了 Deform【变形】菜单组，在介绍装备模块菜单组和动画模块菜单组就不再赘述 Deform【变形】菜单组，各个菜单组的具体命令如下。

3.1　Skeleton【骨架】菜单组

———————————— Joints【关节】————————————

1. Create Joints【创建关节】命令

Create Joints【创建关节】命令的图标为■，Tool Settings【工具设置】面板，如图 3.1 所示。

Symmetry【对称】参数组：在该参数组中主要提供了 Off【禁用】、X_Axis【X 轴】、Y_Axis【Y 轴】和 Z_Axis【Z 轴】4 个选项，默认 Off【禁用】选项。

2. Insert Joints【插入关节】命令

Insert Joints【插入关节】命令的图标为■，该命令没有选项面板。

3. Mirror Joints【镜像关节】

Mirror Joints【镜像关节】命令的图标为■，Mirror Joints Options【镜像关节选项】面板，如图 3.2 所示。

图 3.1　Tool Settings【工具设置】面板

图 3.2　Mirror Joints Options【镜像关节选项】面板

4. Orient Joint【确定关节方向】命令

Orient Joint【确定关节方向】的图标为,Orient Joint Options【确定关节方向选项】面板,如图 3.3 所示。

5. Remove Joints【移除关节】命令

Remove Joints【移除关节】命令的图标为,该命令没有选项面板。

6. Connect Joint【连接关节】命令

Connect Joint【连接关节】的图标为,Connect Joint Options【连接关节选项】面板,如图 3.4 所示。

图 3.3　Orient Joint Options
【确定关节方向选项】面板

图 3.4　Connect Joint Options
【连接关节选项】面板

7. Disconnect Joint【断开关节】命令

Disconnect Joint【断开关节】命令的图标为,该命令没有选项面板。

8. Reboot Skeleton【重定骨架根】命令

Reboot Skeleton【重定骨架根】命令的图标为,该命令没有选项面板。

9. Joint Labeling【关节标签设置】命令组

1）Add Joint Labels【添加关节标签】命令组
Add Joint Labels【添加关节标签】命令组主要包括 31 个命令,如图 3.5 所示。
2）Toggle Selected Labels【切换选定标签】命令
Toggle Selected Labels【切换选定标签】命令没有选项面板。
3）Show All Labels【显示所有标签】命令
Show All Labels【显示所有标签】命令没有选项面板。
4）Hide All Labels【隐藏所有标签】命令
Hide All Labels【隐藏所有标签】命令没有选项面板。
5）Rename Joints From Labels【根据标签重命名关节】命令
Rename Joints From Labels【根据标签重命名关节】命令没有选项面板。

6）Label Based on Joint Names【基于关节名称设置标签】命令

Label Based on Joint Names【基于关节名称设置标签】命令没有选项面板。

────────────────── IK【IK】──────────────────

10. Create IK Handle【创建 IK 控制柄】命令

Create IK Handle【创建 IK 控制柄】命令的图标为 ，Tool Settings【工具设置】面板，如图 3.6 所示。

图 3.5　Add Joint Labels　　　　　　图 3.6　Tool Settings
【添加关节标签】命令组面板　　　　　　【工具设置】面板

Current Solver【当前解算器】参数组主要包括 Rotate-Plane Solver【旋转平面解算器】和 Single-Chain Solver【单链解算器】两个选项，默认 Rotate-Plane Solver【旋转平面解算器】选项。

11. Create IK Spline Handle【创建 IK 样条线控制柄】命令

Create IK Spline Handle【创建 IK 样条线控制柄】的图标为 ，Tool Settings【工具设置】面板，如图 3.7 所示。

Twist Type【扭曲类型】参数组：在该参数组中提供了 Linear【线性】、Ease in【缓入】、Ease out【缓出】和 Ease in out【缓入缓出】4 个选项，默认 Linear【线性】选项。

12. Set Preferred Angle【设置首选角度】命令

Set Preferred Angle【设置首选角度】命令的图标为 ，Set Preferred Angle Options【设置首选角度选项】面板，如图 3.8 所示。

图 3.7　Tool Settings
【工具设置】面板

图 3.8　Set Preferred Angle Options
【设置首选角度选项】面板

13. Assume Preferred Angle【采用首选角度】命令

Assume Preferred Angle【采用首选角度】命令的图标为，Assume Preferred Angle Options【采用首选角度选项】面板，如图 3.9 所示。

14. Enable IK Handle Snap【启用 IK 控制柄捕捉】命令

Enable IK Handle Snap【启用 IK 控制柄捕捉】命令没有选项面板。

15. Enable IK/FK Control【启用 IK/FK 控制】命令

Enable IK/FK Control【启用 IK/FK 控制】命令没有选项面板。

16. Enable Selected IK Handles【启用选定 IK 控制柄】命令

Enable Selected IK Handles【启用选定 IK 控制柄】命令没有选项面板。

17. Disable Selected IK Handles【禁用选定 IK 控制柄】命令

Disable Selected IK Handles【禁用选定 IK 控制柄】命令没有选项面板。

18. Quick Rig【快速装备】命令（新增功能）

Quick Rig【快速装备】命令的图标为，单击该命令，弹出 Quick Rig【快速装备】对话框，如图 3.10 所示。

图 3.9　Set Preferred Angle Options
【设置首选角度选项】面板

图 3.10　Quick Rig
【快速装备】对话框

19. Human IK…【Human IK…】命令

Human IK…【Human IK…】命令的图标为 ，该命令没有选项面板。

3.2 Skin【蒙皮】菜单组

——————————Bind【绑定】——————————

1. Bind Skin【绑定蒙皮】命令

Bind Skin【绑定蒙皮】命令的图标为 ，Bind Skin Options【绑定蒙皮选项】面板，如图 3.11 所示。

（1）Bind to【绑定到】参数组：在该参数组中主要提供了 Joint Hierarchy【关节层次】、Selected Joints【选定关节】和 Object Hierarchy【对象层次】3 个选项，默认 Joint Hierarchy【关节层次】选项。

（2）Bind Method【绑定方法】参数组：在该参数组中主要提供了 Closest Distance【最近距离】、Closest in Hierarchy【在层次中最近】、Heat Map【热量贴图】和 Geodesic Voxel【测地线体素】4 个选项，默认 Closest Distance【最近距离】选项。

（3）Skinning Method【蒙皮方法】参数组：在该参数组中主要提供了 Classic Linear【经典线性】、Dual Quaternion【双四元数】和 Weight Blended【权重已混合】3 个选项，默认 Classic linear【经典线性】选项。

（4）Normalize Weights【规格化权重】参数组：在该参数组中主要提供了 None【无】、Interactive【交互式】和 Post【后期】3 个选项，默认 Interactive【交互式】选项。

（5）Weight Distribution【权重分布】参数组：在该参数组中主要提供了 Distance【距离】和 Neighbors【相邻】2 个选项，默认 Distance【距离】选项。

2. Interactive Bind Skin【交互式绑定蒙皮】命令

Interactive Bind Skin【交互式绑定蒙皮】命令的图标为 ，Interactive Bind Skin Options【交互式绑定蒙皮选项】面板，如图 3.12 所示。

图 3.11 Bind Skin Options 【绑定蒙皮选项】面板

图 3.12 Interactive Bind Skin Options 【交互式绑定蒙皮选项】面板

（1）Bind to【绑定到】参数组：在该参数组中主要提供了 Joint hierarchy【关节层次】、Selected joints【选定关节】和 Object hierarchy【对象层次】3 个选项，默认 Joint hierarchy【关节层次】选项。

（2）Bind method【绑定方法】参数组：在该参数组中主要提供了 Closest distance【最近距离】和 Closest in hierarchy【在层次中最近】2 个选项，默认 Closest distance【最近距离】选项。

（3）Volume Type【体积类型】参数组：在该参数组中主要提供了 Capsule【胶囊】和 Cylinder【圆柱体】2 个选项，默认 Capsule【胶囊】选项。

（4）Skinning method【蒙皮方法】参数组：在该参数组中主要提供了 Classic Linear【经典线性】、Dual Quaternion【双四元数】和 Weight Blended【权重已混合】3 个选项，默认【经典线性】选项。

3. Unbind Skin【取消绑定蒙皮】命令

Unbind Skin【取消绑定蒙皮】命令的图标为，Unbind Skin Options【取消绑定蒙皮选项】面板，如图 3.13 所示。

图 3.13　Unbind Skin Options【取消绑定蒙皮选项】面板

History【历史】参数组：在该参数组中主要提供了 Delete history【删除历史】、Keep history【保持历史】和 Bake History【烘焙历史】3 个选项，默认 Delete history【删除历史】选项。

4. Go to Bind Pose【转到绑定姿势】命令

Go to Bind Pose【转到绑定姿势】命令的图标为，该命令没有选项面板。

──────── Weight Maps【权重贴图】────────

5. Paint Skin Weight【绘制蒙皮权重】命令

Paint Skin Weight【绘制蒙皮权重】命令的图标为，Tool Settings【工具设置】面板，如图 3.14 所示。

（1）Weight Type【权重类型】参数组：在该参数组中主要提供了 Skin Weight【蒙皮权重】和 BQ Blend Weight【DQ 混合权重】2 个选项，默认 Skin Weight【蒙皮权重】选项。

（2）Normalize Weights【规格化权重】参数组：在该参数组中主要提供了 Off【禁用】、Interactive【交互式】和 Post【后期】3 个选项，默认 Off【禁用】选项。

第 3 章 Rigging【装备】模块菜单组

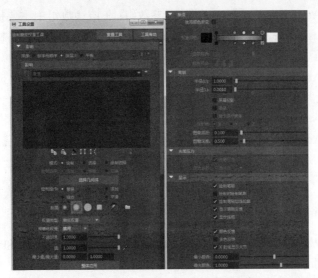

图 3.14 Tool Settings【工具设置】面板

6. Mirror Skin Weights【镜像蒙皮权重】命令

Mirror Skin Weights【镜像蒙皮权重】命令的图标为，Mirror Skin Weights Options【镜像蒙皮权重选项】面板，如图 3.15 所示。

（1）Influence Association 1【影响关联 1】参数组：在该参数组中主要提供了 Closest joint【最近关节】、One to one【一对一】和 Label【标签】3 个选项，默认 Closest joint【最近关节】选项。

（2）Influence Association 2【影响关联 2】参数组：在该参数组中主要提供了 Closest joint【最近关节】、One to one【一对一】和 Label【标签】3 个选项，默认 Closest joint【最近关节】选项。

7. Copy Skin Weights【复制蒙皮权重】命令

Copy Skin Weights【复制蒙皮权重】命令的图标为，Copy Skin Weights Options【复制蒙皮权重选项】面板，如图 3.16 所示。

图 3.15 Mirror Skin Weights Options
【镜像蒙皮权重选项】面板

图 3.16 Copy Skin Weights Options
【复制蒙皮权重选项】面板

（1）Influence Association 1【影响关联 1】参数组：在该参数组中主要提供了 Closest Joint【最近关节】、Closest Bone【最近骨骼】、One to One【一对一】、Label【标签】和 Name【名称】5 个选项，默认 Closest joint【最近关节】选项。

（2）Influence Association 2【影响关联 2】参数组：在该参数组中主要提供了 None【无】、Closest Joint【最近关节】、Closest Bone【最近骨骼】、One to One【一对一】、Label【标签】和 Name【名称】6 个选项，默认 None【无】选项。

（3）Influence Association 3【影响关联 3】参数组：在该参数组中主要提供了 None【无】、Closest Joint【最近关节】、Closest Bone【最近骨骼】、One to One【一对一】、Label【标签】和 Name【名称】6 个选项，默认 None【无】选项。

8. Smooth Skin Weights【平滑蒙皮权重】命令

Smooth Skin Weights【平滑蒙皮权重】命令的图标为，Smooth Skin Weights Options【平滑蒙皮权重选项】面板，如图 3.17 所示。

9. Hammer Skin Weights【锤式蒙皮权重】命令

Hammer Skin Weights【锤式蒙皮权重】命令没有选项面板。

10. Copy Vertex Weights【复制顶点权重】命令

Copy Vertex Weights【复制顶点权重】命令没有选项面板。

11. Paste Vertex Weights【粘贴顶点权重】命令

Paste Vertex Weights【粘贴顶点权重】命令没有选项面板。

12. Prune Small Weights【删减小权重】命令

Prune Small Weights【删减小权重】命令的 Prune Weights Options【删减权重选项】面板，如图 3.18 所示。

图 3.17 Smooth Skin Weights Options
【平滑蒙皮权重选项】面板

图 3.18 Prune Weights Options
【删减权重选项】面板

13. Reset Default Weights【重置默认权重】命令

Reset Default Weights【重置默认权重】命令没有选项面板。

14. Normalize Weights【规格化权重】命令组

Normalize Weights【规格化权重】命令组主要包括 4 个命令，如图 3.19 所示。

15. Export Weight Maps…【导出权重贴图…】命令

Export Weight Maps…【导出权重贴图…】命令的图标为，Export Skin Weight Maps Options【导出蒙皮权重贴图选项】面板，如图 3.20 所示。

图 3.19　Normalize Weights
【规格化权重】命令组面板

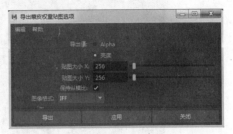

图 3.20　Export Skin Weight Maps Options
【导出蒙皮权重贴图选项】面板

【图像格式】参数组：在该参数组中主要提供了【GIF】、【Soft Image】、【RLA】、【TIFF】、【SGI】、【Alias】、【IFF】、【JPEG】、【EPS】和【Quantel】10 个选项，默认【IFF】选项。

16. Import Weight Maps…【导入权重贴图…】命令

Import Weight Maps…【导入权重贴图…】命令图标为，该命令没有选项面板。

────────────────── Other【其他】──────────────────

17. Interactive Blind Skin Tool【交互式绑定蒙皮工具】命令

Interactive Blind Skin Tool【交互式绑定蒙皮工具】命令的图标为，Tool Settings【工具设置】面板，如图 3.21 所示。

 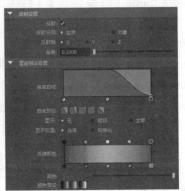

注：由于该面板界面图太长，故被截成两个图。

图 3.21　【交互式绑定蒙皮工具】命令中的 Tool Settings【工具设置】面板

18. Move Skinned Joints【移动蒙皮关节】命令

Move Skinned Joints【移动蒙皮关节】命令的图标为，Tool Settings【工具设置】面板，如图 3.22 所示。

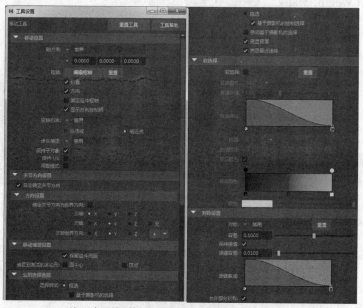

注：由于该面板界面图太长，故被截成两个图。

图 3.22　Tool Settings【工具设置】面板

（1）Transform Constraint【变换约束】参数组：在该参数组中主要提供了 Off【禁用】、Edge【边】和 Surface【曲面】3 个选项，默认 Off【禁用】选项。

（2）Step Snap【步长捕捉】参数组：在该参数组中主要提供了 Off【禁用】、Relative【相对】和 Absolute【绝对】3 个选项，默认 Off【禁用】选项。

（3）Falloff mode【衰减模式】参数组：在该参数组中主要提供了 Volume【体积】、Surface【表面】、Global【全局】和 Object【对象】4 个选项，默认 Volume【体积】选项。

（4）Interpolation【衰减模式】参数组：在该参数组中主要提供了 None【无】、Linear【线性】、Smooth【平滑】和 Spline【样条线】4 个选项，默认 None【无】选项。

（5）Symmetry【对称】参数组：在该参数组中主要提供了 Off【禁用】、Object X【对象 X】、Object Y【对象 Y】、Object Z【对象 Z】、World X【世界 X】、World Y【世界 Y】、World Z【世界 Z】和 Topology【拓扑】8 个选项，默认 Off【禁用】选项。

19. Move Weights to Influences【将权重移动到影响】命令

Move Weights to Influences【将权重移动到影响】命令没有选项面板。

20. Edit Influences【编辑影响】命令组

在 Edit Influences【编辑影响】命令组中，主要包括 4 个命令，如图 3.23 所示。

1）Add Influence【添加影响】命令

Add Influence【添加影响】命令的图标为，Add Influence Options【添加影响选项】面板，如图 3.24 所示。

图 3.23　Edit Influences
【编辑影响】命令组面板

图 3.24　Add Influence Options
【添加影响选项】面板

2）Remove【移除影响】命令

Remove【移除影响】命令的图标为，该命令没有选项面板。

3）Set Max Influences…【设置最大影响…】命令

Set Max Influences…【设置最大影响…】命令的图标为，单击该命令，弹出 Set Max Influences【设置最大影响】面板，如图 3.25 所示。

4）Remove Unused Influences【移除未使用的影响】命令

Remove Unused Influences【移除未使用的影响】命令没有选项面板。

21. Substitute Geometry【替换几何体】命令

Substitute Geometry【替换几何体】命令的图标为，Substitute Geometry Options【替换几何体选项】面板，如图 3.26 所示。

图 3.25　Set Max Influences
【设置最大影响】面板

图 3.26　Substitute Geometry Options
【替换几何体选项】面板

22. Bake Deformation to Skin Weights…【烘焙变形到蒙皮权重…】命令

单击 Bake Deformation to Skin Weights…【烘焙变形到蒙皮权重…】命令，弹出 Bake Deformer Tool【烘焙变形器工具】对话框，如图 3.27 所示。

图 3.27　Bake Deformer Tool【烘焙变形器工具】

3.3　Constrain【约束】菜单组

——————Create【创建】——————

1. Parent【父对象】命令

Parent【父对象】命令的图标为 ，Parent Constraint Options【父约束选项】面板，如图 3.28 所示。

2. Point【点】命令

Point【点】命令的图标为 ，Point Constraint Options【点约束选项】面板，如图 3.29 所示。

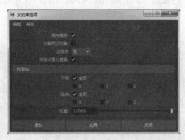

图 3.28　Parent Constraint Options
　　　　【父约束选项】面板

图 3.29　Point Constraint Options
　　　　【点约束选项】面板

3. Orient【方向】命令

Orient【方向】命令的图标为 ，Orient Constraint Options【方向约束选项】面板，如图 3.30 所示。

4. Scale【缩放】命令

Scale【缩放】命令的图标为■，Scale Constraint Options【缩放约束选项】面板，如图 3.31 所示。

图 3.30　Orient Constraint Options
【方向约束选项】面板

图 3.31　Scale Constraint Options
【缩放约束选项】面板

5. Aim【目标】命令

Aim【目标】命令的图标为■，Aim Constraint Options【目标约束选项】面板，如图 3.32 所示。

World up Type【世界上方向类型】参数组：在该参数组中主要提供了 Scene up【场景上方向】、Object up【对象上方向】、Object Rotation up【对象旋转上方向】、Vector【向量】和 None【无】5 个选项，默认 Vector【向量】选项。

6. Pole Vector【极向量】命令

Pole Vector【极向量】命令的图标为■，Pole Vector Options【极向量约束选项】面板，如图 3.33 所示。

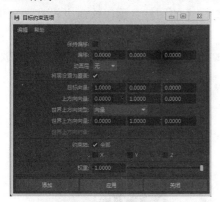

图 3.32　Aim Constraint Options
【目标约束选项】面板

图 3.33　Pole Vector Constraint Options
【极向量约束选项】面板

7. Motion Paths【运动路径】命令组

在 Motion Paths【运动路径】命令组中，主要包括 3 个命令，如图 3.34 所示。

1）Attach to Motion Path【连接到运动路径】命令

Attach to Motion Path【连接到运动路径】命令的图标为■，Attach to Motion Path Options【链接到运动路径选项】面板，如图 3.35 所示。

图 3.34　Motion Paths
【运动路径】命令组面板

图 3.35　Attach to Motion Path Options
【链接到运动路径选项】面板

World up Type【世界上方向类型】参数组：在该参数组中主要提供了 Scene up【场景上方向】、Object up【对象上方向】、Object rotation up【对象旋转上方向】、Vector【向量】和 Normal【法线】5 个选项，默认 Vector【向量】选项。

2）Flow Path Object【流动路径对象】命令

Flow Path Object【流动路径对象】命令的图标为■，Flow Path Object Options【流动路径对象选项】面板，如图 3.36 所示。

3）Set Motion Path Key【设定运动路径关键帧】命令

Set Motion Path Key【设定运动路径关键帧】命令的图标为■，该命令没有选项面板。

8．Closest Point【最近点】命令

Closest Point【最近点】命令的 Closest Point on Options【启用最近点选项】面板，如图 3.37 所示。

图 3.36　Flow Path Object Options
流动路径对象选项】面板

图 3.37　Closest Point on Options
【启用最近点选项】面板

9. Point on Poly【多边形上的点】命令

Point on Poly【多边形上的点】命令的图标为 ，Point On Poly Constraint Options【多边形上的点约束选项】面板，如图 3.38 所示。

10. Geometry【几何体】命令

Geometry【几何体】命令的图标为 ，Geometry Constraint Options【几何体约束选项】面板，如图 7.39 所示。

图 3.38　Point On Poly Constraint Options
【多边形上的点约束选项】面板

图 3.39　Geometry Constraint Options
【几何体约束选项】面板

11. Normal【法线】命令

Normal【法线】命令的图标为 ，Normal Constraint Options【发现约束选项】面板，如图 3.40 所示。

World up type【世界上方向类型】参数组：在该参数组中主要提供了 Scene up【场景上方向】、Object up【对象上方向】、Object rotation up【对象旋转上方向】、Vector【向量】和 Normal【法线】5 个选项，默认 Vector【向量】选项。

12. Tangent【切线】命令

Tangent【切线】命令的图标为 ，Tangent Constraint Options【切线约束选项】面板，如图 3.41 所示。

图 3.40　Normal Constraint Options
【发现约束选项】面板

图 3.41　Tangent Constraint Options
【切线约束选项】面板

World up Type【世界上方向类型】参数组：在该参数组中主要提供了 Scene up【场景上方向】、Object up【对象上方向】、Object Rotation up【对象旋转上方向】、Vector【向量】和 Normal【法线】5 个选项，默认 Vector【向量】选项。

―――――――――――――――――― Modify【修改】――――――――――――――――――

13. Remove Target【移除目标】命令

Remove Target【移除目标】命令的 Remove Target Options【移除目标选项】面板,如图 3.42 所示。

14. Set Rest Position【设置静止位置】命令

Set Rest Position【设置静止位置】命令没有选项面板。

15. Modify Constrained Axis…【修改受约束的轴…】命令

单击 Modify Constrained Axis…【修改受约束的轴…】命令,弹出 Modify Constrained Axis Options【修改受约束的轴选项】面板,如图 3.43 所示。

图 3.42　Remove Target Options
　　　　【移除目标选项】面板

图 3.43　Modify Constrained Axis Options
　　　　【修改受约束的轴选项】面板

3.4　Control【控制】菜单组

―――――――――――――――――― Controller【控制器】――――――――――――――――――

1. Tag As Controller【标记为控制器】命令

Tag As Controller【标记为控制器】命令没有选项面板。

2. Parent Controller【父控制器】命令

Parent Controller【父控制器】命令没有选项面板。

―――――――――――――――――― 【Human IK】――――――――――――――――――

3. Create Control Rig【创建控制装备】命令

单击 Create Control Rig【创建控制装备】命令,弹出 Create Control Rig【创建控制装备】面板,如图 3.44 所示。

Character Set【角色集】

4. Create Character Set【创建角色集】命令

Create Character Set【创建角色集】命令的 Create Character Set Options【创建角色集选项】面板，如图 3.45 所示。

图 3.44　Create Control Rig
【创建控制装备】面板

图 3.45　Create Character Set Options
【创建角色集选项】面板

5. Create Subcharacter Set【创建子角色集】命令

Create Subcharacter Set【创建子角色集】命令的 Create Subcharacter Set Options【创建子角色集选项】面板，如图 3.46 所示。

6. Character Mapper【角色映射器】命令

单击 Character Mapper【角色映射器】命令，弹出 Character Mapper【角色映射器】对话框，如图 3.47 所示。

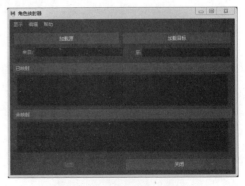

图 3.46　Create Subcharacter Set Options
【创建子角色集选项】面板

图 3.47　Character Mapper
角色映射器】对话框

7. Attribute Editor【属性编辑器】命令

Attribute Editor【属性编辑器】命令没有选项面板。

8. Add to Character Set【添加到角色集】命令

Add to Character Set【添加到角色集】命令没有选线面板。

9. Remove from Character Set【从角色集中移除】命令

Remove from Character Set【从角色集中移除】命令没有选项面板。

10. Merge Character Sets【合并角色集】命令

Merge Character Sets【合并角色集】命令没有选项面板。

11. Select Character Set Node【选择角色集节点】命令

Select Character Set Node【选择角色集节点】命令没有选项面板。

12. Select Character Set Members【选择角色集成员】命令

Select Character Set Members【选择角色集成员】命令没有选项面板。

13. SSET Current Character Set【设定当前角色集】命令组

SSET Current Character Set【设定当前角色集】命令组主要包括 2 个命令，如图 3.48 所示。

14. Redirect【重定向】命令

Redirect【重定向】命令的 Character Redirection Options【角色重定向选项】面板，如图 3.49 所示。

图 3.48　SSET Current Character Set
【设定当前角色集】命令组面板

图 3.49　Character Redirection Options
【角色重定向选项】面板

第 4 章　Animation【动画】模块菜单组

　　动画模块菜单组主要包括 Key【关键帧】、Playback【播放】、Visualize【可视化】、Deform【变形】、Constrain【约束】和【MASH】6 个菜单组。Modeling【建模】模块菜单组、Rigging【装备】模块菜单组和 Animation【动画】模块菜单组中都包含了 Deform【变形】、菜单组，在介绍 Rigging【装备】模块菜单组和 Animation【动画】模块菜单组中，就不再赘述 Deform【变形】菜单组。Rigging【装备】模块菜单包含了 Constrain【约束】菜单组，在此也不再赘述。动画模块菜单组的具体命令如下。

4.1　Key【关键帧】菜单组

────────────Set【集】────────────

1. Set Key【设定关键帧】命令

Set Key【设定关键帧】命令的图标为■，快捷键为 S，Set Key Options【设定关键帧选项】面板，如图 4.1 所示。

2. Set Key on Animated【设置动画关键帧】命令

Set Key on Animated【设置动画关键帧】命令的图标为■，该命令没有选项面板。

3. Set Key on Translate【设置平移关键帧】命令

Set Key on Translate【设置平移关键帧】命令的图标为■，快捷键为 Shift+W，该命令没有选项面板。

4. Set Key on Rotate【设置旋转关键帧】命令

Set Key on Rotate【设置旋转关键帧】命令的图标为■，快捷键为 Shift+E，该命令没有选项面板。

5. Set Key on Scale【设置缩放关键帧】命令

Set Key on Scale【设置缩放关键帧】命令的图标为■，快捷键为 Shift+R，该命令没有选项面板。

6. Set Breakdown Key【设置受控关键点关键帧】命令

Set Breakdown Key【设置受控关键点关键帧】命令的 Set Breakdown Options【设置受控关键点选项】面板，如图 4.2 所示。

图 4.1　Set Key Options
【设定关键帧选项】面板

图 4.2　Set Breakdown Options
【设置受控关键点选项】面板

7. Set Driven Key【设定受驱动关键帧】命令组

在 Set Driven Key【设定受驱动关键帧】命令组中，主要包括 3 个命令，如图 4.3 所示。

8. Set Blend Shape Target Weight Keys【设定融合变形目标权重关键帧】命令

Set Blend Shape Target Weight Keys【设定融合变形目标权重关键帧】命令没有选项面板。

———————————————— Edit【编辑】————————————————

9. Cut【剪切】命令

Cut【剪切】命令的 Cut Keys Options【剪切关键帧选项】面板，如图 4.4 所示。

图 4.3　Set Driven Key
【设定受驱动关键帧】命令组面板

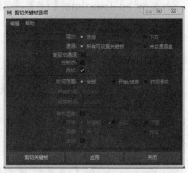

图 4.4　Cut Keys Options
【剪切关键帧选项】面板

10. Copy【复制】命令

Copy【复制】命令的 Copy Keys Options【复制关键帧选项】面板，如图 4.5 所示。

第 4 章 Animation【动画】模块菜单组

11. Paste【粘贴】命令

Paste【粘贴】命令的 Paste Keys Options【粘贴关键帧选项】面板，如图 4.6 所示。

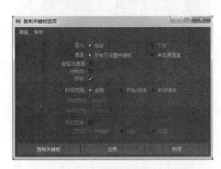

图 4.5　Copy Keys Options
【复制关键帧选项】面板

图 4.6　Paste Keys Options
【粘贴关键帧选项】面板

12. Delete【删除】命令

Delete【删除】命令的 Delete Keys Options【删除关键帧选项】面板，如图 4.7 所示。

13. Scale【缩放】命令

Scale【缩放】命令的 Scale Keys Options【缩放关键帧选项】面板，如图 4.8 所示。

图 4.7　Delete Keys Options
【删除关键帧选项】面板

图 4.8　Scale Keys Options
【缩放关键帧选项】面板

14. Snap【捕捉】命令

Snap【捕捉】命令的 Snap Keys Options【捕捉关键帧选项】面板，如图 4.9 所示。

15. Bake Animation【烘焙动画】命令

Bake Animation【烘焙动画】命令的图标为 ，Animation Options【烘焙模拟选项】面板，如图 4.10 所示。

图 4.9　Snap Keys Options
【捕捉关键帧选项】面板

图 4.10　Animation Options
【烘焙模拟选项】面板

Baked layers【烘焙层】参数组：在该参数组中主要提供了 Keep【保持】、Remove Attributes【移除属性】和 Clear Animation【清除动画】3 个选项，默认 Keep【保持】选项。

16. Hold Current Keys【保持当前关键帧】命令

Hold Current Keys【保持当前关键帧】命令没有选项组。

17. IK/FK Keys【IK/FK 关键帧】命令组

在 IK/FK Keys【IK/FK 关键帧】命令组中，主要包括 4 个命令，如图 4.11 所示。

──────────── Character Set【角色集】────────────

18. Create Character Set【创建角色集】命令

Create Character Set【创建角色集】命令的 Create Character Set Options【创建角色集选项】面板，如图 4.12 所示。

图 4.11　IK/FK Keys
【IK/FK 关键帧】命令组面板

图 4.12　Create Character Set Options
【创建角色集选项】面板

第 4 章 Animation【动画】模块菜单组

19. Create Subcharacter Set【创建子角色集】命令

Create Subcharacter Set【创建子角色集】命令的 Create Subcharacter Set Options【创建子角色集选项】面板，如图 4.13 所示。

20. Character Mapper【角色映射器】命令

单击 Character Mapper【角色映射器】命令，弹出 Character Mapper【角色映射器】面板，如图 4.14 所示。

图 4.13 Create Subcharacter Set Options
【创建子角色集选项】面板

图 4.14 Character Mapper
【角色映射器】面板

21. Attribute Editor【属性编辑器】命令

Attribute Editor【属性编辑器】命令没有选项面板。

22. Add to Character Set【添加到角色集】命令

Add to Character Set【添加到角色集】命令没有选项面板。

23. Remove from Character Set【从角色集中移除】命令

Remove from Character Set【从角色集中移除】命令没有选项面板。

24. Merge Character Set【合并角色集】命令

Merge Character Set【合并角色集】命令没有选项面板。

25. Select Character Set Node【选择角色集节点】命令

Select Character Set Node【选择角色集节点】命令没有选项面板。

26. Select Character Set Members【选择角色集成员】命令

Select Character Set Members【选择角色集成员】命令没有选项面板。

27. Select Current Character Set【设定当前角色集】命令组

在 Select Current Character Set【设定当前角色集】命令组中，主要包括 2 个命令，如图 4.15 所示。

28. Redirect【重定向】命令

Redirect【重定向】命令的 Character Redirect Options【角色重定向选项】面板，如图 4.16 所示。

图 4.15 Select Current Character Set 【设定当前角色集】命令组面板

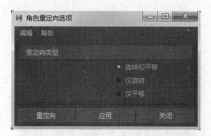

图 4.16 Character Redirect Options 【角色重定向选项】面板

―――――――――――― Time【时间】――――――――――――

29. Scene Time Warp【场景时间扭曲】命令组

在 Scene Time Warp【场景时间扭曲】命令组中，主要包括 4 个命令，如图 4.17 所示。

30. Set Time Code…【设定时间码…】命令

单击 Set Time Code…【设定时间码…】命令，弹出 Set Time code Options【设置时间码选项】对话框，如图 4.18 所示。

图 4.17 Scene Time Warp 【场景时间扭曲】命令组面板

图 4.18 Set Time code Options 【设置时间码选项】对话框

4.2 Play back【播放】菜单组

1. Playblast【播放预览】命令

Playblast【播放预览】命令的图标为 ■，Playblast Options【播放预览选项】面板，如

第 4 章 Animation【动画】模块菜单组

图 4.19 所示。

（1）Format【格式】参数组：在该参数组中主要提供了【avi】和【image】2 个选项，默认【avi】选项。

（2）Encoding【编码】参数组：在该参数组中主要提供了【MS-RLE】、【MS-CRAM】、【MS-YUV】、【IYUV 编码压缩器】、【Toshiba YUV411】和【none】6 个选项，默认【none】选项。

（3）Display Size【显示大小】参数组：在该参数组中主要提供了 From Window【来自窗口】、From Render Settings【来自渲染设置】和 Custom【自定义】3 个选项，默认 From window【来自窗口】选项。

─────────────────── Controls【控制】───────────────────

2. Play【播放】命令

Play【播放】命令的图标为▶，快捷键为 Alt+V，该命令没有选项面板。

3. Backward【向后】命令

Backward【向后】命令的图标为◀，该命令没有选项面板。

4. Stop【停止】命令

Stop【停止】的图标为■，快捷键为 Alt+V，该命令没有选项面板。

5. Next Key【下一关键帧】命令

Next Key【下一关键帧】命令的图标为▶，快捷键为.（点号），该命令没有选项面板。

6. Prev Key【上一关键帧】命令

Prev Key【上一关键帧】命令的图标为◀，快捷键为,（逗号），该命令没有选项面板。

7. Next Frame【下一帧】命令

Next Frame【下一帧】命令的图标为▶，快捷键为"Alt+."，该命令没有选项面板。

8. Prev Frame【上一帧】命令

Prev Frame【上一帧】命令的图标为◀，快捷键为"Alt+,"，该命令没有选项面板。

9. Play Range Start【播放范围开头】命令

Play Range Start【播放范围开头】命令的图标为⏮，快捷键为 Alt+Shift+V，该命令没有选项面板。

10. Play Range End【播放范围末尾】命令

Play Range End【播放范围末尾】命令的图标为 ▶|，该命令没有选项面板。

────────────────────────── Settings【设置】──────────────────────────

11. Set Range to【将范围设定为】命令组

在 Set Range to【将范围设定为】命令组中，主要包括 6 个命令，如图 4.20 所示。

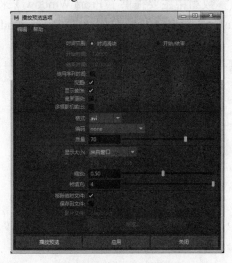

图 4.19　Playblast Options
【播放预览选项】面板

图 4.20　Set Range to
【将范围设定为】命令组面板

12. Playback Speed【播放速度】命令组

在 Playback Speed【播放速度】命令组中，主要包括 3 个命令，如图 4.21 所示。

13. Playback Looping【播放循环】命令组

在 Playback Looping【播放循环】命令组中，主要包括 3 个命令，如图 4.22 所示。

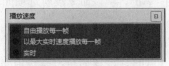

图 4.21　Playback Speed
【播放速度】命令组面板

图 4.22　Playback Speed
【播放速度】命令组面板

14. Eenable Stepped Previes【启用阶跃预览】命令

Eenable Stepped Previes【启用阶跃预览】命令没有选项面板。

第4章 Animation【动画】模块菜单组

---Render【渲染】---

15. Render Current Frame【渲染当前帧】命令

Render Current Frame【渲染当前帧】命令没有选项面板。

16. Display Render Settings…【显示渲染设置…】

单击 Display Render Settings…【显示渲染设置…】命令，弹出【渲染设置】面板。

4.3 Visualize【可视化】菜单组

---Viewport【视口】---

1. Create Editable Motion Trail【创建可编辑的运动轨迹】命令

Create Editable Motion Trail【创建可编辑的运动轨迹】命令的图标为 ，Motion Trail Options【运动轨迹选项】面板，如图 4.23 所示。

2. Create Turntable…【创建转台…】命令

Create Turntable…【创建转台…】命令的图标为 ，Animation Turntable Options【动画转台选项】面板，如图 4.24 所示。

图 4.23 Motion Trail Options 　　图 4.24 Animation Turntable Options
　　【运动轨迹选项】面板　　　　　　　　【动画转台选项】面板

---Ghost【重影】---

3. Ghost Selected【为选定对象生成重影】命令

Ghost Selected【为选定对象生成重影】命令的图标为 ，Ghost Options【重影选项】面板，如图 4.25 所示。

Type of Ghosting【重影类型】参数组：在该参数组中主要提供了 Global Preferences【全局首选项】、Custom Frames【自定义帧】、Custom Frame Steps【自定义帧步数】、Custom Key

Steps【自定义关键帧步数】和 Key Frames【关键帧】5 个选项,默认 Global Preferences【全局首选项】选项。

4. Unghost Selected【取消选定对象的重影】命令

Unghost Selected【取消选定对象的重影】命令的图标为 ▇,Unghost Options【取消重影选项】面板,如图 4.26 所示。

图 4.25　Ghost Options【重影选项】面板

图 4.26　Unghost Options【取消重影选项】面板

5. Unghost All【全部取消重影】命令

Unghost All【全部取消重影】命令的图标为 ▇,该命令没有选项面板。

──────────Snapshot【快照】──────────

6. Create Animation Snapshot【创建动画快照】命令

Create Animation Snapshot【创建动画快照】命令的图标为 ▇,Animation Snapshot Options【动画快照选项】面板,如图 4.27 所示。

7. Update Snapshot【更新快照】命令

Update Snapshot【更新快照】命令的图标为 ▇,该命令没有选项面板。

8. Create Animated Sweep【创建动画扫描】命令

Create Animated Sweep【创建动画扫描】命令的图标为 ▇,Animation Sweep Options【动画扫描选项】面板,如图 4.28 所示。

图 4.27　Animation Snapshot Options
【动画快照选项】面板

图 4.28　Animation Sweep Options
【动画扫描选项】面板

4.4 【MASH】菜单组

MASH 为 Multi Stage Noise Shaping 的缩写，指多级噪声整形技术。

1. Create MASH Network【创建 MASH 网络】命令

Create MASH Network【创建 MASH 网络】命令的 MASH Options【MASH 选项】面板，如图 4.29 所示。

2. MASH Editor【[MASH 编辑器】命令

单击 MASH Editor【[MASH 编辑器】命令，弹出 MASH Editor【MASH 编辑器】面板，如图 4.30 所示。

3. Utilities【工具】命令组

在 Utilities【工具】命令组中，主要包括 12 个命令，如图 4.31 所示。

图 4.29　MASH Options
【MASH 选项】面板

图 4.30　MASH Editor
【MASH 编辑器】面板

图 4.31　Utilities
【工具】命令组面板

第 5 章　FX 模块菜单组

FX 模块菜单组主要包括 nParticle【n 粒子】、Fluids【流体】、nCloth【n 布料】、nHair【n 头发】、nConstraint【n 约束】、nCache【n 缓存】、Fields/Solvers【字段/结算器】、Effects【效果】、【Bifrost】、Boss【凸柱】和【MASH】11 个菜单组。【动画】模块菜单组中都包含了【MASH】菜单组，在此就不再赘述【MASH】菜单组，【FX】模块菜单组的具体命令如下。

5.1　nParticle【n 粒子】菜单组

———Create【创建】———

1. Fill Object【填充对象】命令

Fill Object【填充对象】命令的图标为 ，Fill Object Options【填充对象选项】面板，如图 5.1 所示。

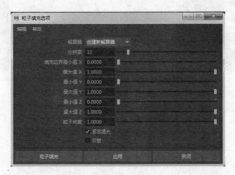

图 5.1　Fill Object Options【填充对象选项】面板

2. Get nParticle Example…【获取 nParticle 示例…】命令

Get nParticle Example…【获取 nParticle 示例…】命令的图标为 ，单击该命令弹出 Content Browser【内容浏览器】对话框，如图 5.2 所示。在该对话框中双击需要的效果即可加载到场景中。

3. Goal【目标】命令

Goal【目标】命令的图标为 ，Goal Options【目标选项】面板，如图 5.3 所示。

第 5 章 FX 模块菜单组

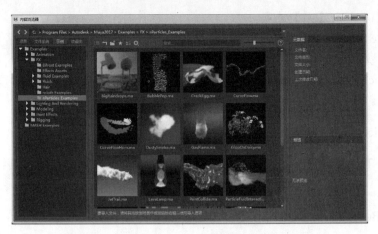

图 5.2 Content Browser【内容浏览器】对话框

4. Instancer【实例化器】命令

Instancer【实例化器】命令的图标为 ，Particle Instancer Options【实例化器选项】面板，如图 5.4 所示。

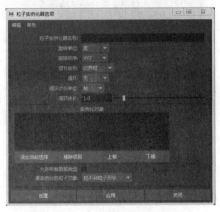

图 5.3 Goal Options
【目标选项】面板

图 5.4 Particle Instancer Options
【实例化器选项】面板

（1）Rotation Units【旋转单位】参数组：在该参数组中主要提供了 Degrees【度】和 Radians【弧度】2 个选项，默认 Degrees【度】选项。

（2）Rotation Order【旋转顺序】参数组：在该参数组中主要提供了【xyz】、【xzy】、【yxz】、【yzx】、【zxy】和【zyx】6 个选项，默认【xyz】选项。

（3）Level of Detail【细节级别】参数组：在该参数组中主要提供了 Geometry【几何体】、Bounding Box【边界框】和 Bounding Boxes【边界框】3 个选项，默认 Bounding Box【边界框】选项。

（4）Cycle【循环】参数组：在该参数组中主要提供了 None【无】和 Sequential【顺序】

2个选项，默认 None【无】选项。

（5）Cycle Step Units【循环步长单位】参数组：在该参数组中主要提供了 Frames【帧】和 Seconds【秒】2个选项，默认 None【无】选项。

（6）Particle Object to Instance【要实例化的粒子对象】参数组：在该参数组中只有 No Particle Shapes Found【找不到粒子形状】参数。

5. nParticle Tool【nParticle 工具】命令

nParticle Tool【nParticle 工具】命令的图标为 ，Tool Settings【工具设置】面板如图 5.5 所示。

6. Soft Body【柔体】命令

Soft Body【柔体】命令的图标为 ，Soft Options【软性选项】面板，如图 5.6 所示。

图 5.5　Tool Settings【工具设置】面板　　　　图 5.6　Soft Options【软性选项】面板

Creation Options【创建选项】参数组：在该参数组中主要包括 Make Soft【生成柔体】、Duplicate，Make Copy Soft【复制，将副本生成柔体】和 Duplicate，Make Original Soft【复制，将原始生成柔体】3个选项，默认 Make Soft【生成柔体】选项。

7. Paint Soft Body Weights Tool【绘制柔体权重工具】命令

Paint Soft Body Weights Tool【绘制柔体权重工具】命令的图标为 ，Tool Settings【工具设置】面板，如图 5.7 所示。

（1）Vector Index【向量索引】参数组：在该参数组中主要包括【x/R】、【Y/G】和【z/B】3个选项，默认【x/R】选项。

（2）Import value【导入值】参数组：在该参数组中主要包括 Luminance【亮度】、Alpha【Alpha】、Red【红】、Green【绿】和 Blue【蓝】5个选项，默认 Luminance【亮度】选项。

图 5.7 Tool Settings【工具设置】面板

（3）Image format【导出值】参数组：在该参数组中主要包括 Luminance【亮度】、【Alpha】、【RGB】和【RGBA】4 个选项，默认【亮度】选项。

（4）Image format【图像格式】参数组：在该参数组中主要包括【GIF】、【Soft Image】、【RLA】、【TIFF】、【SGI】、Alias【锯齿】、【IFF】、【JPEG】和【EPS】9 个选项，默认【IFF】选项。

8. Springs【弹簧】命令

Springs【弹簧】命令的图标为 ，Spring Options【弹簧选项】面板，如图 5.8 所示。

Creation method【创建方法】参数组：在该参数组中主要包括 Min Max【最小值/最大值】、All【全部】和 Wire frame【线框】3 个选项，默认 Min Max【最小值/最大值】选项。

9. Create Options【创建选项】命令组

在 Create Options【创建选项】命令组中，主要包括 5 个命令，如图 5.9 所示。

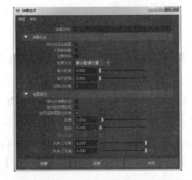

图 5.8 Spring Options【弹簧选项】面板　　图 5.9 Create Options【创建选项】命令组面板

———————————————— Emit【发射】————————————————

10. Create Emitter【创建发射器】命令

Create Emitter【创建发射器】命令的图标为▨，Emitter Options（Create）【发射器选项（创建）】面板，如图5.10所示。

图5.10　Emitter Options（Create）【发射器选项（创建）】面板

（1）Emitter Type【发射器类型】参数组：在该参数组中主要包括Omni【泛向】、Directional【方向】和Volume【体积】3个选项，默认Omni【泛向】选项。

（2）Cycle Emission【循环发射】参数组：在该参数组中主要包括None（Time Random off）【无（禁用Time Random）】和Frame（Time Random on）【帧（启用Time Random）】2个选项，默认None（Time Random off）【无（禁用Time Random）】选项。

（3）Volume Shape【体积形状】参数组：在该参数组中主要包括Cube【立方体】、Sphere【球体】、Cylinder【圆柱体】、Cone【圆锥体】和Torus【圆环】5个选项，默认Cube【立方体】选项。

11. Emit from Object【从对象发射】命令

Emit from Object【从对象发射】命令的图标为▨，Emitter Options（Emit form Object）【发射器选项（从对象发射）】面板，如图5.11所示。

（1）Emitter type【发射器类型】参数组：在该参数组中主要包括Omni【泛向】、Directional【方向】、Surface【表面】和Curve【曲线】4个选项，默认Omni【泛向】选项。

（2）Cycle emission【循环发射】参数组：在该参数组中主要包括None（Time Random off）【无（禁用Time Random）】和Frame（Time Random on）【帧（启用Time Random）】2个选项，默认None（Time Random off）【无（禁用Time Random）】选项。

第 5 章 FX 模块菜单组

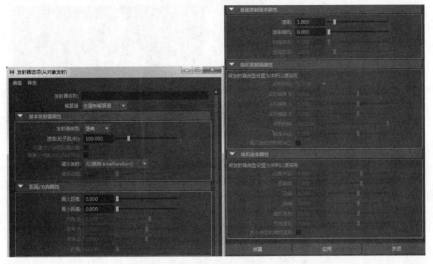

图 5.11　Emitter Options（Emit form Object）【发射器选项（从对象发射）】面板

（3）Volume Shape【体积形状】参数组：在该参数组中主要包括 Cube【立方体】、Sphere【球体】、Cylinder【圆柱体】、Cone【圆锥体】和 Torus【圆环】5 个选项，默认 Cube【立方体】选项。

12．Per-point Emission Rates【逐点发射速率】命令

Per-point Emission Rates【逐点发射速率】命令的图标为，该命令没有选项面板。

13．Use Selected Emitter【使用选定发射器】命令

Use Selected Emitter【使用选定发射器】命令的图标为，该命令没有选项面板。

————————————————Editors【编辑器】————————————————

14．Particle Collision Event Editor【粒子碰撞事件编辑器】命令

Particle Collision Event Editor【粒子碰撞事件编辑器】命令图标为，该命令没有选项面板。

15．Sprite Wizard…【精灵导向…】命令

————————————————Legacy Particles【旧版粒子】————————————————

16．Create Emitter【创建发射器】命令

Create Emitter【创建发射器】命令的图标为，Emitter Options（Create）【发射器选项（创建）】面板，如图 5.12 所示。

209

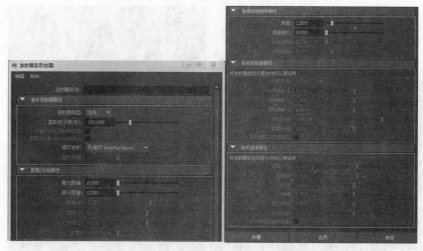

图 5.12 Create Emitter【发射器选项（创建）】面板

（1）Emitter type【发射器类型】参数组：在该参数组中主要包括 Omni【泛向】、Directional【方向】、Surface【表面】和 Curve【曲线】4 个选项，默认 Omni【泛向】选项。

（2）Cycle Emission【循环发射】参数组：在该参数组中主要包括 None（Time Random off）【无（禁用 Time Random）】和 Frame（Time Random on）【帧（启用 Time Random）】2 个选项，默认 None（Time Random off）【无（禁用 Time Random）】选项。

（3）Volume Shape【体积形状】参数组：在该参数组中主要包括 Cube【立方体】、Sphere【球体】、Cylinder【圆柱体】、Cone【圆锥体】和 Torus【圆环】5 个选项，默认 Cube【立方体】选项。

17. Create Springs【创建弹簧】命令

Create Springs【创建弹簧】命令的图标为 ，Spring Options【弹簧选项】面板，如图 5.13 所示。

图 5.13 Spring Options【弹簧选项】面板

Creation method【创建方法】参数组：在该参数组中主要包括 Min Max【最小值/最大值】、All【全部】和 Wire frame【线框】3 个选项，默认 Min Max【最小值/最大值】选项。

18. Emit from Object【从对象发射】命令

Emit from Object【从对象发射】命令的图标为 ，Emit Options（Emit from Object）【发射器选项（从对象发射）】面板，如图 5.14 所示。

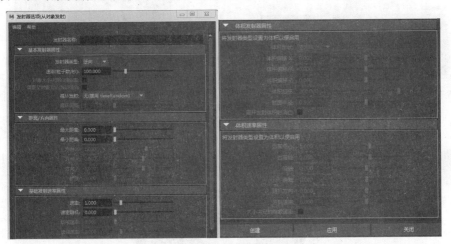

图 5.14 Emit Options（Emit from Object）【发射器选项（从对象发射）】面板

（1）Emitter type【发射器类型】参数组：在该参数组中主要包括 Omni【泛向】、Directional【方向】、Surface【表面】和 Curve【曲线】4 个选项，默认 Omni【泛向】选项。

（2）Cycle Emission【循环发射】参数组：在该参数组中主要包括 None（Time Random off）【无（禁用 Time Random）】和 Frame（Time Random on）【帧（启用 Time Random）】2 个选项，默认 None（Time Random off）【无（禁用 Time Random）】选项。

（3）Volume Shape【体积形状】参数组：在该参数组中主要包括 Cube【立方体】、Sphere【球体】、Cylinder【圆柱体】、Cone【圆锥体】和 Torus【圆环】5 个选项，默认 Cube【立方体】选项。

19. Make Collide【使碰撞】命令

Make Collide【使碰撞】命令的图标为 ，Collision Options【碰撞选项】面板，如图 5.15 所示。

20. Particle Tool【粒子工具】命令

Particle Tool【粒子工具】命令的图标为 ，Tool Settings【工具设置】面板，如图 5.16 所示。

图 5.15　Collision Options【碰撞选项】面板　　　图 5.16　Tool Settings【工具设置】面板

21. Soft Body【柔体】命令

Soft Body【柔体】命令的图标为 ，Soft Options【软性选项】面板，如图 5.17 所示。

图 5.17　Soft Options【软性选项】面板

Creation Options【创建选项】参数组：在该参数组中主要包括 Make Soft【生成柔体】、Duplicate，Make Copy Soft【复制，将副本生成柔体】和 Duplicate，Make Original Soft【复制，将原始生成柔体】3 个选项，默认 Make Soft【生成柔体】选项。

5.2　Fluids【流体】菜单组

———————————————Create【创建】———————————————

1. 3D Container【3D 容器】命令

3D Container【3D 容器】命令的图标为 ，Create 3D Container with Emitter Options【创建具有发射器的 3D 容器选项】面板，如图 5.18 所示。

（1）Emitter Type【发射器类型】参数组：在该参数组中主要包括 Omni【泛向】和 Volume【体积】2 个选项，默认 Omni【泛向】选项。

（2）Cycle Emission【循环发射】参数组：在该参数组中主要包括 None（Time Random off）【无（禁用 Time Random）】和 Frame（Time Random on）【帧（启用 Time Random）】2 个选项，默认 None（Time Random off）【无（禁用 Time Random）】选项。

第 5 章 FX 模块菜单组

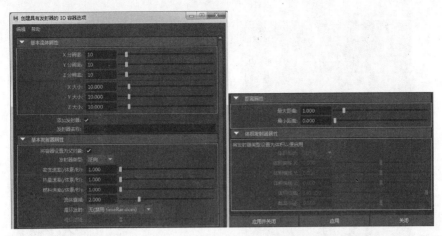

图 5.18 Create 3D Container with Emitter Options【创建具有发射器的 3D 容器选项】面板

（3）Volume shape【体积形状】参数组：在该参数组中主要包括 Cube【立方体】、Sphere【球体】、Cylinder【圆柱体】、Cone【圆锥体】和 Torus【圆环】5 个选项，默认 Cube【立方体】选项。

2. 2D Container【2D 容器】命令

2D Container【2D 容器】命令的图标为 ，Create 2D Container with Emitter Options【创建具有发射器的 2D 容器选项】面板，如图 5.19 所示。

（1）Emitter type【发射器类型】参数组：在该参数组中主要包括 Omni【泛向】和 Volume【体积】2 个选项，默认 Omni【泛向】选项。

（2）Cycle emission【循环发射】参数组：在该参数组中主要包括 None（Time Random off）【无（禁用 Time Random）】和 Frame（Time Random on）【帧（启用 Time Random）】2 个选项，默认 None（Time Random off）【无（禁用 Time Random）】选项。

（3）Volume Shape【体积形状】参数组：在该参数组中主要包括 Cube【立方体】、Sphere【球体】、Cylinder【圆柱体】、Cone【圆锥体】和 Torus【圆环】5 个选项，默认 Cube【立方体】选项。

3. Add/Edit Contents【添加/编辑内容】命令组

在 Add/Edit Contents【添加/编辑内容】命令组中，主要包括 6 个命令，如图 5.20 所示。

1）Emitter【发射器】命令

Emitter【发射器】命令的图标为 ，Emitter Options【发射器选项】面板，如图 5.21 所示。

（1）Emitter Type【发射器类型】参数组：在该参数组中主要包括 Omni【泛向】和 Volume【体积】2 个选项，默认 Omni【泛向】选项。

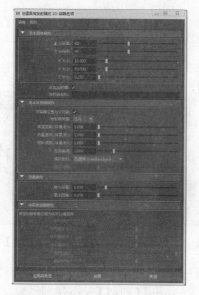

图 5.19 Create 2D Container with Emitter Options
【创建具有发射器的 2D 容器选项】面板

图 5.20 Add/Edit Contents
【添加/编辑内容】命令组面板

（2）Cycle Emission【循环发射】参数组：在该参数组中主要包括 None（Time Random off）【无（禁用 Time Random）】和 Frame（Time Random on）【帧（启用 Time Random）】2 个选项，默认 None（Time Random off）【无（禁用 Time Random）】选项。

（3）Volume Shape【体积形状】参数组：在该参数组中主要包括 Cube【立方体】、Sphere【球体】、Cylinder【圆柱体】、Cone【圆锥体】和 Torus【圆环】5 个选项，默认 Cube【立方体】选项。

2）Emitter from Object【从对象发射】命令

Emitter from Object【从对象发射】命令的图标为 ，Emit from Object Options【从对象发射选项】面板，如图 5.22 所示。

图 5.21 Emitter Options
【发射器选项】面板

图 5.22 Emit from Object Options
【从对象发射选项】面板

（1）Emitter Type【发射器类型】参数组：在该参数组中主要包括 Omni【泛向】、Surface【表面】和 Curve【曲线】3 个选项，默认 Surface【表面】选项。

（2）Cycle Emission【循环发射】参数组：None（Time Random off）【无（禁用 Time Random）】和 Frame（Time Random on）【帧（启用 Time Random）】2 个选项，默认 None（Time Random off）【无（禁用 Time Random）】选项。

3）Gradients【渐变】命令

Gradients【渐变】命令的 Fluid Gradients Options【流体渐变选项】面板，如图 5.23 所示。

图 5.23　Fluid Gradients Options【流体渐变选项】面板

（1）Density【密度】参数组：在该参数组中主要包括 Constant【恒定】、X gradient【X 渐变】、Y gradient【Y 渐变】、Z gradient【Z 渐变】、-X gradient【-X 渐变】、-Y gradient【-Y 渐变】、-Z gradient【-Z 渐变】和 Center gradient【中心渐变】8 个选项，默认 Y gradient【Y 渐变】选项。

（2）Velocity【速度】参数组：在该参数组中主要包括 Constant【恒定】、X gradient【X 渐变】、Y gradient【Y 渐变】、Z gradient【Z 渐变】、-X gradient【-X 渐变】、-Y gradient【-Y 渐变】、-Z gradient【-Z 渐变】和 Center gradient【中心渐变】8 个选项，默认 Constant【恒定】选项。

（3）Temperature【温度】参数组：在该参数组中主要包括 Constant【恒定】、X gradient【X 渐变】、Y gradient【Y 渐变】、Z gradient【Z 渐变】、-X gradient【-X 渐变】、-Y gradient【-Y 渐变】、-Z gradient【-Z 渐变】和 Center gradient【中心渐变】8 个选项，默认 Constant【恒定】选项。

（4）Fuel【燃料】参数组：在该参数组中主要包括 Constant【恒定】、X gradient【X 渐变】、Y gradient【Y 渐变】、Z gradient【Z 渐变】、-X gradient【-X 渐变】、-Y gradient【-Y 渐变】、-Z gradient【-Z 渐变】和 Center gradient【中心渐变】8 个选项，默认 Constant【恒定】选项。

4）Paint Fluids Tool【绘制流体工具】命令

Paint Fluids Tool【绘制流体工具】命令的图标为，Tool Settings【工具设置】面板，如图 5.24 所示。

（1）Paint Able Attributes【可绘制属性】参数组：在该参数组中主要包括 Density【密度】、Density and Color【密度和颜色】、Density and Fuel【密度和燃料】、Velocity【速度】、Temperature【温度】、Fuel【燃料】、Color【颜色】和 Falloff【衰减】8 个选项，默认 Density【密度】选项。

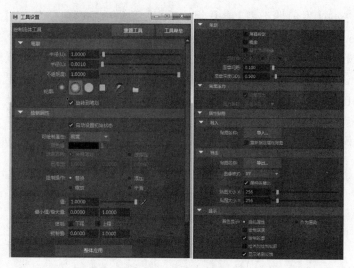

图 5.24　Tool Settings【工具设置】面板

（2）Image format【图像格式】参数组：在该参数组中主要包括【GIF】、【Soft Image】、【RLA】、【TIFF】、【SGI】、Alias【锯齿】、【IFF】、【JPEG】和【EPS】9 个选项，默认【IFF】选项。

5）With Curve【连同曲线】命令

With Curve【连同曲线】命令的 Set Fluid Contents with Curve Options【使用曲线设置流体内容选项】面板，如图 5.25 所示。

6）Initial States…【初始状态…】命令

Initial States…【初始状态…】命令的图标为▥，Initial States Options【初始状态选项】面板，如图 5.26 所示。

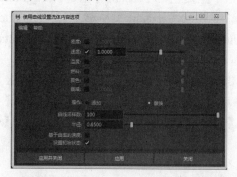

图 5.25　Set Fluid Contents with Curve Options
　　　　　【使用曲线设置流体内容选项】面板

图 5.26　Initial States Options
　　　　　【初始状态选项】面板

4. Get Example…【获取示例…】命令组

在 Get Example…【获取示例…】命令组中，主要包括 2 个命令，如图 7.27 所示。

5. Ocean【海洋】命令

Ocean【海洋】命令的图标为 ，Create Ocean【创建海洋】面板，如图 7.28 所示。

图 5.27　Get Example…【获取示例…】命令组面板

图 5.28　Create Ocean【创建海洋】面板

6. Pond【池塘】命令

Pond【池塘】命令的 Create Pond【创建池塘】面板，如图 5.29 所示。

------------------------Edit【编辑】------------------------

7. Extend Fluid【扩展流体】命令

Extend Fluid【扩展流体】命令的图标为 ■，Extend Fluid Options【扩展流体选项】面板，如图 5.30 所示。

图 5.29　Create Pond
　　　　【创建池塘】面板

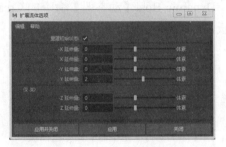

图 5.30　Extend Fluid Options
　　　　【扩展流体选项】面板

8. Edit Fluid Resolution【编辑流体分辨率】命令

Edit Fluid Resolution【编辑流体分辨率】命令的图标为 ■，Edit Fluid Resolution Options【编辑流体分辨率选项】面板，如图 5.31 所示。

9. Make Collide【使碰撞】命令

Make Collide【使碰撞】命令的图标为 ■，Make Collide Options【使碰撞选项】面板，如图 5.32 所示。

10. Make Motion Field【生成运动场】命令

Make Motion Field【生成运动场】命令没有选项面板。

图 5.31　Edit Fluid Resolution Options
【编辑流体分辨率选项】面板

图 5.32　Make Collide Options
【使碰撞选项】面板

——————————Ocean/Pond【海洋/池塘】——————————

11．Add Dynamic Locator【添加动力学定位器】命令组

在 Add Dynamic Locator【添加动力学定位器】命令组中，主要包括了 4 个命令，如图 5.33 所示。

1）Surface【曲面】命令

Surface【曲面】命令的图标为 ，该命令没有选项面板。

2）Dynamic Boat【动态船】命令

Dynamic Boat【动态船】命令的 Create Dynamic Boat Locator【创建动力学船定位器】面板，如图 5.34 所示。

图 5.33　Add Dynamic Locator
【添加动力学定位器】命令组面板

图 5.34　Create Dynamic Boat Locator
【创建动力学船定位器】面板

3）Dynamic Simple【动态简单】命令

Dynamic Simple【动态简单】命令的 Add Dynamic Simple Locator【添加动力学简单定位器】面板，如图 5.35 所示。

4）Dynamic Surface【动态曲面】命令

Dynamic Surface【动态曲面】命令的 Create Dynamic Surface Locator【创建动力学表面定位器】面板，如图 5.36 所示。

图 5.35　Add Dynamic Simple Locator
【添加动力学简单定位器】面板

图 5.36　Create Dynamic Surface Locator
【创建动力学表面定位器】面板

12．Add Preview Plane【添加预览平面】命令

Add Preview Plane【添加预览平面】命令的图标为 ，该命令没有选项面板。

第 5 章 FX 模块菜单组

13. Create Boat【创建船】命令组

在 Create Boat【创建船】命令组中，主要包括了 3 个命令，如图 5.37 所示。

1）Float Selected Object（s）【漂浮选定对象】命令

Float Selected Object（s）【漂浮选定对象】命令的 Float Selected Objects【漂浮选定对象】面板

图 5.37　Create Boat【创建船】命令组面板　　图 5.38　Float Selected Objects【漂浮选定对象】面板

2）Make Boat【生成船】命令

Make Boat【生成船】命令的 Make Boats【生成船】面板，如图 5.39 所示。

3）Make Motor Boat【生成摩托艇】命令

Make Motor Boat【生成摩托艇】命令的图标为 ，Make Motor Boats【生成摩托艇】面板，如图 5.40 所示。

图 5.39　Make Boats【生成船】面板　　图 5.40　Make Motor Boats【生成摩托艇】面板

14. Create Wake【创建尾迹】命令

Create Wake【创建尾迹】命令的图标为 ，Create Wake【创建尾迹】面板，如图 5.40 所示。

5.3　nCloth【n 布料】菜单组

──────── Create【创建】 ────────

1. Create Passive Collider【创建被动碰撞对象】命令

Create Passive Collider【创建被动碰撞对象】命令的图标为 ，Make Collide Options【使撞选项】面板，如图 5.41 所示。

图 5.40　Create Wake【创建尾迹】面板　　图 5.41　Make Collide Options【使碰撞选项】面板

2. Get nCloth【创建 nCloth】命令

Get nCloth【创建 nCloth】命令的图标为■，Create nCloth Options【创建 nCloth 选项】面板，如图 5.42 所示。

3. Get nCloth Example…【获取 nCloth 示例…】命令

单击 Get nCloth Example…【获取 nCloth 示例…】命令，弹出 Content Browser【内容浏览器】对话框，如图 5.423 所示。在该对话框中双击需要的效果即可加载到场景中。

图 5.42 Create nCloth Options
【创建 nCloth 选项】面板

图 5.43 Content Browser
【内容浏览器】对话框

---------------------------------Edit【编辑】---------------------------------

4. Delete History【删除历史】命令

Delete History【删除历史】命令的 Delete nCloth History Options【删除 nCloth 历史选项】面板，如图 5.44 所示。

5. Display Current Mesh【显示当前网络】命令

Display Current Mesh【显示当前网络】命令的图标为■，该命令没有选项面板。

6. Display Input Mesh【显示输入网络】命令

Display Input Mesh【显示输入网络】命令的图标为■，该命令没有选项面板。

7. Remove nCloth【移除 nCloth】命令

Remove nCloth【移除 nCloth】命令的图标为■，该命令没有选项面板。

---------------------------------Properties【特性】---------------------------------

8. Convert nCloth【转化 nCloth 输出空间】命令组

在 Convert nCloth【转化 nCloth 输出空间】命令组中，主要包括 2 个命令，如图 5.45 所示。

图 5.44　Delete nCloth History Options　　　　图 5.45　Convert nCloth
【删除 nCloth 历史选项】面板　　　　　　　【转化 nCloth 输出空间】命令组面板

9. Rest Shape【静止形状】命令组

在 Rest Shape【静止形状】命令组中，主要包括 3 个命令，如图 5.46 所示。

———————————— Maps【贴图】————————————

10. Paint Texture Properties【绘制纹理特性】命令组

在 Paint Texture Properties【绘制纹理特性】命令组中，主要包括 21 个命令，Paint Texture Properties【绘制纹理特性】命令组，如图 5.47 所示。

11. Paint Vertex Properties【绘制顶点特性】命令组

在 Paint Vertex Properties【绘制顶点特性】命令组中，主要包括 21 个命令，Paint Vertex Properties【绘制顶点特性】命令组面板，如图 5.48 所示。

图 5.46　Rest Shape　　　图 5.47　Paint Texture Properties　　图 5.48　Paint Vertex Properties
【静止形状】命令组面板　　【绘制纹理特性】命令组面板　　　【绘制顶点特性】命令组面板

Paint Vertex Properties【绘制顶点特性】命令组中的所有命令的 Tool Settings【工具设置】面板基本相同，只有 nCloth Attributes【nCloth 属性】卷展栏中的 Paint Attributes【绘制属性】选项参数不同。选择不同的命令，则对应不同的 Paint Attributes【绘制属性】参数选项。

Paint Vertex Properties【绘制顶点特性】命令组命令的 Tool Settings【工具设置】面板，如图 5.49 所示。

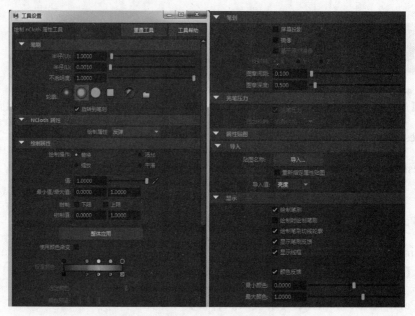

图 5.49 Tool Settings【工具设置】面板

（1）Paint Attributes【绘制属性】参数组：在该参数组中，主要包括 Collide Strength【碰撞强度】、Thickness【厚度】、Bounce【反弹】、Friction【摩擦力】、Stickiness【黏滞】、Field Magnitude【场幅值】、Mass【质量】、Stretch【拉伸】、Compression【压缩】、Bend【弯曲】、Bend Angle Drop off【弯曲角度衰减】、Restitution Angle【恢复角度】、Damp【阻尼】、Rigidity【刚性】、Deform【变形】、Input Attract【输入吸引】、Rest Length Scale【静止长度比例】、Wrinkle【褶皱】、Lift【升力】、Drag【阻力】和 Tangential Drag【切向阻力】21 个选项，默认选项，选择不同的命令有所不同。

（2）Import Value【导入值】参数组：在该参数组中，主要包括 Luminance【亮度】、Alpha【Alpha】、Red【红】、Green【绿】和 Blue【蓝】5 个选项，默认 Luminance【亮度】选项。

12. Convert Texture to Vertex Map【将纹理贴图转化为顶点贴图】命令组

在 Convert Texture to Vertex Map【将纹理贴图转化为顶点贴图】命令组中，主要包括 21 个命令，如图 5.50 所示。

13. Convert Vertex to Texture Map【将顶点贴图转换为纹理贴图】命令组

在 Convert Vertex to Texture Map【将顶点贴图转换为纹理贴图】命令组中，主要包括 21 个命令，如图 5.51 所示。

第5章 FX模块菜单组

5.4 nHair【n头发】菜单组

——Create【创建】——

1. Create Hair【创建头发】命令

Create Hair【创建头发】命令的图标为，Create Hair Options【创建头发选项】面板，如图 5.52 所示。

图 5.50 Convert Texture to Vertex Map【将纹理贴图转化为顶点贴图】面板

图 5.51 Convert Vertex to Texture Map【将顶点贴图转换为纹理贴图】面板

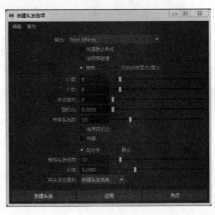

图 5.52 Create Hair Options【创建头发选项】面板

Output【输出】参数组：在该参数组中，主要包括 Paint Effects【Paint Effects】、NURBS curves【NURBS 曲线】和 Paint Effects and NURBS curves【Paint Effects 和 NURBS 曲线】3 个选项，默认 Paint Effects【Paint Effects】选项。

2. Get Hair Example…【获取头发示例…】命令

Get Hair Example…【获取头发示例…】命令的图标为，单击 Get Hair Example…【获取头发示例…】命令，弹出 Content Browser【内容浏览器】对话框，如图 5.53 所示。

3. Paint Hair Follicles【绘制毛囊】命令

Paint Hair Follicles【绘制毛囊】命令的图标为，Paint Hair Follicles【绘制毛囊设置】面板，如图 5.54 所示。

Output【输出】参数组：在该参数组中，主要包括 Paint Effects【Paint Effects】、NURBS curves【NURBS 曲线】和 Paint Effects and NURBS curves【Paint Effects 和 NURBS 曲线】3 个选项，默认 Paint Effects【Paint Effects】选项。

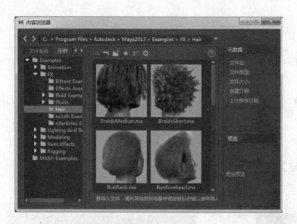

图 5.53 Content Browser
【内容浏览器】对话框

图 5.54 Paint Hair Follicles
【绘制毛囊设置】面板

4. Paint Hair Textures【绘制头发纹理】命令组

在 Paint Hair Textures【绘制头发纹理】命令组中，主要包括 3 个命令，如图 5.55 所示。

———————————— Edit【编辑】————————————

5. Assign Hair System【指定头发系统】命令组

在 Assign Hair System【指定头发系统】命令组中，只包括 New Hair System【新建头发系统】命令。

New Hair System【新建头发系统】命令。

New Hair System【新建头发系统】的图标为■，该命令没有选项面板。

6. Convert Selection【转化当前选择】命令组

在 Convert Selection【转化当前选择】命令组中，主要包括 9 个命令，如图 5.56 所示。

7. Display【显示】命令组

在 Display【显示】命令组中，主要包括 6 个命令，如图 5.57 所示。

图 5.55 Paint Hair Textures
【绘制头发纹理】命令组面板

图 5.56 Convert Selection
【转化当前选择】命令组面板

图 5.57 Display
【显示】命令组面板

8. Make Selected Curves Dynamic【动力学化选定曲线】命令

Make Selected Curves Dynamic【动力学化选定曲线】命令的图标为 ，Make Curves Dynamic Options【使曲线动力学化选项】面板，如图 5.58 所示。

Output【输出】参数组：在该参数组中，主要包括 Paint Effects【Paint Effects】、NURBS curves【NURBS 曲线】和 Paint Effects and NURBS curves【Paint Effects 和 NURBS 曲线】3 个选项，默认 NURBS curves【NURBS 曲线】选项。

9. Modify Curves【修改曲线】命令组

在 Modify Curves【修改曲线】命令组中，主要包括 7 个命令，如图 5.59 所示。

1）Lock Length【锁定长度】命令

Lock Length【锁定长度】命令的图标为 ，快捷键为 L，该命令没有选项面板。

2）Unlock Length【解除锁定长度】命令

Unlock Length【解除锁定长度】命令的图标为 ，快捷键为 L，该命令没有选项面板。

3）Straighten【拉直】命令

Straighten【拉直】命令的图标为 ，Straighten Curves Options【拉直曲线选项】面板，如图 5.60 所示。

图 5.58 Make Curves Dynamic Options【使曲线动力学化选项】面板

图 5.59 Modify Curves 【修改曲线】命令组面板

图 5.60 Straighten Curves Options 【拉直曲线选项】面板

4）Smooth【平滑】命令

Smooth【平滑】命令的图标为 ，Smooth Curves Options【平滑曲线选项】面板，如图 5.61 所示。

5）Curl【卷曲】命令

Curl【卷曲】命令的图标为 ，Curl Curves Options【卷曲曲线选项】面板，如图 5.62 所示。

图 5.61 Smooth Curves Options 【平滑曲线选项】面板

图 5.62 Curl Curves Options 【卷曲曲线选项】面板

6）Bend【弯曲】命令

Bend【弯曲】命令的图标为■，Bend Curves Options【弯曲曲线选项】面板，如图 5.63 所示。

7）Scale Curvature【缩放曲率】命令

Scale Curvature【缩放曲率】命令的图标为■，Scale Curvature Options【缩放曲率选项】面板，如图 5.64 所示。

图 5.63　Bend Curves Options
【弯曲曲线选项】面板

图 5.64　Scale Curvature Options
【缩放曲率选项】面板

10．Set Rest Position【设置静止位置】命令组

在 Set Rest Position【设置静止位置】命令组中，主要包括 2 个命令，如图 5.65 所示。

11．Set Start Position【设置开始位置】命令组

在 Set Start Position【设置开始位置】命令组中，主要包括 2 个命令，如图 5.66 所示。

图 5.65　Set Rest Position
【设置静止位置】命令组面板

图 5.66　Set Start Position
【设置开始位置】命令组面板

———————————— Tools【工具】————————————

12．Add Output Curves to Hair【将输出曲线添加到头发】命令

Add Output Curves to Hair【将输出曲线添加到头发】命令没有选项面板。

13．Assign Paint Effects Brush to Hair【将 Paint Effects 笔刷指定给头发】命令

Assign Paint Effects Brush to Hair【将 Paint Effects 笔刷指定给头发】命令的图标为■，该命令没有选项面板。

14．Add Paint Effects Outputs Output to Hair【将 Paint Effects 输出添加到头发】命令

Add Paint Effects Outputs Output to Hair【将 Paint Effects 输出添加到头发】命令，没有选项面板。

第 5 章 FX 模块菜单组

15. Randomize Follicles【随机化毛囊】命令

Randomize Follicles【随机化毛囊】命令的 Randomize Follicles Options【随机化毛囊选项】面板，如图 5.67 所示。

16. Scale Hair Tool【缩放头发工具】命令

Scale Hair Tool【缩放头发工具】命令的图标为■，该命令没有选项面板。

17. Transplant Hair【移植头发】命令

Transplant Hair【移植头发】命令的图标为■，Transplant Hair Options【移植头发选项】面板，如图 5.68 所示。

图 5.67　Randomize Follicles Options
【随机化毛囊选项】面板

图 5.68　Transplant Hair Options
【移植头发选项】面板

———————— Delete【删除】————————

18. Delete Hair【删除头发】命令

Delete Hair【删除头发】命令的图标为■，该命令没有选项面板。

19. Delete Entire Hair System【删除整个头发系统】命令

Delete Entire Hair System【删除整个头发系统】命令的图标为■，该命令没有选项面板。

5.5　nConstraint【n 约束】菜单组

———————— Create【创建】————————

1. Component【组件】命令

Component【组件】命令的图标为■，Create Component nConstraint Options【创建组件 nConstraint 选项】面板，如图 5.69 所示。

（1）Constraint Type【约束类型】参数组：在该参数组中，主要包括 Stretch【拉伸】和 Bend【弯曲】2 个选项，默认 Stretch【拉伸】选项。

（2）Component【组件】参数组：在该参数组中，主要包括 Edges【边】、Crosslinks【交叉链接】和 Edges and Cross links【边和交叉链接】3 个选项，默认 Edges【边】选项。

2. Component to Component【组件到组件】命令

Component to Component【组件到组件】命令的图标为▣,Create Component to Component Constraint Option Box【创建组件到组件约束选项框】面板,如图 5.70 所示。

图 5.69　Create Component nConstraint Options
　　　【创建组件 nConstraint 选项】面板

图 5.70　Create Component to Component Constraint
　　　Option Box【创建组件到组件约束选项框】面板

3. Force Field【力场】命令

Force Field【力场】命令的图标为▣,Create Force Field Constraint Option Box【创建力场约束选项框】面板,如图 5.71 所示。

4. Point to Surface【点到曲面】命令

Point to Surface【点到曲面】命令的图标为▣,Create Point to Surface Constraint Option Box【创建点到曲面约束选项框】面板,如图 5.72 所示。

图 5.71　Create Force Field Constraint Option Box
　　　【创建力场约束选项框】面板

图 5.72　Create Point to Surface Constraint
　　　Option Box【创建点到曲面约束选项框】面板

5. Slide on Surface【在曲面上滑动】命令

Slide on Surface【在曲面上滑动】命令的图标为▣,Create Slide on Surface Constraint Option Box【创建在曲面上滑动约束选项框】面板,如图 5.73 所示。

6. Tearable Surface【可撕裂曲面】命令

Tearable Surface【可撕裂曲面】命令的图标为▣,Create Tearable Surface Constraint Option Box【创建可撕裂曲面约束选项框】面板,如图 5.74 所示。

图 5.73　Create Slide on Surface Constraint Option
　　　Box【创建在曲面上滑动约束选项框】面板

图 5.74　Create Tearable Surface Constraint Option
　　　Box【创建可撕裂曲面约束选项框】面板

第 5 章　FX 模块菜单组

7. Transform Constraint【变换约束】命令

Transform Constraint【变换约束】命令的图标为■，Create Transform Constraint Option Box【创建变换约束选项框】面板，如图 5.75 所示。

————————————————————Edit【编辑】————————————————————

8. Attract to Matching Mesh【吸引到匹配网络】命令

Attract to Matching Mesh【吸引到匹配网络】命令的图标为■，Create Attract to Matching Mesh Constraint Option Box【创建吸引到匹配网络约束选项框】面板，如图 5.76 所示。

图 5.75　Create Transform Constraint Option Box【创建变换约束选项框】面板

图 5.76　Create Attract to Matching Mesh Constraint Option Box【创建吸引到匹配网络约束选项框】面板

9. Disable Collision【禁用碰撞】命令

Disable Collision【禁用碰撞】命令的图标为■，Create Disable Collision Constraint Option Box【创建禁用碰撞约束选项框】面板，如图 5.77 所示。

10. Exclude Collide Pairs【排除碰撞】命令

Exclude Collide Pairs【排除碰撞】命令的图标为■，Create Exclude Collision Pair Constraint Option Box【创建禁用碰撞约束选项框】面板，如图 5.78 所示。

图 5.77　Create Disable Collision Constraint Option Box【创建禁用碰撞约束选项框】面板

图 5.78　Create Exclude Collision Pair Constraint Option Box【创建禁用碰撞约束选项框】面板

11. Remove Dynamic Constraint【移除动态约束】命令

Remove Dynamic Constraint【移除动态约束】命令的图标为■，该命令没有选项面板。

12. Weld Adjacent Borders【焊接相邻边界】命令

Weld Adjacent Borders【焊接相邻边界】命令的图标为■，Create Exclude Collision Pair Constraint Option Box【创建焊接相邻边界约束选项框】面板，如图 5.79 所示。

———————————Membership【成员身份】———————————

13. nConstraint Membership Tool【nConstraint 成员身份工具】命令

nConstraint Membership Tool【nConstraint 成员身份工具】命令的图标为■，该命令没有选项面板。

14. Add Members【添加成员】命令

Add Members【添加成员】命令的图标为■，该命令没有选项面板。

15. Remove Members【移除成员】命令

Remove Members【移除成员】命令的图标为■，该命令没有选项面板。

16. Replace Members【替换成员】命令

Replace Members【替换成员】命令的图标为■，该命令没有选项面板。

17. Select Members【选择成员】命令

Select Members【选择成员】命令的图标为■，该命令没有选项面板。

———————————Maps【贴图】———————————

18. Paint Properties by Vertex Map【通过顶点贴图绘制特性】命令组

在 Paint Properties by Vertex Map【通过顶点贴图绘制特性】命令组中，主要包括 3 个命令，如图 5.80 所示。

图 5.79 Create Exclude Collision Pair Constraint Option Box【创建焊接相邻边界约束选项框】面板

图 5.80 Paint Properties by Vertex Map【通过顶点贴图绘制特性】命令组面板

19. Paint Properties by Texture Map【通过纹理贴图绘制特性】命令组

在 Paint Properties by Texture Map【通过纹理贴图绘制特性】命令组中，主要包括 3 个命令，如图 5.81 所示。

20. Convert Texture to Vertex Map【将纹理贴图转化为顶点贴图】命令组

在 Convert Texture to Vertex Map【将纹理贴图转化为顶点贴图】命令组中，主要包括 3 个命令，如图 5.82 所示。

第 5 章　FX 模块菜单组

图 5.81　Paint Properties by Texture Map　　　图 5.82　Convert Texture to Vertex Map
【通过纹理贴图绘制特性】命令组面板　　　　　【将纹理贴图转化为顶点贴图】命令组面板

21．Convert Vertex to Texture Map【将顶点贴图转化为纹理贴图】命令组

在 Convert Vertex to Texture Map【将顶点贴图转化为纹理贴图】命令组中，主要包括 3 个命令，如图 5.83 所示。

5.6　nCache【n 缓存】菜单组

———————————— Create【创建】————————————

1．Create New Cache【创建新缓存】命令组

在 Create New Cache【创建新缓存】命令组中，主要包括 2 个命令，如图 5.84 所示。

图 5.83　Convert Vertex to Texture Map　　　　图 5.84　Create New Cache
【将顶点贴图转化为纹理贴图】命令组面板　　　　【创建新缓存】命令组面板

1）nObject【n 对象】命令

nObject【n 对象】命令的图标为■，Create nCache Options【创建 nCache 选项】面板，如图 5.85 所示。

Cache format【缓存格式】参数组：在该参数组中，主要包括【Max】和【Min】2 个选项，默认【Max】选项。

2）Maya Fluid【Maya 流体】命令

Maya Fluid【Maya 流体】命令的图标为■，Create Fluid nCache Options【创建流体 nCache 选项】面板，如图 5.86 所示。

Cache format【缓存格式】参数组：在该参数组中，主要包括【Max】和【Min】2 个选项，默认【Max】选项。

2．Merge Caches【合并缓存】命令组

在 Merge Caches【合并缓存】命令组中，主要包括 2 个命令，如图 5.87 所示。

1）nObject【n 对象】命令

nObject【n 对象】命令的图标为■，Merge nCache Options【合并 nCache 选项】面板，如图 5.88 所示。

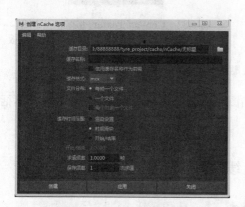

图 5.85　Create nCache Options
【创建 nCache 选项】面板

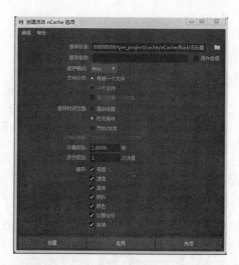

图 5.86　Create Fluid nCache Options
【创建流体 nCache 选项】面板

图 5.87　Merge Caches
【合并缓存】命令组面板

图 5.88　Merge nCache Options
【合并 nCache 选项】面板

Cache format【缓存格式】参数组：在该参数组中，主要包括【Max】和【Min】2 个选项，默认【Max】选项。

2）Maya Fluid【Maya 流体】命令

Maya Fluid【Maya 流体】命令的图标为，Merge Fluid nCache Options【合并流体 nCache 选项】面板，如图 5.89 所示。

Cache format【缓存格式】参数组：在该参数组中，主要包括【Max】和【Min】2 个选项，默认【Max】选项。

3. Replace Caches【替换缓存】命令组

在 Replace Caches【替换缓存】命令组中，主要包括 2 个命令，如图 5.90 所示。

第 5 章 FX 模块菜单组

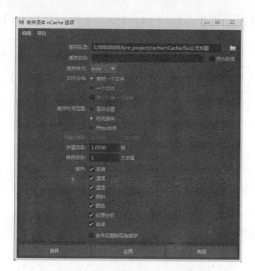

图 5.89 Merge Fluid nCache Options
【合并流体 nCache 选项】面板

图 5.90 Replace Caches
【替换缓存】命令组面板

1）nObject【n 对象】命令

nObject【n 对象】命令的图标为，Replace nCache Options【替换 nCache 选项】面板，如图 5.91 所示。

Cache format【缓存格式】参数组：在该参数组中，主要包括【Max】和【Min】2 个选项，默认【Max】选项。

2）Maya Fluid【Maya 流体】命令

Maya Fluid【Maya 流体】命令的图标为，Replace Fluid nCache Options【替换流体 nCache 缓存】，如图 5.92 所示。

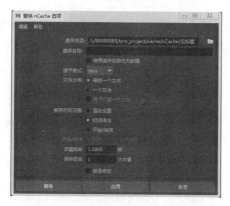

图 5.91 Replace nCache Options
【替换 nCache 选项】面板

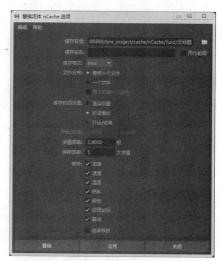

图 5.92 Replace Fluid nCache Options
【替换流体 nCache 缓存】面板

Cache format【缓存格式】参数组：在该参数组中，主要包括 Max【Max】和 Min【Min】2 个选项，默认 Max【Max】选项。

4. Enable All Caches on Selected【启用选定对象上的所有缓存】命令

Enable All Caches on Selected【启用选定对象上的所有缓存】命令的图标为▇，该命令没有选项面板。

5. Disable All Caches on Selected【禁用选定对象上的所有缓存】命令

Disable All Caches on Selected【禁用选定对象上的所有缓存】命令的图标为▇，该命令没有选项面板。

——————————————Edit【编辑】——————————————

6. Append to Cache【附加到缓存】命令

Append to Cache【附加到缓存】命令的图标为▇，该命令没有选项面板。

7. Attach Cache…【附加缓存…】命令

Attach Cache…【附加缓存…】命令的图标为▇，该命令没有选项面板。

8. Delete Cache【删除缓存】命令

Delete Cache【删除缓存】命令的图标为▇，Delete nCache Options【删除 nCache 选项】面板，如图 5.93 所示。

——————————————Frame【框显】——————————————

9. Replace Cache Frame【替换缓存】命令

Replace Cache Frame【替换缓存】命令的图标为▇，该命令没有选项面板。

10. Delete Cache Frame【删除缓存】命令

Delete Cache Frame【删除缓存】命令的图标为▇，Delete nCache Frames Options【删除 nCache 帧选项】面板，如图 5.94 所示。

图 5.93　Delete nCache Options【删除 nCache 选项】面板

图 5.94　Delete nCache Frames Options【删除 nCache 帧选项】面板

———————————— nCloth【n 布料】————————————

11. Paint Cache Weights Tool【绘制缓存权重工具】命令

Paint Cache Weights Tool【绘制缓存权重工具】命令的图标为 ，Tool Settings【工具设置】面板，如图 5.95 所示。

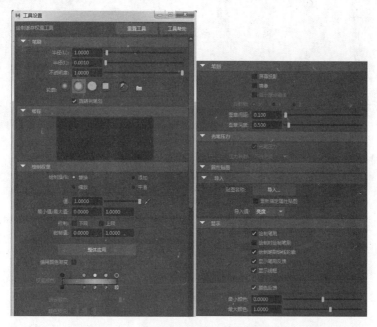

图 5.95　Tool Settings【工具设置】面板

Luminance【导入值】参数组：在该参数组中，主要包括 Luminance【亮度】、【Alpha】、Red【红】、Green【绿】和 Blue【蓝】5 个选项，默认 Luminance【亮度】选项。

12. Transfer Cache To Input Mesh【将缓存传递到输入网络】命令

Transfer Cache To Input Mesh【将缓存传递到输入网络】命令的图标为 ，该命令没有选项面板。

———————————— Legacy Cache【旧版缓存】————————————

13. Create Particle Disk Cache【创建粒子磁盘缓存】命令

Create Particle Disk Cache【创建粒子磁盘缓存】命令的图标为 ，Create Particle Disk Cache Options【创建粒子存盘缓存选项】面板，如图 5.96 所示。

14. Edit Over sampling or Cache Settings…【编辑过采样或缓存设置…】命令

单击 Edit Over sampling or Cache Settings…【编辑过采样或缓存设置…】命令，弹出 Attribute Editor【属性编辑器】面板，如图 5.97 所示。

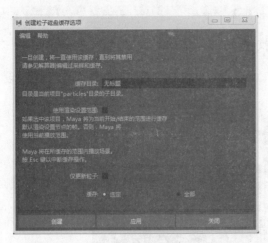

图 5.96　Create Particle Disk Cache Options
【创建粒子存盘缓存选项】面板

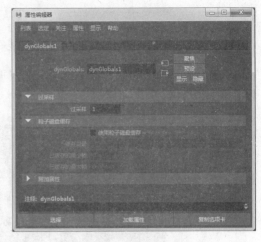

图 5.97　Attribute Editor
【属性编辑器】面板

15. Memory Caching【内存缓存】命令组

在 Memory Caching【内存缓存】命令组中，主要包括 3 个命令，如图 5.98 所示。

图 5.98　Memory Caching【内存缓存】命令组面板

5.7　Fields/Solvers【字段/结算器】菜单组

――――――――――――Create【创建】――――――――――――

1. Air【空气】命令

Air【空气】命令的图标为，Air Options【空气选项】命令组面板如图 5.99 所示。

Volume shape【体积形状】参数组：在该参数组中，主要包括 None【无】、Cube【立方体】、Sphere【球体】、Cylinder【圆柱体】、Cone【圆锥体】和 Torus【圆环】6 个选项，默认 None【无】选项。

2. Drag【阻力】命令

Drag【阻力】命令图标为，Drag Options【阻力选项】命令组面板如图 5.100 所示。

Volume shape【体积形状】参数组：在该参数组中，主要包括 None【无】、Cube【立方体】、Sphere【球体】、Cylinder【圆柱体】、Cone【圆锥体】和 Torus【圆环】6 个选项，默认 None【无】选项。

第 5 章　FX 模块菜单组

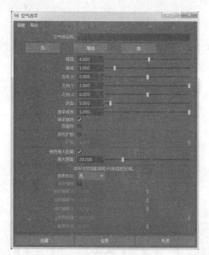

图 5.99　Air Options【空气选项】面板

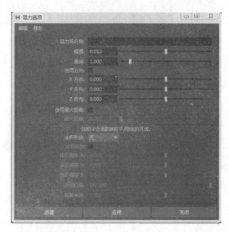

图 5.100　Drag Options【阻力选项】面板

3. Gravity【重力】命令

Gravity【重力】命令的图标为 ，Gravity Options【重力选项】面板如图 5.101 所示。

Volume shape【体积形状】参数组：在该参数组中，主要包括 None【无】、Cube【立方体】、Sphere【球体】、Cylinder【圆柱体】、Cone【圆锥体】和 Torus【圆环】6 个选项，默认 None【无】选项。

4. Newton【牛顿】命令

Newton【牛顿】命令的图标为 ，Newton Options【重力选项】面板，如图 5.102 所示。

图 5.101　Gravity Options【重力选项】面板

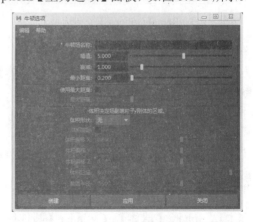

图 5.102　Newton Options【重力选项】面板

Volume shape【体积形状】参数组：在该参数组中，主要包括 None【无】、Cube【立方体】、Sphere【球体】、Cylinder【圆柱体】、Cone【圆锥体】和 Torus【圆环】6 个选项，默认 None【无】选项。

237

5. Radial【径向】命令

Radial【径向】命令的图标为▓，Radial Options【径向选项】面板，如图 5.103 所示。
Volume shape【体积形状】参数组：在该参数组中，主要包括 None【无】、Cube【立方体】、Sphere【球体】、Cylinder【圆柱体】、Cone【圆锥体】和 Torus【圆环】6 个选项，默认 None【无】选项。

6. Turbulence【湍流】命令

Turbulence【湍流】命令的图标为▓，Turbulence Options【湍流选项】面板，如图 5.104 所示。

图 5.103　Radial Options【径向选项】面板

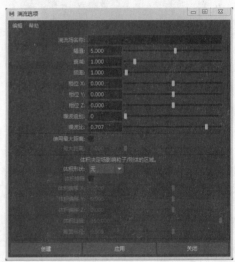

图 5.104　Turbulence Options【湍流选项】面板

Volume shape【体积形状】参数组：在该参数组中，主要包括 None【无】、Cube【立方体】、Sphere【球体】、Cylinder【圆柱体】、Cone【圆锥体】和 Torus【圆环】6 个选项，默认 None【无】选项。

7. Uniform【统一】命令

Uniform【统一】命令的图标为▓，Uniform Options【一致选项】面板，如图 5.105 所示。
Volume shape【体积形状】参数组：在该参数组中，主要包括 None【无】、Cube【立方体】、Sphere【球体】、Cylinder【圆柱体】、Cone【圆锥体】和 Torus【圆环】6 个选项，默认 None【无】选项。

8. Vortex【漩涡】命令

Vortex【漩涡】命令的图标为▓，Vortex Options【漩涡选项】面板，如图 5.106 所示。

第 5 章　FX 模块菜单组

图 5.105　Uniform Options【一致选项】面板

图 5.106　Vortex Options【漩涡选项】面板

Volume shape【体积形状】参数组：在该参数组中，主要包括 None【无】、Cube【立方体】、Sphere【球体】、Cylinder【圆柱体】、Cone【圆锥体】和 Torus【圆环】6 个选项，默认 None【无】选项。

————————————Volume【体积】————————————

9. Volume Axis【体积轴】命令

Volume Axis【体积轴】命令的图标为，Volume Axis Options【体积轴选项】面板，如图 5.107 所示。

Volume shape【体积形状】参数组：在该参数组中，主要包括 None【无】、Cube【立方体】、Sphere【球体】、Cylinder【圆柱体】、Cone【圆锥体】和 Torus【圆环】6 个选项，默认 Cube【立方体】选项。

10. Volume Curve【体积曲线】命令

Volume Curve【体积曲线】命令没有选项面板。

————————————Connect【连接】————————————

11. Use Selected as source【使用选定对象作为源】命令

Use Selected as source【使用选定对象作为源】命令的图标为，该命令没有选项面板。

12. Assign Selected【指定给选定对象】命令

Assign Selected【指定给选定对象】命令的图标为，该命令没有选项图标。

————————————Solvers【解算器】————————————

13. Connect to Time【连接到时间】命令

Connect to Time【连接到时间】命令没有选项面板。

239

14. Initial State【初始状态】命令组

在 Initial State【初始状态】命令组中，主要包括 8 个命令，如图 5.108 所示。

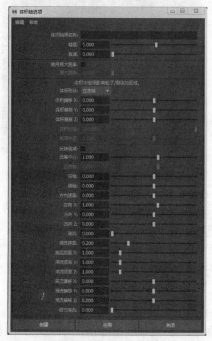

图 5.107 Volume Axis Options【体积轴选项】面板　　图 5.108 Initial State【初始状态】命令组面板

1）Set for Selected【为选定对象设定】命令

Set for Selected【为选定对象设定】命令没有选项面板。

2）Set for All Dynamic【为所有动力学对象设定】命令

Set for All Dynamic【为所有动力学对象设定】命令没有选项面板。

3）Clear Initial State【清除初始状态】命令

Clear Initial State【清除初始状态】命令没有选项面板。

──────────── Nucleus【网络】 ────────────

4）Set From Mesh【从网络设定】命令

Set From Mesh【从网络设定】命令没有选项面板。

5）Relax Initial State【松弛初始状态】命令

Relax Initial State【松弛初始状态】命令的 Relax Initial State Options【松弛初始状态选项】面板，如图 5.109 所示。

6）Resolve Interpenetration【解决穿透】命令

Resolve Interpenetration【解决穿透】命令的 Resolve Interpenetration Options【解决穿透选项】面板，如图 5.110 所示。

图 5.109　Relax Initial State Options
【松弛初始状态选项】面板

图 5.110　Resolve Interpenetration Options
【解决穿透选项】面板

———————— Fluids【流体】————————

7）Set Initial State【设置初始状态】命令

Set Initial State【设置初始状态】命令的图标为，Set Initial State Options【设置初始状态选项】面板，如图 5.111 所示。

8）Save State As Example…【将状态另存为示例…】命令

Save State As Example…【将状态另存为示例…】命令没有选项面板。

15．Interactive Playback【交互式播放】命令

Interactive Playback【交互式播放】命令的图标为，该命令没有选项面板。

———————— Nucleus【Nucleus】————————

16．Assign Solver【指定解算器】命令组

在 Assign Solver【指定解算器】命令组中，只包括一个命令，如图 5.112 所示。

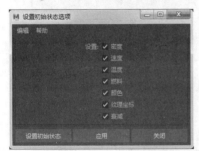

图 5.111　Set Initial State Options
【设置初始状态选项】面板

图 5.112　Assign Solver
【指定解算器】命令组面板

———————— Legacy Rigid Bodies【旧版刚体】————————

17．Create Active Rigid Body【创建主动刚体】命令

Create Active Rigid Body【创建主动刚体】命令的图标为，Rigid Options【刚性选项】命令组面板，如图 5.113 所示。

Stand in【替代对象】参数组：在该参数组中，主要包括 None【无】、Cube【立方体】、和 Sphere【球体】3 个选项，默认 None【无】选项。

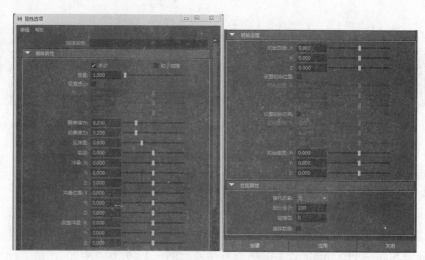

图 5.113 Rigid Options【刚性选项】面板

18. Create Passive Rigid Body【创建被动刚体】命令

Create Passive Rigid Body【创建被动刚体】命令的图标为 ，Rigid Options【刚性选项】面板，如图 5.114 所示。

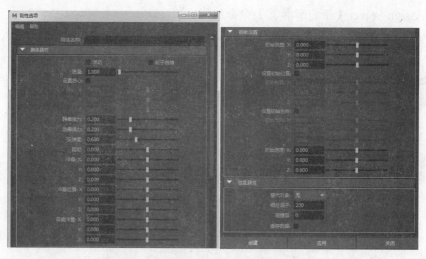

图 5.114 Rigid Options【刚性选项】面板

Stand in【替代对象】参数组：在该参数组中，主要包括 None【无】、Cube【立方体】、和 Sphere【球体】3 个选项，默认 None【无】选项。

19. Create Nail Constraint【创建钉子约束】命令

Create Nail Constraint【创建钉子约束】命令的图标为 ，Constraint Options【约束选项】面板，如图 5.115 所示。

Constraint Type【约束类型】参数组：在该参数组中，主要包括 Nail【钉子】、Pin【固定】、Hinge【铰链】、Spring【弹簧】和 Barrier【屏障】5 个选项，默认 Nail【钉子】选项。

20. Create Pin Constraint【创建固定约束】命令

Create Pin Constraint【创建固定约束】命令的图标为，Constraint Options【约束选项】面板，如图 5.116 所示。

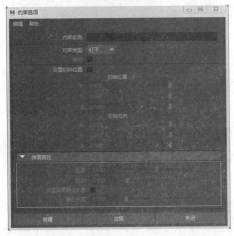

图 5.115 【创建钉子约束】命令中的
Constraint Options【约束选项】面板

图 5.116 【创建固定约束】命令中的
Constraint Options【约束选项】面板

Constraint Type【约束类型】参数组：在该参数组中，主要包括 Nail【钉子】、Pin【固定】、Hinge【铰链】、Spring【弹簧】和 Barrier【屏障】5 个选项，默认 Pin【固定】选项。

21. Create Hinge Constraint【创建铰链约束】命令

Create Hinge Constraint【创建铰链约束】命令的图标为，Constraint Options【约束选项】面板，如图 5.117 所示。

Constraint type【约束类型】参数组：在该参数组中，主要包括 Nail【钉子】、Pin【固定】、Hinge【铰链】、Spring【弹簧】和 Barrier【屏障】5 个选项，默认 Hinge【铰链】选项。

22. Create Spring Constraint【创建弹簧约束】命令

Create Spring Constraint【创建弹簧约束】命令的图标为，Constraint Options【约束选项】面板，如图 5.118 所示。

Constraint type【约束类型】参数组：在该参数组中，主要包括 Nail【钉子】、Pin【固定】、Hinge【铰链】、Spring【弹簧】和 Barrier【屏障】5 个选项，默认 Spring【弹簧】选项。

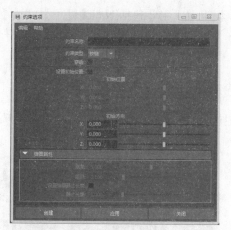
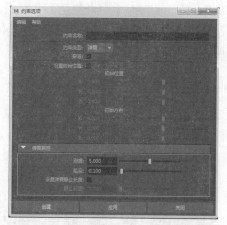

图 5.117 【创建铰链约束】命令中的 Constraint Options【约束选项】面板

图 5.118 【创建弹簧约束】命令中的 Constraint Options【约束选项】面板

23. Create Barrier Constraint【创建屏障约束】命令

Create Barrier Constraint【创建屏障约束】命令的图标为 ，Constraint Options【约束选项】面板，如图 5.119 所示。

Constraint type【约束类型】参数组：在该参数组中，主要包括 Nail【钉子】、Pin【固定】、Hinge【铰链】、Spring【弹簧】和 Barrier【屏障】5 个选项，默认 Barrier【屏障】选项。

24. Set Active Key【设定主动关键帧】命令

Set Active Key【设定主动关键帧】命令没有选项面板。

25. Set Passive Key【设定被动关键帧】命令

Set Passive Key【设定被动关键帧】命令没有选项面板。

26. Break Rigid Body Connections【断开刚体连接】命令

Break Rigid Body Connections【断开刚体连接】命令没有选项面板。

27. Set Rigid Body Interpenetration【设定刚体穿透】命令

Set Rigid Body Interpenetration【设定刚体穿透】命令没有选项面板。

28. Set Rigid Body Collision【设定刚体碰撞】命令

Set Rigid Body Collision【设定刚体碰撞】命令没有选项面板。

29. Rigid Body Solver Attributes【刚体解算器属性】命令

Rigid Body Solver Attributes【刚体解算器属性】命令没有选项面板。

30. Current Rigid Body Solver【当前刚体解算器】命令

Current Rigid Body Solver【当前刚体解算器】命令没有选项面板。

31. Create Rigid Body Solver【创建刚体解算器】命令

Create Rigid Body Solver【创建刚体解算器】命令没有选项面板。

5.8 Effects【效果】菜单组

————Assets【资源】————

1. Get Effects Asset…【获取效果资源…】命令

Get Effects Asset…【获取效果资源…】命令的图标为 ，单击该命令，弹出 Content Browser【内容浏览器】对话框，如图 5.120 所示。然后，双击需要的效果选项即可。

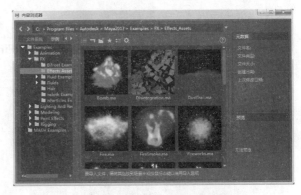

图 5.119　Constraint Options　　　　图 5.120　Content Browser
　　【约束选项】面板　　　　　　　　　【内容浏览器】对话框

2. Apply Effect【应用效果】命令

Apply Effect【应用效果】命令没有选项面板。

3. Collide With Effect【带效果碰撞】命令

Collide With Effect【带效果碰撞】命令没有选项面板。

————Create【创建】————

4. Fire【火】命令

Fire【火】命令的图标为 ，Create Fire Effect Options【创建火效果选项】面板，如图 5.121 所示。

图 5.121　Create Fire Effect Options【创建火效果选项】面板

Fire Emitter Type【火发射器类型】参数组：在该参数组中，主要包括 Omni-directional Point【泛向粒子】、Directional Point【定向粒子】、Surface【表面】和 Curve【曲线】4 个选项，默认 Surface【表面】选项。

5. Fireworks【火焰】命令

Fireworks【火焰】命令的图标为，Create Fireworks Effect Options【创建焰火效果选项】面板，如图 5.122 所示。

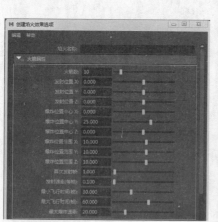
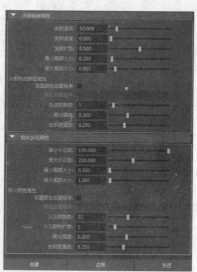

图 5.122　Create Fireworks Effect Options【创建焰火效果选项】面板

6. Flow【流】命令组

在 Flow【流】命令组中，主要包括 3 个命令，如图 5.123 所示。

第 5 章　FX 模块菜单组

图 5.123　Flow【流】命令组面板

1）Create Curve Flow【创建曲线流】命令

Create Curve Flow【创建曲线流】命令的图标为 ，Create Flow Effect Options【创建流效果选项】命令组面板，如图 5.124 所示。

2）Create Surface Flow【创建曲面流】命令

Create Surface Flow【创建曲面流】命令的图标为 ，Create Surface Flow Effect Options【创建曲面流效果选项】面板，如图 5.125 所示。

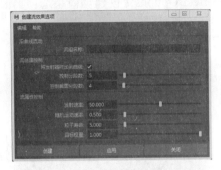

图 5.124　Create Flow Effect Options
【创建流效果选项】命令组面板

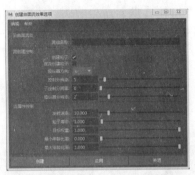

图 5.125　Create Surface Flow Effect Options
【创建曲面流效果选项】面板

3）Delete Selected Flow【删除选定流】命令

Delete Selected Flow【删除选定流】命令的图标为 ，Delete Surface Flow Effect Options【删除曲面流效果选项】面板，如图 5.126 所示。

7．Lightning【闪电】命令

Lightning【闪电】命令的图标为 ，Create Lightning Effect Options【创建闪电效果选项】面板，如图 5.127 所示。

图 5.126　Delete Surface Flow Effect Options
【删除曲面流效果选项】面板

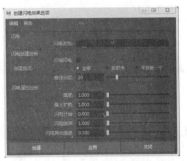

图 5.127　Create Lightning Effect Options
【创建闪电效果选项】面板

247

8. Shatter【破碎】命令

Shatter【破碎】命令的图标为 ![icon]，Create Shatter Effect Options【创建破碎效果选项】面板，如图 5.128 所示。

Post operation【后期操作】参数组：在该参数组中，主要包括 Shapes【形状】、Rigid Bodies with Collisions off【碰撞为禁用的刚体】、Soft Bodies with Goals【具有目标的柔体】、Soft Bodies with Lattice and Goals【具有晶格和目标的柔体】和 Sets【集】5 个选项，默认 Shapes【形状】选项。

9. Smoke【烟】命令

Smoke【烟】命令的图标为 ![icon]，Create Smoke Effect Options【创建烟效果选项】面板，如图 5.129 所示。

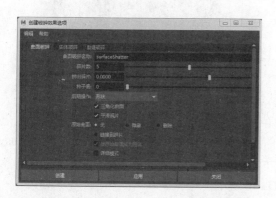

图 5.128　Create Shatter Effect Options 【创建破碎效果选项】面板

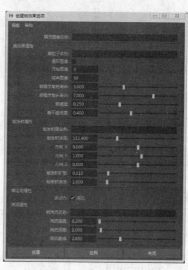

图 5.129　Create Smoke Effect Options 【创建烟效果选项】面板

5.9　Bifrost【Bifrost】菜单组

────────── Create【创建】──────────

1. Liguid【液体】命令

Liguid【液体】命令没有选项面板。

2. Aero【阿诺德】命令

Aero【阿诺德】命令没有选项面板。

3. Get Example...【获取示例…】命令

单击 Get Example...【获取示例…】命令，弹出 Content Browser【内容浏览器】面板，如图 5.130 所示，双击需要的示例，即可载入场景中。

———————————————————— Add【添加】————————————————————

4. Emitter【发射器】命令

Emitter【发射器】命令没有选项面板。

5. Collider【碰撞对象】命令

Collider【碰撞对象】命令没有选项面板。

6. Foam【泡沫】命令

Foam【泡沫】命令没有选项面板。

7. Foam Mask【泡沫遮罩】命令

Foam Mask【泡沫遮罩】命令没有选项面板。

8. Kill plane【终结平面】命令

Kill plane【终结平面】命令没有选项面板。

9. Adaptive Camera【自适应摄影机】命令

Adaptive Camera【自适应摄影机】命令没有选项面板。

10. Adaptive Mesh【自适应网络】命令

Adaptive Mesh【自适应网络】命令没有选项面板。

11. Guide【导向】命令

Guide【导向】命令没有选项面板。

12. Emission Region【发射区域】命令

Emission Region【发射区域】命令没有选项面板。

13. Motion Field【运动场】命令

Motion Field【运动场】命令没有选项面板。

14. Kill Field【禁用场】命令

―――――――――――――――Remove【移除】―――――――――――――――

15. Compute【移除…】命令组

在 Compute【移除…】命令组中，主要包括 8 个命令，如图 5.131 所示。

图 5.130 Content Browser【内容浏览器】面板　　　图 5.131 Compute【移除…】命令组面板

―――――――――――――――Compute【计算】―――――――――――――――

16. Compute and Cache to Disk…【计算并缓存到磁盘…】命令

Compute and Cache to Disk…【计算并缓存到磁盘…】命令的 Bifrost Compute and Cache Options【Bifrost 计算和缓存选项】面板，如图 5.132 所示。

Compression Format【压缩格式】参数组：在该参数组中，主要包括 Simple（Lossless，Least Compression）【简单（无损最小压缩）】、Float（Lossless）【浮点（无损）】、和 Quantization（Lossy）【量化（有损）】3 个选项，默认 Simple（Lossless，Least Compression）【简单（无损最小压缩）】选项。

17. Flush Scratch Cache【清空暂时缓存】命令

Flush Scratch Cache【清空暂时缓存】命令，没有选项面板。

18. Stop Background Processing【停止后台处理】命令

Stop Background Processing【停止后台处理】命令没有选项面板。

―――――――――――――――Initial State【初始状态】―――――――――――――――

19. 设置初始状态…【Set Intial State…】命令

设置初始状态…【Set Intial State…】命令的 BIF Set Intial State…【BIF 设置初始状态】面板，如图 5.133 所示。

20. Clear Initial State【清除初始状态】命令

Clear Initial State【清除初始状态】命令没有选项面板。

第 5 章　FX 模块菜单组

图 5.132　Bifrost Compute and Cache Options
【Bifrost 计算和缓存选项】面板

图 5.133　BIF Set Intial State…
【BIF 设置初始状态】

──────Options【选项】──────

21. Bifrost Options…【Bifrost 选项…】命令

单击 Bifrost Options…【Bifrost 选项…】命令，弹出 Bifrost Options【Bifrost 选项】面板，如图 5.134 所示。

22. Display Bifrost HUD【显示 Bifrost HUD】命令

Display Bifrost HUD【显示 Bifrost HUD】命令没有选项面板。

5.10　Boss【凸柱】菜单组

Boss Editor【凸柱编辑器】命令

单击 Boss Editor【凸柱编辑器】命令，弹出 Boss Ripple/Wave Generator【凸柱波纹/信号发生器】对话框，如图 5.135 所示。

图 5.134　Bifrost Options
【Bifrost 选项】面板

图 5.135　凸柱波纹/信号发生器
【Boss Ripple/Wave Generator】对话框

第 6 章 Rendering【渲染】模块菜单组

渲染模块菜单组主要包括 Lighting/Shading【照明/着色】、Texturing【纹理】、Render【渲染】、Toon【卡通】和 Stereo【立体】5 个菜单组。渲染模块菜单组的具体命令介绍如下。

6.1 Lighting/Shading【照明/着色】菜单组

───────────── Materials【材质】─────────────

1. Material Attributes【材质属性】命令

Material Attributes【材质属性】命令没有选项面板。

2. Assign New Material…【指定新材质…】命令

Assign New Material…【指定新材质…】命令没有选项面板。

3. Assign Favorite Material【指定收藏材质】命令组

在 Assign Favorite Material【指定收藏材质】命令组中,主要包括 9 个命令,如图 6.1 所示。

4. Assign Existing Maps【指定现有材质】命令组

在 Assign Existing Maps【指定现有材质】命令组中,主要包括 2 个命令,如图 6.2 所示。

图 6.1　Assign Favorite Material
【指定收藏材质】命令组面板

图 6.2　Assign Existing Maps
【指定现有材质】命令组面板

───────────── Baking【烘焙】─────────────

5. Transfer Maps…【传递贴图…】命令

Transfer Maps…【传递贴图…】命令没有选项面板。

第 6 章　Rendering【渲染】模块菜单组

────────────────Light Linking【灯光连接】────────────────

6. Light Linking Editor【灯光连接编辑器】命令组

在 Light Linking Editor【灯光连接编辑器】命令组中，主要包括 2 个命令，如图 6.2 所示。

图 6.3　Light Linking Editor【灯光连接编辑器】命令组面板

7. Make Light Links【生成灯光连接】命令

Make Light Links【生成灯光连接】命令没有选项面板。

8. Break Light Links【断开灯光连接】命令

Break Light Links【断开灯光连接】命令没有选项面板。

9. Select Objects Illuminated by Light【选择灯光照明的对象】命令

Select Objects Illuminated by Light【选择灯光照明的对象】命令没有选项面板。

10. Select Lights Illuminating Object【选择照明对象的灯光】命令

Select Lights Illuminating Object【选择照明对象的灯光】命令没有选项面板。

11. Make Shadow Links【生成阴影连接】命令

Make Shadow Links【生成阴影连接】命令没有选项面板。

12. Break Shadow Links【断开阴影连接】命令

Break Shadow Links【断开阴影连接】命令没有选项面板。

13. Select Object Shadowed by Light【选择灯光使其产生阴影的对象】命令

Select Object Shadowed by Light【选择灯光使其产生阴影的对象】命令没有选项面板。

14. Select Lights Shadowing Object【选择使对象产生阴影的灯光】命令

Select Lights Shadowing Object【选择使对象产生阴影的灯光】命令没有选项面板。

6.2　Texturing【纹理】菜单组

────────────────Painting【绘制】────────────────

1. 3D Paint Tool【3D 绘制工具】命令

3D Paint Tool【3D 绘制工具】命令的图标为，Tool Settings【工具设置】面板，如图 6.4 所示。

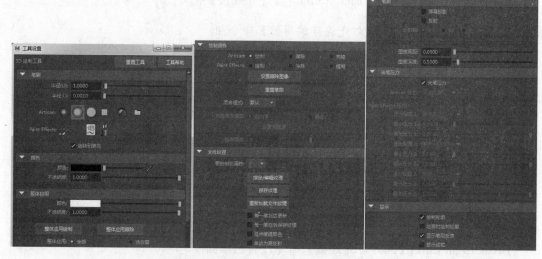

图 6.4 Tool Settings【工具设置】面板

―――――――――― PSD ――――――――――

2. Create PSD Network…【创建 PSD 网络…】命令

Create PSD Network…【创建 PSD 网络…】命令的图标为 ,该命令没有选项面板。

3. Edit PSD Network…【编辑 PSD 网络…】命令

Edit PSD Network…【编辑 PSD 网络…】命令的图标为 ,该命令没有选项面板。

4. Update PSD Networks【更新 PSD 网络】命令

Update PSD Networks【更新 PSD 网络】命令的图标为 ,该命令没有选项面板。

―――――――――― Projection【投影】――――――――――

5. Create Texture Reference Object【创建纹理引用对象】命令

Create Texture Reference Object【创建纹理引用对象】命令的图标为 ,该命令没有选项面板。

6. Delete Texture Reference Object【删除纹理引用对象】命令

Delete Texture Reference Object【删除纹理引用对象】命令的图标为 ,该命令没有选项面板。

7. Select Texture Reference Object【选择纹理引用对象】命令

Select Texture Reference Object【选择纹理引用对象】命令的图标为 ,该命令没有选项面板。

第6章 Rendering【渲染】模块菜单组

———————【Nurbs】———————

8. NURBS Texture Placement Tool【NURBS 纹理放置工具】命令

NURBS Texture Placement Tool【NURBS 纹理放置工具】命令的图标为 ，Tool Settings【工具设置】面板，如图 6.5 所示。

6.3 Render【渲染】菜单组

———————Render Settings【渲染设置】———————

1. Render Settings…【渲染设置…】命令

Render Settings…【渲染设置…】命令没有选项面板。

2. Render Using【使用以下渲染器渲染】命令组

在 Render Using【使用以下渲染器渲染】命令组中，主要包括 5 个命令，如图 6.6 所示。

图 6.5　Tool Settings【工具设置】面板　　图 6.6　Render Using【使用以下渲染器渲染】命令组面板

3. Test Resolution【测试分辨率】命令组

在 Test Resolution【测试分辨率】命令组中，要包括 9 个选项，如图 6.7 所示。

4. Set NURBS Tessellation【设置 NURBS 细分】命令

Set NURBS Tessellation【设置 NURBS 细分】命令的 Set NURBS Tessellation Options【设置 NURBS 细分选项】面板，如图 6.8 所示

图 6.7　Test Resolution 【测试分辨率】命令组面板　　图 6.8　Test Resolution 【测试分辨率】命令组面板

Curvature tolerance【曲率容差】参数组：在该参数组中主要包括 Highest【最高质量】、High Quality【高质量】、Medium Quality【中等质量】、Low Quality【低质量】和 No Curvature check【不检查曲率】5 个选项，默认 Medium Quality【中等质量】选项。

5. Run Render Diagnostics【运行渲染诊断】命令

Run Render Diagnostics【运行渲染诊断】命令没有选项面板。

6. Export Pre-Compositing【导出预合成】命令

Export Pre-Compositing【导出预合成】命令的图标为■，Export Pre-Compositing【导出预合成】面板，如图 6.9 所示。

图 6.9 Export Pre-Compositing【导出预合成】面板

──────────────── Rendering【渲染】────────────────

7. Render Current Frame【渲染当前帧】命令

Render Current Frame【渲染当前帧】命令没有选项面板。

8. Redo Previous Render【重做上一次渲染】命令

Redo Previous Render【重做上一次渲染】命令没有选项面板。

9. IPR Render Current Frame【IPR 渲染当前帧】命令

IPR Render Current Frame【IPR 渲染当前帧】命令没有选项面板。

10. Redo Previous IPR Render【重做上一次 IPR 渲染】命令

Redo Previous IPR Render【重做上一次 IPR 渲染】命令没有选项面板。

11. Render Sequence【渲染序列】命令（新增功能）

Render Sequence【渲染序列】命令的 Render Sequence（Current Frame）【渲染序列（当前帧）】面板，如图 6.10 所示。

第 6 章 Rendering【渲染】模块菜单组

---Batch Sequence【批渲染】---

12. Batch Render【批渲染】命令

Batch Render【批渲染】命令没有选项面板。

13. Cancel Batch Render【取消批渲染】命令

Cancel Batch Render【取消批渲染】命令没有选项面板。

14. Show Batch Render【显示批渲染】命令

Show Batch Render【显示批渲染】命令没有选项面板。

15. Create Backburner Job…【创建 Backburner 作业…】命令

Create Backburner Job…【创建 Backburner 作业…】命令没有选项面板。

6.4 Toon【卡通】菜单组

---Shading【着色】---

1. Assign Fill Shader【指定填充着色器】命令组

在 Assign Fill Shader【指定填充着色器】命令组中，主要包括 7 个命令，如图 11 所示。

图 6.10 Render Sequence（Current Frame）
【渲染序列（当前帧）】面板

图 6.11 Assign Fill Shader
【指定填充着色器】命令组面板

2. Assign Outline【指定轮廓】命令组

在 Assign Outline【指定轮廓】命令组中，主要包括 4 个命令，如图 6.12 所示。

3. Create Modifier【创建修改器】命令

Create Modifier【创建修改器】命令的图标为 ，该命令没有选项面板。

4. Set Camera Background Color【设置摄影机背景色】命令组

在 Set Camera Background Color【设置摄影机背景色】命令组中，主要包括 4 个命令，如图 6.13 所示。

图 6.12　Assign Outline
【指定轮廓】命令组面板

图 6.13　Set Camera Background Color
【设置摄影机背景色】命令组面板

──────── Geometry【几何体】────────

5. Reverse Surfaces【反转表面】命令

Reverse Surfaces【反转表面】命令的图标为 ，该命令没有选项面板。

6. Convert Toon To Polygons【将卡通线转化为多边形】命令

Convert Toon To Polygons【将卡通线转化为多边形】命令的图标为 ，Convert Toon To Polygons Options【将卡通线转化为多边形选项】面板，如图 6.14 所示。

──────── Painting【绘制】────────

7. Paint Line Attributes【绘制线属性】命令组

在 Paint Line Attributes【绘制线属性】命令组中，主要包括 6 个命令，如图 6.15 所示。

图 6.14　Convert Toon To Polygons Options
　　　【将卡通线转化为多边形选项】面板

图 6.15　Paint Line Attributes
【绘制线属性】命令组面板

8. Assign Paint Effects Brush to Toon Lines【将 Paint Effects 笔刷指定给卡通线】命令

Assign Paint Effects Brush to Toon Lines【将 Paint Effects 笔刷指定给卡通线】命令的图标为 ，该命令没有选项面板。

──────── Library【库】────────

9. Get Toon Example…【获取卡通示例…】命令

Get Toon Example…【获取卡通示例…】命令的图标为 ，该命令没有选项面板。

第 6 章 Rendering【渲染】模块菜单组

6.5 Stereo【立体】菜单组

———————————Cameras【摄影机】———————————

1. Editors【编辑器】命令组

在 Editors【编辑器】命令组中，主要包括 2 个命令，如图 6.16 所示。

2. Create【创建】命令组

在 Create【创建】命令组中，主要包括 2 个命令，如图 6.17 所示。

图 6.16　Editors【编辑器】命令组面板 　　图 6.17　Create【创建】命令组面板

———————————Links【连接】———————————

3. Make Links【创建连接】命令

Make Links【创建连接】命令的没有选项面板。

4. Break Links【断开连接】命令

Break Links【断开连接】命令没有选项面板。

第 7 章　View【视图】模块菜单组

视图模块菜单组主要包括 View【视图】、Shading【着色】、Lighting【照明】、Show【显示】、Renderer【渲染器】和 Panels【面板】6 个菜单组。视图模块菜单组的具体命令介绍如下。

7.1　View【视图】菜单组

1. Select Camera【选择摄影机】命令

Select Camera【选择摄影机】命令没有选项面板。

2. Lock Camera【锁定摄影机】命令

Lock Camera【锁定摄影机】命令没有选项面板。

3. Create Camera From View【从视图创建摄影机】命令

Create Camera From View【从视图创建摄影机】命令的快捷键为 Ctrl+Shift+C，该命令没有选项面板。

4. Cycle Through Cameras【在摄影机之间循环切换】命令

Cycle Through Cameras【在摄影机之间循环切换】命令没有选项面板。

5. Undo View Change【撤销视图更改】命令

Undo View Change【撤销视图更改】命令的快捷键为 Alt+Z，该命令没有选项面板。

6. Redo View Change【重做视图更改】命令

Redo View Change【重做视图更改】命令的快捷键为 Alt+Y，该命令没有选项面板。

7. Default View【默认视图】命令

Default View【默认视图】命令的快捷键为 Alt+Home，该命令没有选项面板。

8. View Along Axis【沿轴查看】命令组

在 View Along Axis【沿轴查看】命令组，主要包括 6 个命令，如图 7.1 所示。

第 7 章 View【视图】模块菜单组

9. Lock at Selection【注视当前选择】命令

Lock at Selection【注视当前选择】命令没有选项面板。

10. Center View of Selection【当前选择的中心视图】命令

Center View of Selection【当前选择的中心视图】命令没有选项面板。

11. Frame All【框显全部】命令

Frame All【框显全部】命令没有选项面板。

12. Frame Selection【框显当前选择】命令

Frame Selection【框显当前选择】命令的快捷键为 F，该命令没有选项面板。

13. Frame Selection with Children【框显当前选择（包含子对象）】命令

Frame Selection with Children【框显当前选择（包含子对象）】命令的快捷键为 Ctrl+F，该命令没有选项面板。

14. Align Camera To Polygon【将摄影机与多边形对齐】命令

Align Camera To Polygon【将摄影机与多边形对齐】命令没有选项面板。

15. Predefined【预定义书签】命令组

在 Predefined【预定义书签】命令组中，主要包括 7 个命令，如图 7.2 所示。

16. Bookmarks【书签】命令组

在 Bookmarks【书签】命令组中只包含 Edit Bookmarks…【编辑书签…】命令。

17. Camera Settings【摄影机设置】命令组

在 Camera Settings【摄影机设置】命令组中，主要包括 16 个命令，如图 7.3 所示。

18. Camera Attribute Editor…【摄影机属性编辑器…】命令

Camera Attribute Editor…【摄影机属性编辑器…】命令没有选项面板。

19. Camera Tools【摄影机工具】命令组

在 Camera Tools【摄影机工具】命令组中，主要包括 11 个命令，如图 7.4 所示。

20. Image Plane【图像平面】命令组

在 Image Plane【图像平面】命令组中，主要包括 3 个命令，如图 7.5 所示。

图 7.1　View Along Axis　　　图 7.2　Predefined　　　图 7.3　Camera Settings
【沿轴查看】命令组面板　　　【预定义书签】命令组面板　　　【摄影机设置】命令组面板

21. View Sequence Time【查看序列时间】命令

View Sequence Time【查看序列时间】命令没有选项面板。

7.2　Shading【着色】菜单组

1. Wireframe【线框】命令

Wireframe【线框】命令没有选项面板。

2. Smooth Shade All【对所有项目进行平滑着色处理】命令

Smooth Shade All【对所有项目进行平滑着色处理】命令没有选项面板。

3. Smooth Shade Selected Items【对选定项目进行平滑着色处理】命令

Smooth Shade Selected Items【对选定项目进行平滑着色处理】命令没有选项面板。

4. Flat Shade All【对所有项目进行平面着色】命令

Flat Shade All【对所有项目进行平面着色】命令没有选项面板。

5. Flat Shade Selected Items【对选定项目进行平面着色】命令

Flat Shade Selected Items【对选定项目进行平面着色】命令没有选项面板。

6. Bounding Box【边界框】命令

Bounding Box【边界框】命令没有选项面板。

7. Points【点】命令

Points【点】命令没有选项面板。

8. Use Default Material【使用默认材质】命令

Use Default Material【使用默认材质】命令没有选项面板。

9. Wireframe on Shade【着色对象上的线框】命令

Wireframe on Shade【着色对象上的线框】命令没有选项面板。

10. X-Ray【X 射线显示】命令

X-Ray【X 射线显示】命令没有选项面板。

11. X-Ray Joints【X 射线显示关节】命令

X-Ray Joints【X 射线显示关节】命令没有选项面板。

12. X-Ray Active Components【X 射线显示活动组件】命令

X-Ray Active Components【X 射线显示活动组件】命令没有选项面板。

13. Object Transparency Sorting【对象透明度排序】命令

Object Transparency Sorting【对象透明度排序】命令没有选项面板。

14. Polygon Transparency Sorting【多边形透明度排序】命令

Polygon Transparency Sorting【多边形透明度排序】命令没有选项面板。

15. Interactive Shading【交互式着色】命令

Interactive Shading【交互式着色】命令没有选项面板。

16. Backface Culling【背面消隐】命令

Backface Culling【背面消隐】命令没有选项面板。

17. Smooth Wireframe【平滑线框】命令

Smooth Wireframe【平滑线框】命令没有选项面板。

18. Hardware Texturing【硬件纹理】命令

Hardware Texturing【硬件纹理】命令没有选项面板。

19. Hardware Fog【硬件雾】命令

Hardware Fog【硬件雾】命令没有选项面板。

20. Depth of Field【景深】命令

Depth of Field【景深】命令没有选项面板。

21. Apply Current to All【将当前样式应用于所有对象】命令

Apply Current to All【将当前样式应用于所有对象】命令没有选项面板。

7.3 Lighting【照明】

1. Use Default Lighting【使用默认照明】命令

Use Default Lighting【使用默认照明】命令没有选项面板。

2. Use All Lights【使用所有灯光】命令

Use All Lights【使用所有灯光】命令的快捷键为数字键 7，该命令没有选项面板。

3. Use Selected Lights【使用选定灯光】命令

Use Selected Lights【使用选定灯光】命令没有选项面板。

4. Use Flat Lights【使用平面照明】命令

Use Flat Lights【使用平面照明】命令没有选项面板。

5. Use No Lights【不使用灯光】命令

Use No Lights【不使用灯光】命令没有选项面板。

6. Two Sided Lighting【双面照明】命令

Two Sided Lighting【双面照明】命令没有选项面板。

7. Shadows【阴影】命令

Shadows【阴影】命令没有选项面板。

7.4 Show【显示】菜单组

1. Isolate Select【隔离选择】命令组

在 Isolate Select【隔离选择】命令组中，主要包括 7 个命令，如图 7.6 所示。

图 7.4 Camera Tools　　　　图 7.5 Image Plane　　　　图 7.6 Isolate Select
【摄影机工具】命令组面板　　【图像平面】命令组面板　　【隔离选择】命令组面板

────────Bookmark Current Objects【为当前对象建立书签】────────

2. All【全部】命令

All【全部】命令没有选项面板。

3. None【无】命令

None【无】命令没有选项面板。

4. NURBS Curves【NURBS 曲线】命令

NURBS Curves【NURBS 曲线】命令没有选项面板。

5. NURBS Surfaces【NURBS 曲面】命令

NURBS Surfaces【NURBS 曲面】命令没有选项面板。

6. NURBS CV【UNRBS CV】命令

NURBS CV【UNRBS CV】命令没有选项面板。

7. NURBS Hulls【NURBS 壳线】命令

NURBS Hulls【NURBS 壳线】命令没有选项面板。

8. Polygons【多边形】命令

Polygons【多边形】命令没有选项面板。

9. Subdiv Surfaces【细分曲面】命令

Subdiv Surfaces【细分曲面】命令没有选项面板。

10. Planes【平面】命令

Planes【平面】命令没有选项面板。

11. Lights【灯光】命令

Lights【灯光】命令没有选项面板。

12. Cameras【摄影机】命令

Cameras【摄影机】命令没有选项面板。

13. Image Planes【图像平面】命令

Image Planes【图像平面】命令没有选项面板。

14. Joints【关节】命令

Joints【关节】命令没有选项面板。

15. Ik Handles【IK 控制柄】命令

Ik Handles【IK 控制柄】命令没有选项面板。

16. Deformers【变形器】命令

Deformers【变形器】命令没有选项面板。

17. Dynamics【动力学】命令

Dynamics【动力学】命令没有选项面板。

18. Particle Instancers【粒子实例化器】命令

Particle Instancers【粒子实例化器】命令没有选项面板。

19. Fluids【流体】命令

Fluids【流体】命令没有选项面板。

20. Hair Systems【头发系统】命令

Hair Systems【头发系统】命令没有选项面板。

21. Follicles【毛囊】命令

Follicles【毛囊】命令没有选项面板。

22. nCloth【n 布料】命令

nCloth【n 布料】命令没有选项面板。

23. nParticle【n粒子】命令

nParticle【n粒子】命令没有选项面板。

24. nRigid【n刚体】命令

nRigid【n刚体】命令没有选项面板。

25. Dynamic Constraints【动态约束】命令

Dynamic Constraints【动态约束】命令没有选项面板。

26. Locators【定位器】命令

Locators【定位器】命令没有选项面板。

27. Dimensions【尺度】命令

Dimensions【尺度】命令没有选项面板。

28. Pivots【枢轴】命令

Pivots【枢轴】命令没有选项面板。

29. Handles【控制柄】命令

Handles【控制柄】命令没有选项面板。

30. Texture Placements【纹理放置】命令

Texture Placements【纹理放置】命令没有选项面板。

31. Strokes【笔划】命令

Strokes【笔划】命令没有选项面板。

32. Motion Trails【运动轨迹】命令

Motion Trails【运动轨迹】命令没有选项面板。

33. Plugin Shapes【插件形状】命令

Plugin Shapes【插件形状】命令没有选项面板。

34. Clip Ghosts【片段重影】命令

Clip Ghosts【片段重影】命令没有选项面板。

35. Grease Pencil【油性铅笔】命令

Grease Pencil【油性铅笔】命令没有选项面板。

36. GPU Cache【GPU 缓存】命令

GPU Cache【GPU 缓存】命令没有选项面板。

37. Manipulators【操纵器】命令

Manipulators【操纵器】命令没有选项面板。

38. Grid【栅格】命令

Grid【栅格】命令没有选项面板。

39.【HUD】命令

【HUD】命令没有选项面板。

40. Hold-Outs【透底】命令

Hold-Outs【透底】命令没有选项面板。

41. Selection Highlighting【选择亮显】命令

Selection Highlighting【选择亮显】命令没有选项面板。

42. Playblast Display【播放预览显示】命令组

在 Playblast Display【播放预览显示】命令组中，主要包括 41 个命令，如图 7.7 所示。

图 7.7　Playblast Display【播放预览显示】命令组面板

7.5 Renderer【渲染器】菜单组

1. Viewport 2.0【Viewport 2.0】命令

Viewport 2.0【Viewport 2.0】命令没有选项面板。

2. Legacy Default Viewport【旧版默认视口】命令

Legacy Default Viewport【旧版默认视口】命令没有选项面板。

3. Legacy High Quality Viewport【旧版高质量视口】命令

Legacy High Quality Viewport【旧版高质量视口】命令没有选项面板。

7.6 Panels【面板】菜单组

1. Perspective【透视】命令组

在 Perspective【透视】命令组中，主要包括 2 个命令，如图 7.8 所示。

2. Stereo【立体】命令组

在 Stereo【立体】命令组中，主要包括 2 个命令，如图 7.9 所示。

3. Orthographic【正交】命令组

Orthographic【正交】命令组中，主要包括 4 个命令，如图 7.10 所示。
New【前】命令组中，主要包括 6 个命令，如图 7.11 所示。

图 7.8　Perspective　　　　图 7.9　Stereo　　　　图 7.10　Orthographic
【透视】命令组面板　　　　【立体】命令组面板　　　　【正交】命令组面板

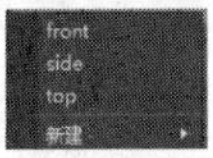

4. Look Through Selected【沿选定对象观看】命令

Look Through Selected【沿选定对象观看】命令的 Look Through Selected Options【沿选定对象观看选项】面板，如图 7.12 所示。

5. Panel【面板】命令组

在 Panel【面板】命令组中，主要包括 22 个命令，如图 7.13 所示。

图 7.11　New　　　　　图 7.12　Look Through Selected　　　图 7.13　Panel
【前】命令组面板　　　【沿选定对象观象】命令组面板　　　【面板】命令组面板

6. Hypergraph Panel【Hypergraph 面板】命令组

在 Hypergraph Panel【Hypergraph 面板】命令组中，主要包括 3 个命令，如图 7.13 所示。

7. Layouts【布局】命令组

在 Layouts【布局】命令组中，主要包括 10 个命令，如图 7.15 所示。

8. Saved Layouts【保存的布局】命令组

在 Saved Layouts【保存的布局】命令组中，主要包括 20 个命令，如图 7.16 所示。

图 7.14　Hypergraph Panel　　　图 7.15　Layouts　　　图 7.16　Saved Layouts
【Hypergraph 面板】命令组面板　　【布局】命令组面板　　【保存的布局】命令组面板

9. Tear off…【撕下…】命令

Tear off…【撕下…】命令没有选项面板。

10. Tear off Copy…【撕下副本…】命令

Tear off Copy…【撕下副本…】命令没有选项面板

11. Panel Editor…【面板编辑器…】命令

Panel Editor…【面板编辑器…】命令没有选项面板。

第 8 章　工具属性及工具架图标

本章主要介绍 Maya 软件中的工具图标和工具架图标。

8.1　Maya 工具图标

Maya 软件包括 Select Tool【选择】、 Lasso Tool【套索】、 Paint Selection Tool【绘制选择】、 Move Tool【移动】、 Rotate Tool【旋转】和 Scale Tool【缩放】6 个工具图标。

1. Select Tool【选择】工具图标

功能：选择场景和编辑器窗口中的对象和组建。
Select Tool【选择】工具属性面板，如图 8.1 所示。

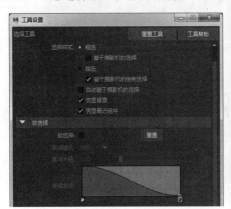 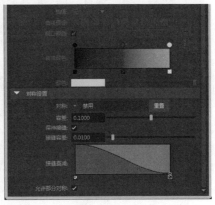

图 8.1　Select（选择）Tool Settings【工具设置】

2. Lasso Tool【套索】工具图标

功能：选择场景，在其对象和组件周围通过自由绘制形状方式来选择这些对象和组件。
Lasso Tool【套索】工具属性面板，如图 8.2 所示。

3. Paint Selection Tool【绘制选择】工具图标

功能：通过绘制选择场景中的组件。
Paint Selection Tool【绘制选择】工具属性面板，如图 8.3 所示。

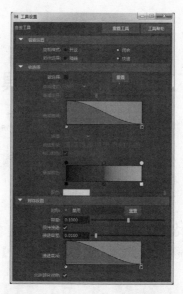

图 8.2　Lasso（套索）Tool Settings【工具设置】

图 8.3　Paint Selection（绘制选择 Tool Settings）【工具设置】

4. Move Tool【移动】工具图标

功能：通过拖动变换操纵器移动选定对象或组件。
Move Tool【移动】工具属性面板，如图 8.4 所示。

5. Rotate Tool【旋转】工具图标

功能：通过拖动旋转操纵器旋转选定对象或组件。
Rotate Tool【旋转】工具属性面板，如图 8.5 所示。

第 8 章 工具属性及工具架图标

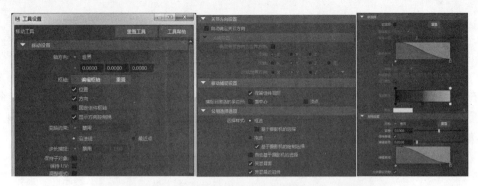

图 8.4 Move（移动）Tool Settings【工具设置】

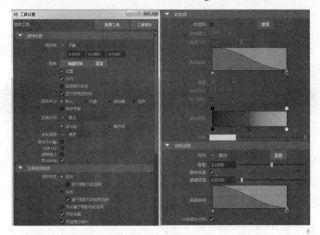

图 8.5 Rotate（旋转）Tool Settings【工具设置】

6. Scale Tool【缩放】工具图标

功能：通过拖动缩放操纵器缩放选定对象或组件。

Scale Tool【缩放】工具属性面板，如图 8.6 所示。

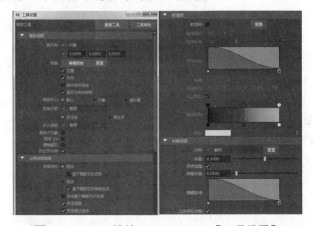

图 8.6 Scale（缩放）Tool Settings【工具设置】

8.2 Shelf Tabs【工具架】图标

在 Maya 2018 中，为了让用户工作方便，将一些常用的命令以工具架的形式，分 14 类放置在工作界面的工具栏下方。用户只要单击相应的图标，即可执行该命令，而不需要通过菜单栏进行操作。工具架的分类标签如图 8.7 所示。

图 8.7　工具架分类标签

1. Curves/Surfaces【曲线/曲面】图标

在 Curves/Surfaces【曲线/曲面】工具架中主要包括了如下 38 个命令图标，各图标名称如图 8.8 所示。图标名称和功能介绍如下。

图 8.8　Curves/Surfaces【曲线/曲面】图标

（1）NURBS Circle【NURBS 圆形】图标

功能：在上栅格上创建 NURBS 圆形。

（2）NURBS Square【NURBS 方形】图标

功能：在栅格上创建 NURBS 方形。

（3）EP Curve Tool【EP 曲线工具】图标

功能：通过指定编辑点在栅格或激活的面上创建曲线。

（4）【Pencil Curve Tool（铅笔曲线工具）】图标

功能：在栅格或激活的曲面上创建曲线。

（5）Three Point Circular Arc【三点圆弧】图标

功能：通过指定三个点在栅格上创建圆弧。

（6）Attach Curves【附加曲线】

功能：将两条曲线附加为一条曲线

（7）Detach Curves【分离曲线】

功能：将曲线沿选择的曲线点处分离为两段曲线。

（8）Insert Knot【插入结】

功能：沿确定的曲线点处插入控制点。

（9）Extend Curve【延伸曲线】

功能：延伸选择曲线或曲面上的曲线。

（10）Offset Curve【偏移曲线】

功能：对选择的曲线或曲面上的曲线、等参线或修剪边进行偏移。

(11) Rebuild Curve【重建曲线】

功能：对选择的曲线或曲面上的曲线进行重建。

(12) Add Points Tool【添加点工具】

功能：给选择的曲线添加点。

(13) Curve Editing Tool【曲线编辑工具】

功能：对选择的曲线或曲面上的曲线进行编辑。

(14) Bezier Curve Tool【Bezier 曲线工具】

功能：创建 Bezier 曲线。

(15) NURBS Sphere【NURBS 球体】

功能：在栅格上创建 NURBS 球体。

(16) NURBS Cube【NURBS 立方体】

功能：在栅格上创建 NURBS 立方体。

(17) NURBS Cylinder【NURBS 圆柱体】

功能：在栅格上创建 NURBS 圆柱体。

(18) NURBS Cone【NURBS 圆锥体】

功能：在栅格上创建 NURBS 圆锥体。

(19) NURBS Plane【NURBS 平面】

功能：在栅格上创建 NURBS 平面。

(20) NURBS Torus【NURBS 圆环】

功能：在栅格上创建 NURBS 圆环。

(21) Revolve【旋转】

功能：将选择的曲线、等参线或修剪边旋转成面。

(22) Loft【放样】

功能：将选择的曲线、等参线或修剪边放样成面。

(23) Planar【平面】

功能：将选择曲线、等参线或修剪边共同形成一个平面。

(24) Extrude【挤出】

功能：选择曲线、等参线或修剪边，最后选择路径。单击该命令，将沿路径挤出平面。

(25) Birail 1 Tool【双轨成形 1 工具】

功能：选择剖面线，再选择两条轨道线，即可形成一个曲面。

(26) Bevel Plus【倒角+】

功能：对选择的曲线、等参线、修剪边或 CoS 进行倒角处理。

(27) Project Curve on Surface【在曲面上投影曲线】

功能：将选择的曲线、等参线或修剪边投射到目标曲面上。

(28) Intersect Surface【曲面相交】

功能：选择两个或多个曲面，最后一个曲面与其他曲面相交。

（29）Trim Tool【修剪工具】

功能：对选择的曲线和曲面进行修剪。

（30）Untrim Surfaces【取消修剪曲面】

功能：取消已修剪的曲面修剪效果。

（31）Attach Surface【附加曲面】

功能：将选择的两个面进行附加操作。

（32）Detach Surfaces【分离曲面】

功能：将平面沿选择等参线位置进行分离曲面操作。

（33）Open/Close Surfaces【开放/闭合曲面】

功能：对选择的曲面或等参线进行开放或闭合操作。

（34）Insert Isoparms【插入等参线】

功能：对选择的曲面插入等参线。

（35）Extend Surfaces【延伸曲面】

功能：对选择的曲面进行曲面延伸。

（36）Rebuild Surfaces【重建几何体工具】

功能：对选择曲面进行重建。

（37）Sculpt Geometry Tool【雕刻几何体工具】

功能：对选择的几何体对象进行雕刻。

（38）Surface Editing Tool【曲面编辑工具】

功能：对选择的曲面进行编辑。

2．Polygons【多边形】图标

在 Polygons【多边形】工具架中主要包括 32 个命令图标，各图标如图 8.9 所示。图标名称和功能介绍如下。

图 8.9　Polygons【多边形】图标

（1）Polygons Sphere【多边形球体】

功能：在栅格上创建多边形球体。

（2）Polygons Cube【多边形立方体】

功能：在栅格上创建多边形立方体。

（3）Polygons Cylinder【多边形圆柱体】

功能：在栅格上创建多边形圆柱体。

（4）Polygons Cone【多边形圆锥体】

功能：在栅格上创建多边形圆锥体。

(5) Polygons Plane【多边形平面】

功能：在栅格上创建多边形平面。

(6) Polygons Torus【多边形圆环】

功能：在栅格上创建多边形圆环。

(7) Polygons Pyramid【多边形棱锥】

功能：在栅格上创建多边形棱锥。

(8) Polygons Pipe【多边形管道】

功能：在栅格上创建多边形管道。

(9) Polygons Type【多边形类型】

功能：在栅格上创建多边形文本。

(10) 【SVG】

功能：使用剪贴板中的扩展向量图形或导入的 SVG 文件创建多边形。

(11) Combine【结合】

功能：将选定多边形对象组合成单个对象，以便进行执行合并或面修剪之类的操作。

(12) Separate【分离】

功能：将选定多边形对象壳或任意选定面的壳，从对象中分离为单独对象。

(13) Mirror【镜像】

功能：沿某个轴镜像几何体。

(14) Smooth【平滑】

功能：向选定多边形对象添加多边形以对其进行平滑处理。

(15) Subdiv Proxy【细分曲面代理】

功能：向选定代理对象添加多边形。

(16) Reduce【减少】

功能：减少选定对象的多边形组件数量。

(17) Multi-Cut Tool【多切割工具】

功能：在多边形上切割、切片和插入边。

(18) Extrude【挤出】

功能：挤出选定组件。

(19) Bridge【桥接】

功能：在两组边或面之间创建桥接面。

(20) Bevel【倒角组件】

功能：沿选定边或面创建倒角。

(21) Merge【合并】

功能：基于当前选择合并点/边界边。

(22) Target Weld Tool【目标焊接工具】

功能：将两个边或顶点合并为一个组件。

（23） Flip Triangle Edge【翻转三角形边】

功能：翻转两个三角形之间的边。

（24） Collapse【收拢】

功能：收拢选定边或面。

（25） Extract【提取】

功能：从对应壳提取当前选定的面并显示操作器，以便调整其偏移。

（26） Planar Mapping【平面映射】

功能：创建与选定平明平行的投影。

（27） Cylindrical Mapping【圆柱形映射】

功能：在选定面上创建圆柱形投影。

（28） Spherical Mapping【球形映射】

功能：在选定面上创建球形投影。

（29） Automatic Projection【自动投影】

功能：选择要自动映射的面。

（30） Create Contour Stretch Mapping【创建轮廓拉伸贴图】

功能：创建轮廓拉伸贴图。

（31） UV Editor…【UV 编辑器…】

功能：纹理坐标映射视图。

（32） 3D UV Grab Tool【3D UV 抓取工具】

功能：抓取 3D 视口中的 UV。

3. Sculpting【雕刻】图标

在 Sculpting【雕刻】工具架中主要包括 26 个命令图标，各图标名称如图 8.10 所示，名称和功能如下。

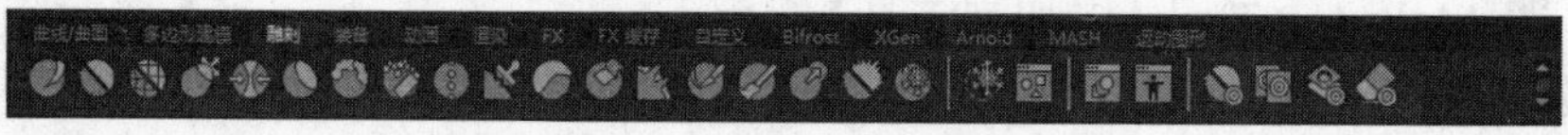

图 8.10 Sculpting【雕刻】图标

（1） Sculpting Tool【雕刻工具】

功能：抬起曲面。

（2） Smooth Tool【平滑工具】

功能：平均分配曲面细节。

（3） Relax Tool【松弛工具】

功能：使网络的曲面平滑，而不影响其形状。

（4） Grab Tool【抓取工具】

功能：沿曲面以任意方向拖动单个顶点。

第 8 章 工具属性及工具架图标

（5）Pinch Tool【收缩工具】

功能：锐化软边。

（6）Flatten Tool【展平工具】

功能：展平曲面。

（7）Foamy Tool【泡沫工具】

功能：柔和抬起曲面。

（8）Spray Tool【喷射工具】

功能：在曲面上随机喷射图章压印。

（9）Repeat Tool【重复工具】

功能：在曲面上压印图案。

（10）Imprint Tool【压印工具】

功能：压印曲面上图章的单个副本。

（11）Wax Tool【上蜡工具】

功能：构建曲面。

（12）ScrapeTool【刮擦工具】

功能：最小化或移除曲面上的凸起区域。

（13）Fill Tool【填充工具】

功能：填充曲面上的凹陷部分。

（14）Knife Tool【刀工具】

功能：将精细笔划切割为曲面。

（15）Smear Tool【涂抹工具】

功能：沿笔划方向拉动曲面。

（16）Bulge Tool【凸起工具】

功能：使曲面上的区域膨胀。

（17）Amplify Tool【放大工具】

功能：增强曲面细节。

（18）Freeze Tool【冻结工具】

功能：绘制曲面区域以防止进一步修改。

（19）Convert to Frozen【转化为冻结】

功能：将修改区域转化为冻结。

（20）Open Content Browser for Sculpting Base Meshes【浏览器打开按钮】

功能：打开内容浏览器以雕刻基础网络。

（21）Shape Editor【形变编辑器】

功能：创建、编辑和管理形变。

（22）Pose Editor【姿势编辑器】

功能：创建、编辑和管理姿势空间变形。

（23）Smooth Target Tool【平滑目标工具】

功能：在编辑模式下仅对当前融合变形目标平均分配曲面细节。

（24）Clone Target Tool【克隆目标工具】

功能：在编辑模式下将在融合变形目标源中所做的编辑复制到当前目标。

（25）Mask Target Tool【遮罩目标工具】

功能：在网络上绘制区域，以便把对当前融合变形目标所做的编辑隐藏在编辑模式下。

（26）Erase Target Tool【擦除目标工具】

功能：移除在编辑模式下对当前融合变形目标进行的编辑。

4. Rigging【装备】图标

在 Rigging【装备】工具架中主要包括 18 个命令图标，如图 8.11 所示，各图标名称和功能介绍如下。

图 8.11　Rigging【装备】图标

（1）Create Locator【创建定位器】

功能：在栅格上创建定位器对象。

（2）Create Joints【创建关节】

功能：单击以放置关节，单击现有关节以添加骨架，单击/拖动以定位关节。

（3）Create IK Handle【创建 IK 控制柄】

功能：在关键上创建 IK 控制柄。

（4）Bind Skin【绑定蒙皮】

功能：平滑绑定蒙皮。

（5）Quick Rig【快速装备】

功能：快速装备角色。

（6）Human IK【角色 IK】

功能：显示角色控制窗口。

（7）Paint Skin Weights【绘制蒙皮权重】

功能：在平滑绑定蒙皮上绘制权重。

（8）Blend Shape【融合变形】

功能：【融合变形】使一个基础物体与多个目标物体进行混合，将一个物体的形状以平滑过渡的方式变形为另一物体的形状。

（9）Create Lattice【创建晶格】

功能：为选定对象创建晶格，【创建晶格】常用于变形结构复杂的对象。

（10）Create Cluster【创建簇】

功能：为选定对象创建簇。【簇变形器】可以同时控制一组变形对象上的点，通过不同影响力变形"簇"有效作用区域的可变形对象。

（11） Parent Constraint【父约束】

功能：创建父约束。【父约束】可以将一个对象的位移和旋转关联到其他对象上，一个被约束对象的运动也能被多个目标对象的平均位置约束。

（12） Point Constraint【点约束】

功能：创建点约束。【点约束】可以使一个对象跟随另一个对象的位置移动，或使一个对象跟随多个对象的平均位置移动。

（13） Orient Constraint【方向约束】

功能：创建方向约束。【方向约束】可以将一个对象的方向与另一个或更多其他对象的方向匹配，常用来制作多个对象的同步变换方向的动画。

（14） Scale Constraint【缩放约束】

功能：创建缩放约束。【缩放约束】可以将一个对象的位移和旋转关联到其他对象上，一个被约束对象的运动也能被多个目标的平均位置约束。

（15） Aim Constraint【目标约束】

功能：创建目标约束。【目标约束】可以约束一个对象的方向，使被约束的对象始终瞄准目标对象。

（16） Pole Vector Constraint【极向量约束】

功能：创建极向量约束。【极向量约束】可以使 IK 旋转平面手柄的极向量终点跟随一个对象或多个对象的平均位置移动。

（17） Set Key【设置关键帧】

功能：为选择的对象创建关键帧。

（18） Set Driven Key【设置受驱动关键帧】

功能：打开 Set Driven Key【设置受驱动关键帧】对话框。

5. Animation【动画】图标

在 Animation【动画】工具架中主要包括 16 个命令图标，如图 8.12 所示。其中， Parent Constraint【父约束】、 Point Constraint【点约束）、 Orient Constraint【方向约束】、 Scale Constraint【缩放约束】、 Aim Constraint【目标约束】和 Pole Vector Constraint【极向量约束】6 个图标在 Rigging【装备】图标工具架中已经详细介绍。其他图标名称和功能介绍如下。

图 8.12　Animation【动画】图标

（1）Play blast【播放预览】

功能：通过屏幕捕获预览动画。

（2）Motion Trail【运动轨迹】

功能：对随时间运动对象生成运动轨迹。

（3）Ghosting【重影】

功能：使运动对象产生重影。

（4）No Ghosting【无重影】

功能：取消运动对象的重影。

（5）Bake Animation【烘焙动画】

功能：将现有动画烘焙为关键帧。

（6）Set Key【设置关键帧】

功能：为运动对象设置关键帧。

（7）Set Key on Animated【设置动画关键帧】

功能：为已设置动画的通道设置关键帧。

（8）Set Key on Translate【设置平移关键帧】

功能：为动画设置平移关键帧。

（9）Set Key on Rotate【设置旋转关键帧】

功能：为动画设置旋转关键帧。

（10）Set Key on Scale【设置缩放关键帧】

功能：为动画设置缩放关键帧。

6. Rendering【渲染】图标

在 Rendering【渲染】工具架中主要包括 23 个命令图标，如图 8.13 所示。图标名称和功能如下。

图 8.13　Rendering【渲染】图标

（1）Ambient Light【环境光】

功能：在栅格或激活的曲面上创建环境光。

（2）Directional Light【平行光】

功能：在栅格或激活的曲面上创建平行光。

（3）Point Light【点光源】

功能：在栅格或激活的曲面上创建点光源。

（4）Spot Light【聚光灯】

功能：在栅格或激活的曲面上创建聚光灯。

（5） Area Light【区域光】

功能：在栅格或激活的曲面上创建区域光。

（6） Volume Light【体积光】

功能：在栅格或激活的曲面上创建体积光。

（7） Create Camera【创建摄影机】

功能：在栅格上创建摄影机。

（8） Edit Material Attributes【编辑材质属性】

功能：打开【着色组件属性编辑器】属性面板。

（9） Anisotropic Material【各向异性材质】

功能：为给选择对象赋予各向异性材质。

（10） Blinn Material【Blinn 材质】

功能：为选择对象赋予 Blinn 材质。

（11） Lambert Material【Lambert 材质】

功能：为选择对象赋予 Lambert 材质。

（12） Phong Material【Phong 材质】

功能：为选择对象赋予 Phong 材质。

（13） Phong E Material【Phong E 材质】

功能：为选择对象赋予 Phong E 材质。

（14） Layered Material【分层材质】

功能：为选择对象赋予分层材质。

（15） Ramp Material【渐变材质】

功能：为选择对象赋予渐变材质。

（16） Shading Map【着色贴图】

功能：为选择对象赋予着色贴图。

（17） Surface Material【表面材质】

功能：为选择对象赋予表面材质。

（18） Use Background【使用背景】

功能：为选择对象赋予背景材质。

（19） Render Diagnostics【渲染诊断】

功能：在脚本编辑器中显示渲染帧断。

（20） Batch Render【批渲染】

功能：将当前场景导出到文件并在后台渲染该文件。

（21） Cancel Batch Render【取消批渲染】

功能：取消当前正在进行批渲染的文件。

（22） Show Batch Render【显示批渲染】

功能：显示批渲染图像。

（23）3D Paint【3D 绘制工具】

功能：对场景中的对象绘制文件纹理。

7.【FX】图标

在【FX】工具架中主要包括如下 29 个命令图标，如图 8.14 所示。图标名称和功能介绍如下。

图 8.14　【FX】图标

（1）Emitter【发射器】

功能：在栅格或激活的对象上创建 nParticle 发射器。

（2）Add Emitter【添加发射器】

功能：为选择对象添加 nParticle 发射器。

（3）3D Fluid Container with Emitter【具有发射器的 3D 流体容器】

功能：创建发射器和 3D 流体容器。

（4）2D Fluid Container with Emitter【具有发射器的 2D 流体容器】

功能：创建发射器和 2D 流体容器。

（5）Emit Fluid from Object【从对象发射流体】

功能：给选择对象或曲面添加流体。

（6）Make Collide【使碰撞】

功能：选择流体和几何体对象单击 Make Collide【使碰撞】图标，使流体和几何体产生碰撞关联。

（7）Create Hair【创建头发】

功能：为选择曲面、面或对象创建头发。

（8）Paint hair tool options【绘制头发工具选项】

功能：打开 Paint Hair Follicles Settings【绘制毛囊设置】和 Tool Setting【工具设置】面板。

（9）Make Selected Curves Dynamic【动力学化选定曲线】

功能：利用选定曲线创建动力学化曲线。

（10）Current Position【当前位置】

功能：显示选定头发系统或所有头发系统的当前位置曲线或 pfxHair。

（11）【Start Position（开始位置）】

功能：显示选定头发系统或所有头发系统的开始位置曲线。

（12）Rest Position【静止位置】

功能：显示选定头发系统或所有头发系统的静止位置曲线。

（13）To Hair System【到头发系统】

功能：选择与当前选择相关的头发系统节点。

（14）To Follicles【到毛囊】

功能：选择与当前选择相关的毛囊。

（15）To Start Curves【到开始曲线】

功能：选择与当前选择相关的开始曲线。

（16）To Rest Curves【到静止曲线】

功能：选择与当前选择相关的静止曲线。

（17）Scale Hair Tool【缩放头发工具】

功能：选择头发（直接或间接），然后向左或向右拖动进行缩放。

（18）To Start Curve End CV【到开始曲线末端 CV】

功能：选择与当前选择相关的开始曲线末端 CV。

（19）To Rest Curve End CV【到静止曲线末端 CV】

功能：选择与当前选择相关的静止曲线末端 CV。

（20）From Current【来自当前】

功能：将选定头发的开始位置设置为当前位置。

（21）Current and Start【当前和开始】

功能：显示选定头发系统或所有头发系统的当前位置曲线和开始位置曲线。

（22）From Current【来自当前】

功能：将选定头发的静止位置设置为当前位置。

（23）Current and Rest【当前和静止】

功能：显示选定头发系统或所有头发系统的当前位置曲线或静止位置曲线。

（24）Create an nCloth【创建 nCloth】

功能：给选定的网络创建 nCloth。

（25）Make the selected mesh（es）collide with nCloth and nParth【创建被动碰撞对象】

功能：使选定网络与 nCloth 和 nParticle 发生碰撞。

（26）Remove nCloth【移除 nCloth】

功能：对选定的 nCloth 或 nRigid 网络，移除 nCloth 或 nRigid 网络。

（27）Display the input mesh【显示输入网络】

功能：显示输入网络。

（28）Display the Current mesh【显示当前网络】

功能：显示当前网络。

（29）Start Interactive playback【启动交互式播放】

功能：启动或停止交互式播放。

8. FX Caching【FX 缓存】图标

在 FX Caching【FX 缓存】工具架中主要包括 10 个命令图标，如图 8.15 所示。图标名称和功能介绍如下。

图 8.15　FX Caching【FX 缓存】图标

（1）　Set Initial State【设置初始状态】

功能：将流体的当前状态设置为其初始条件。

（2）　Clear Initial State【清除初始状态】

功能：移除以前设置的初始条件。

（3）　Create Cache【创建缓存】

功能：预备运行以创建播放和渲染缓存。

（4）　Delete Cache【删除缓存】

功能：删除选定 nCloth 网络、nParticle 或流体对象上的缓存。

（5）　Given a Selected nCloth Mesh, nHair or nPariticle, Create a Cache【创建新的 nObject】

功能：给选定的 nCloth 网络、nHair 或 nParticle 创建缓存。

（6）　Delete cache on selected nCloth mesh, nHair or nParticle【删除缓存】

功能：删除选定的 nCloth 网络、nHair 或 nParticle 上的缓存。

（7）　Enable all caches on selected nCloth, nCloth or nParticle objects【启用选定对象上的所有缓存】

功能：启用选定的 nCloth 网络、nHair 或 nParticle 对象的所有缓存。

（8）　Disable all caches on selected nCloth, nHair or nParticle objects【禁用选定对象上的所有缓存】

功能：禁用选定的 nCloth 网络、nHair 或 nParticle 对象的所有缓存。

（9）　Given a selected nCloth or nParticle, merge existing caches into a new cache【合并到新缓存】

功能：将选定的 nCloth 或 nParticle 缓存合并到新缓存中。

（10）　Replace frame（s）in nCache【替换 nCache 中的帧】

功能：替换 nCache 中的帧。

9.【Arnold】图标

在【Arnold】工具架中主要包括 11 个命令图标，如图 8.16 所示。图标名称和功能介绍如下。

图 8.16　【Arnold】图标

（1）　Flush Texture Cache【清空纹理缓存】

功能：清空纹理缓存中的数据。

（2）Flush Background Cache【清空后台缓存】

功能：清除场景后台所有缓存数据。

（3）Flush Quad Caches【清空四缓存】

功能：清空四缓存中的数据。

（4）Flush All Caches【清空所有缓存】

功能：清空所有缓存数据。

（5）TX Manager【TX 管理】

功能：方便用户对 TX 文件进行统一管理。

（6）Create Area Light【创建面光源】

功能：在网络上创建面光源。

（7）Create SkyDome Light【创建穹顶天光】

功能：场景中创建穹顶天光。

（8）Create Physical Sky【创建物理灯光】

功能：在场景中创建物理灯光。

（9）Create Mesh Light【创建网络灯光】

功能：在场景中创建物理灯光。

（10）Create Photometric Light【创建光度学灯光】

功能：在场景中创建光度学灯光。

（11）Arnold Render View【Arnold 渲染视图】

功能：为场景提供实时演染效果。

10.【Bifrost】图标

在【Bifrost】工具架中主要包括 8 个命令图标，如图 8.17 所示。图标名称和功能介绍如下。

图 8.17 【Bifrost】图标

（1）Liquid【液体】

功能：创建一个空 Bifrost 液体容器，若选择一个或多个多边形网络对象并单击该图标，则以选择的对象创建液体发射器。

（2）AERO【为 Authentic（真实）、Energetic（动感）、Reflective（反射）及 Open（开阔）四个单词的缩写】

功能：选择一个或多个多边形网络对象为 Aero 发射器。

（3）Emitter【发射器】

功能：选择一个或多个多边网络、Bifrost Liquid 或 Bifrost AERO 对象作为发射器。

（4）Collider【碰撞对象】

功能：选择一个或多个多边形网络、Bifrost Liquid 或 Bifrost AERO 对象作为碰撞对象。

(5) Foam【泡沫】

功能：选择 Bifrost Liquid 对象以将泡沫添加到模拟。

(6) Guide【导向】

功能：选择一个或多个多边形网络或 Bifrost Liquid 对象作为导向网络。

(7) Emission Region【发射区域】

功能：选择一个或多个多边形网络或 Bifrost Liquid 对象作为发射区域。发射区域仅用于导向模拟。

(8) Field【场】

功能：给选定的一个或多个液体或 AERO 容器添加场。

11. MASH Network【MASH 网络】图标

在 MASH Network【MASH 网络】工具架中主要包括 16 个命令图标，如图 8.18 所示。图标名称和功能介绍如下。

图 8.18　MASH Network【MASH 网络】图标

(1) Create MASH Network【创建 MASH 网络】

功能：创建 MASH 网络。

(2) MASH Editor【MASH 编辑器】

功能：打开 MASH 编辑器。

(3) Connect MASH to Type【将 MASH 连接到类型】

功能：将 MASH 连接到类型。

(4) Switch MASH Geometry Type【切换 MASH 几何体类型】

功能：切换到 MASH 几何体类型。

(5) Cache MASH Network【缓存 MASH 网络】

功能：缓存 MASH 网络。

(6) Add Trails to Particles【向粒子添加轨迹】

功能：给运动粒子添加轨迹。

(7) Create Mesh from points【从点创建网络】

功能：从点创建网络。

(8) Create MASH Points Node【创建 MASH 点节点】

功能：创建 MAHS 点节点。

(9) Polygon Type【多边形文字】

功能：在栅格上创建多边形文字。

(10) SVG【SVG】

功能：使用剪贴板中的可扩展向量图形或导入的 SVG 文件创建多边形。

(11) Curve Warp【曲线扭曲】

功能：沿路经变形，使选择的对象沿 NURBS 曲线路径变形。

(12) Create MASH Deformer【创建 MASH 变形器】

功能：先选择一个 Waiter，然后选择要变形的对象，单击图标【创建 MASH 变形器】，即可给选择对象添加变形器。

(13) Create Jjggle Deformer【创建抖动变形器】

功能：为选择对象创建抖动变形器。

(14) Create Blend Deformer【创建融合变形器】

功能：为选择对象创建融合变形器。

(15) Texture Deformer【纹理变形】

功能：为选择对象添加纹理变形。

(16) Import Alembic…【导入 Alembic…】

功能：将 Alembic 文件导入当前场景中。

12. Motion Graphics【动态图像】图标

在 Motion Graphics【动态图像】工具架中主要包括 25 个命令图标，如图 8.19 所示。图标名称和功能介绍如下。

图 8.19 Motion Graphics【动态图像】

(1) Create MASH Network【创建 MASH 网络】

功能：给选定对象添加 MASH 网络。

(2) Connect MASH to Type【将 MASH 连接到类型】

功能：将 MASH 连接到选定的 MASH_Waiter 和类型变换对象。

(3) Create MASH Deformer【创建 MASH 变形器】

功能：给对象添加 MASH 变形器。

(4) Polygon Sphere【多边形球体】

功能：在栅格上创建多边形几何体。

(5) Polygon Type【多边形文本】

功能：在栅格上创建多边形文本。

(6) SVG【SVG】

功能：使用剪贴板中的可扩展向量图形或导入的 SVG 文件创建多边形。

(7) Bevel【倒角】

功能：沿选定边或面创建倒角。

(8) Bridge【桥接】

功能：在两组边或面之间创建桥接面。

（9）Extrude【挤出】

功能：对选定的组件进行挤出。

（10）Add Divisions【添加细分】

功能：对选定边或面进行细分。

（11）EP Curve Tool【EP 曲线工具】

功能：在栅格或对象上创建选定类型的曲线。

（12）Bend【弯曲】

功能：给选定对象添加弯曲变形器，对对象进行弯曲变形处理。

（13）Texture Deformer【纹理变形器】

功能：给选定对象添加纹理变形器，对对象进行纹理变形处理。

（14）Blend Shapes【融合变形】

功能：选择需要融合变形的两个或多个对象，最后选择基础形状，单击该命令即可进行融合变形。

（15）Curve Warp【曲线扭曲】

功能：使选择的对象沿曲线进行扭曲变形。

（16）Camera【摄影机】

功能：在栅格上创建摄影机。

（17）Ambient Light【环境光】

功能：在栅格或激活的曲面上创建灯光。

（18）3D Container【3D 流体容器】

功能：创建 3D 流体容器。

（19）Emitter【发射器】

功能：在栅格或激活的对象上创建 n 粒子发射器。

（20）Emit from Object【从对象发射粒子】

功能：创建从对象发射 n 粒子。

（21）nParticle Tool【n 粒子工具】

功能：在栅格或激活的对象上创建 n 粒子。

（22）Radial【创建径向场】

功能：给选择的对象创建场特效。

（23）Fill Object【填充对象】

功能：给选择的网络填充 n 粒子。

（24）Soft Body【柔体】

功能：给选择的对象创建柔体。

（25）Set Key【设置关键帧】

功能：给选择的对象设置关键帧。

13.【XGen】图标

XGen 是以任意数量的随机或均匀放置的基本体填充多边形网络曲面的几何体实例化

第 8 章 工具属性及工具架图标

器。使用 XGen 在曲面上创建任意基本体，例如：头发、毛发、羽毛、鳞片和岩石等。在【XGen】工具架中主要包括 30 个命令图标，如图 8.20 所示。图标名称和功能介绍如下。

图 8.20 【XGen】图标

（1）Open the XGen Window【打开 XGen 窗口】

功能：打开 XGen 窗口。

（2）Update the XGen Preview【更新 XGen 预览】

功能：对 XGen 预览进行更新。

（3）Clear the XGen Preview【清除 XGen 预览】

功能：对 XGen 预览进行清除。

（4）Create a new XGen Description【创建新的 XGen 描述】

功能：打开 Create a new XGen Description【创建 XGen 描述】对话框，共用户进行设置 XGen 描述和创建。

（5）Patches bindings menu【面片绑定菜单】

功能：单击该图标，弹出下拉菜单。在菜单中提供【添加选定面】、【替换为选定面】、【移除选定面】、【基于贴图绑定面…】和【选择边界面】5 个选项选择，单击任意选项，即可执行相应命令。

（6）Add or Move Guides【添加或移动导向】

功能：对选择的导向进行添加或移动。

（7）Hide/Show the Guides of the current XGen Description【隐藏/显示当前 XGen 描述的导向】

功能：对当前 XGen 描述的导向进行隐藏或显示。

（8）Turn the selectability of Guides on and off【启用和禁用导向的可选性】

功能：对导向的可选性进行启用或禁用。

（9）Mirror Selected Guides across the X-axis【绕 X 轴镜像选定导向】

功能：对选定导向以 X 轴进行镜像。

（10）Match XGen patch visibility to geometry visibility【将 XGen 面片可见性匹配到几何体可见性】

功能：将 XGen 面片可见性匹配到几何体可见性。

（11）Flip Selected between patches and their associate geometry【翻转选定面片及其关联的几何体】

功能：对选定面片及其关联的几何体进行翻转。

（12）Select Primitives【选择基本体】

功能：使用鼠标中键对基本体进行隔离。

（13）Cull Selected Primitives【消隐选定基本体】

功能：在"生成器"选项卡中对选择的多个基本体进行消隐。

（14）Sculpt guides【雕刻导向】

功能：对导向进行雕刻。

（15）Convert Primitives to polygons【将基本体转化为多边形】

功能：将选定的基本体转换为多边形。

（16）Open the Interactive Groom Editor【打开交互式修饰编辑器】

功能：单击该图标，打开 Open the Interactive Groom Editor【打开交互式修饰编辑器】共读者操作。

（17）Create Hair and fur using interactive groom splines【使用交互式修饰样条线创建头发和毛发】

功能：使用交互式修饰样条线创建头发和毛发。

（18）Density Tool【密度工具】

功能：对选定的头发数量进行增加或减少。

（19）Place Tool【放置工具】

功能：添加单根头发。

（20）Length Tool【长度工具】

功能：对头发进行加长或缩短。

（21）Cut Tool【切割工具】

功能：对头发进行修剪。

（22）Width Tool【框度工具】

功能：对头发的框度进行缩放。

（23）Comb Tool【梳理工具】

功能：按笔刷笔划的方向更改头发的方向和弯曲度。

（24）Grab Tool【抓取工具】

功能：沿任意方向拉动选择的头发。

（25）Smooth Tool【平滑工具】

功能：拉直头发或平滑不同尺寸和形状的头发之间的过渡。

（26）Noise Tool【噪波工具】

功能：绘制 3D 噪波以变化头发方向，让头发看起来更自然。

（27）Clump Tool【成束工具】

功能：拉动头发以将头发扎成束。

（28）Part Tool【分离工具】

功能：将头发彼此分离。

（29）Freeze Tool【冻结工具】

功能：锁定头发，使其无法修改。

（30）Select Tool【选择工具】

功能：通过绘制选择头发。

第 9 章　界面与快捷菜单

本章主要介绍 Maya 2018 的界面布局模式和快捷菜单。

9.1　Maya 2018 界面布局模式的分类

Maya 2018 界面布局模式主要有【Maya Classic（Maya 经典）】、【Maya-Standard（建模-标准）】、【Maya-Expert（建模-专家）】、【Sculpting（刻）】、【Pose Sculpting（姿势雕刻）】、【UV Editing（UV 编辑）】、【XGen（XGen）】、【XGen-Interactive Groom（XGen-交互式修饰）】、【Rigging（装备）】和【Animation（动画）】10 种。

1.【Maya Classic（Maya 经典）】模式

【Maya Classic（Maya 经典）】模式界面，如图 9.1 所示。

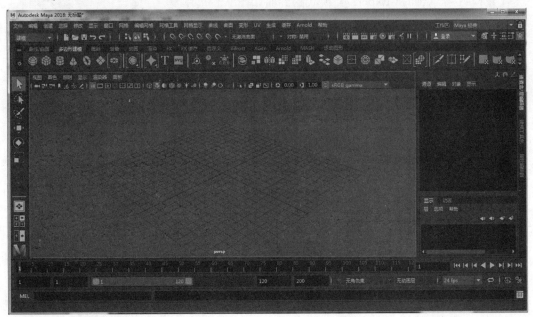

图 9.1　【Maya Classic（Maya 经典）】模式

2.【建模-标准（Maya-Standard）】

【建模-标准（Maya-Standard）】模式的界面如图 9.2 所示。

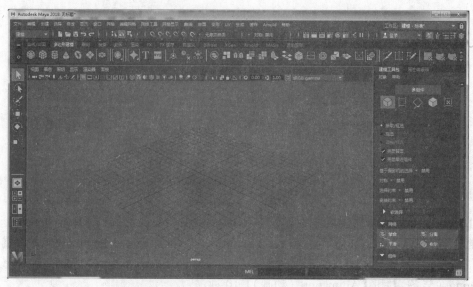

图 9.2 【Maya-Standard（建模-标准）】

3.【建模-专家（Maya-Expert）】

【建模-专家（Maya-Expert）】模式的界面如图 9.3 所示。

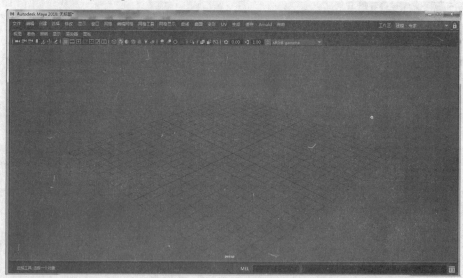

图 9.3 【Maya-Expert（建模-专家）】

4.【Sculpting（雕刻）】

【Sculpting（雕刻）】模式的界面如图 9.4 所示。

第 9 章　界面与快捷菜单

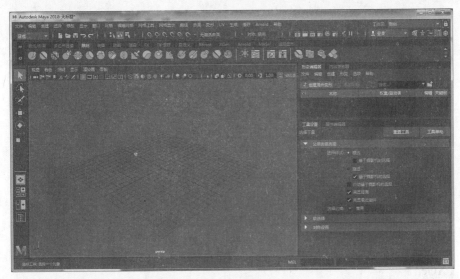

图 9.4　【Sculpting（雕刻）】

5.【Pose Sculpting（姿势雕刻）】

【Pose Sculpting（姿势雕刻）】模式的界面如图 9.5 所示。

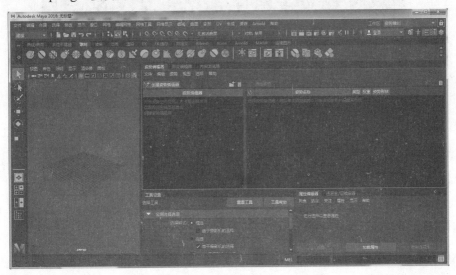

图 9.5　【Pose Sculpting（姿势雕刻）】

6.【UV Editing（UV 编辑）】

【UV Editing（UV 编辑）】模式的界面如图 9.6 所示。

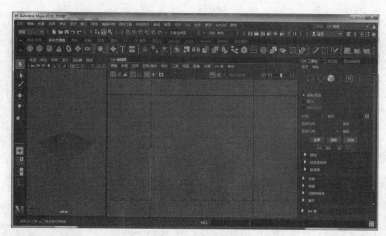

图 9.6 【UV 编辑】

7.【XGen】

【XGen】模式的界面如图 9.7 所示。

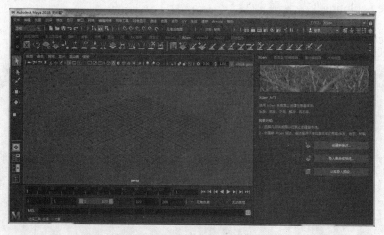

图 9.7 【XGen】模式的界面

8.【XGen-Interactive Groom（XGen-交互式修饰）】

【XGen-Interactive Groom（XGen-交互式修饰）】的界面如图 9.8 所示。

9.【装备（Rigging）】

【装备（Rigging）】模式的界面如图 9.9 所示。

10.【Animation（动画）】

【Animation（动画）】模式的界面如图 9.10 所示。

第 9 章 界面与快捷菜单

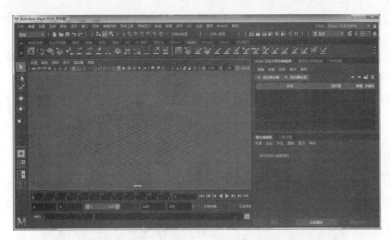

图 9.8 【XGen-Interactive Groom（XGen-交互式修饰）】

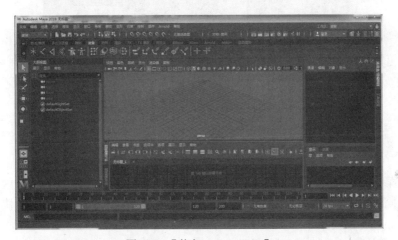

图 9.9 【装备（Rigging）】

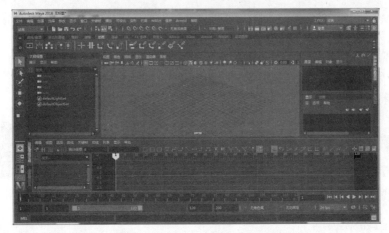

图 9.10 【Animation（动画）】

9.2 快捷菜单介绍

为提高工作效率，Maya 2018 为用户提供了标记快捷菜单、右键开捷菜单和热键快捷菜单等。

1. 标记菜单

标记菜单包括了 Maya 2018 中所有的菜单命令，将鼠标放在工作区中的任意位置，单击空格键，即可调出该菜单，如图 9.11 所示。

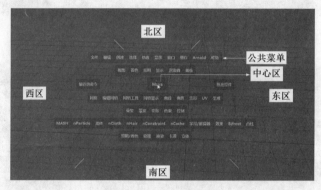

图 9.11 【标记菜单】

在标记菜单中分【中心区】、【东区】、【西区】、【南区】和【北区】5 个区。在按住空格键不放的同时，将鼠标移到不同的区域，单击鼠标左键，即可弹出不同的快捷菜单。

1)【中心区】快捷菜单

【中心区】快捷菜单如图 9.12 所示。

2)【东区】快捷菜单

【东区】快捷菜单如图 9.13 所示。

3)【西区】快捷菜单

【西区】快捷菜单如图 9.14 所示。

图 9.12 【中心区】快捷菜单

图 9.13 【东区】快捷菜单

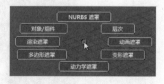
图 9.14 【西区】快捷菜单

4)【南区】快捷菜单

【南区】快捷菜单如图 9.15 所示。

第 9 章 界面与快捷菜单

5)【北区】快捷菜单

【北区】快捷菜单如图 9.16 所示。

2. 右键菜单

右键快捷菜单使用起来非常方便,种类也非常多,在不同的对象及不同状态下打开的快捷菜单也不相同。下面介绍几种常用的右键快捷菜单。

(1) 按住 Shift,同时键单击鼠标右键,并按住鼠标右键不放。此时,会弹出一个【创建多边形对象】的快捷菜单,如图 9.17 所示。

图 9.15 【南区】快捷菜单　　图 9.16 【北区】快捷菜单　　图 9.17 【创建多边形对象】快捷菜单

(2) 在创建的多边形对象上按住 Shift 键和鼠标右键。此时,弹出多边形的一些编辑命令,如图 9.18 所示。

(3) 直接在创建的多边形上按住鼠标右键,此时,弹出一个显示多边形元素的快捷菜单,如图 9.19 所示。

(4) 在创建的 NURBS 曲线上,按住 Shift 键的同时,按住鼠标右键不放,则弹出一些曲线编辑的命令,如图 9.20 所示。

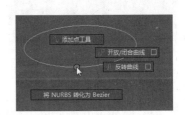

图 9.18 【多边形的一些编辑命令】快捷菜单　　图 9.19 【多边形元素显示】快捷菜单　　图 9.20 【曲线命令显示】快捷菜单

（5）直接在创建的 NURBS 曲线上，按住鼠标右键，则弹出 NURBS 曲线的元素切换快捷菜单，如图 9.21 所示。

（6）在 NURBS 对象上，直接按住鼠标右键，则弹出显示 NURBS 对象元素切换命令的快捷俄菜单，如图 9.22 所示。

图 9.21 【NURBS 曲线的元素切换命令】快捷菜单　　图 9.22 【NURBS 对象元素切换命令】快捷菜单

3. 热键快捷菜单

在 Maya2018 中，用户需要掌握常用的热键快捷菜单。

（1）按住 A 键并单击鼠标左键，弹出【控制对象的输入和输出节点的选择】快捷菜单，如图 9.23 所示。

（2）按住 Q 键并单击鼠标左键，弹出【选择遮罩命令切换】的快捷菜单，如图 9.24 所示。

（3）按住 O 键并单击鼠标左键，弹出【多边形各种元素的选择和编辑菜单命令切换】的快捷菜单，如图 9.25 所示。

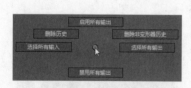  

图 9.23 【控制对象的输入和输出　　图 9.24 【选择遮罩命令　　图 9.25 【多边形各种元素的选择
　　　　节点的选择】快捷菜单　　　　　　　切换】快捷菜单　　　　　和编辑菜单命令切换】快捷菜单

（4）按 W/E/R 键并单击鼠标左键，弹出【各种坐标方向的选择菜单命令切换】的快捷菜单，如图 9.26 所示。

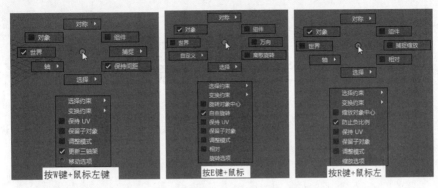

图9.26 【各种坐标方向的选择菜单命令切换】快捷菜单

第 10 章　创建对象的属性

本章主要介绍 Maya 2018 中创建对象的属性，主要包括多边形对象、NURBS、体积对象、摄影机和类型等对象属性。

10.1　Polygon【多边形】对象公共属性面板

所谓 Polygon【多边形】公共属性面板，是指 Maya 2018 中所有 Polygon【多边形】对象都有的属性。在场景中单选 Polygon【多边形】对象，按键盘上的"Ctrl+A"组合键，即可在界面的右侧显示多边形公共属性面板如图 10.1 所示。

在【多边形公共属性】面板中主要包括了 Transform Attributes【变换属性】、Pivots【枢轴】、Limit Information【限制信息】、Display【显示】、Node Behavior【节点行为】、【UUID】和 Extra Attributes【附加属性】7 个卷展开栏。

1. Transform Attributes【变换属性】卷展栏

Transform Attributes【变换属性】卷展栏参数主要用来调节对象的位置、旋转、斜切和旋转轴，详细参数如图 10.2 所示。

图 10.1　Polygon Attributes
【多边形公共属性】面板

图 10.2　Transform Attributes
【变换属性】卷展栏

在 Rotate Order【旋转顺序】参数组中提供【zyz】、【yzx】、【zxy】、【xzy】、【yxz】和【zyx】6 种旋转顺序选择。

2. Pivots【枢轴】卷展栏

Pivots【枢轴】卷展栏参数主要用来调节置枢轴、局部空间和世界空间相，详细参数如图 10.3 所示。

3. Limit Information【限制信息】

Limit Information【限制信息】卷展栏参数主要用来调节平移、旋转和缩放的最大、当

第 10 章 创建对象的属性

前和最小值范围，详细参数如图 10.4 所示。

图 10.3 Pivots【枢轴】卷展栏

图 10.4 Limit Information【限制信息】卷展栏

4．Display【显示】卷展栏参数

Display【显示】卷展栏参数主要用来调节选择控制柄、超纵器、重影信息、边界框信息、绘制覆盖和选择等相关参数，详细参数如图 10.5 所示。

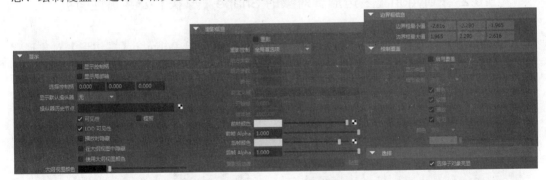

图 10.5 Display【显示】卷展栏

（1）Show Manip Default【显示默认操纵柄】参数组：在该参数组中提供了 None【无】、Translate【转换】、Rotate【旋转】、Scale【比例】、Transform【变换】、Global Default【全局默认】、Smart【智能】和 Specified【以指定】8 种操纵柄方式。

（2）Ghosting Control【重影控制】参数组：在该参数组中提供了 Global Prefs【全局首选项】、Global Frames【自定义帧】、Custom Frame Steps【自定义帧步数】、Custom Key Steps【自定义关键帧步数】和 Key Frames【关键帧】5 种控制重影的方式。

5．Node Behavior【节点行为】卷展栏参数

Node Behavior【节点行为】卷展栏参数主要用来调节缓存、冻结和节点状态等参数。详细参数如图 10.6 所示。

在 Node State【节点状态】参数组中提供 Normal【正常】、Has No Effect【无效果】、

Blocking【阻塞】、Waiting-Normal【等待-正常】、Waiting-Has No Effect【等待-无效果】和 Waiting-Blocking【等待-阻塞】6 种状态供选择。

6.【UUID】参数卷展栏参数

【UUID】参数卷展栏中没有相关参数，只显示了【UUID】的序列号，如图 10.7 所示。

图 10.6　Node Behavior【节点行为】卷展栏

图 10.7　【UUID】卷展栏

7. Extra Attributes【附加属性】参数卷展栏参数

Extra Attributes【附加属性】卷展栏参数在正常情况下没有任何参数可设置，只有在通过 Load Attributes【加载属性】按钮加载属性之后才可进行相关参数设置，如图 10.8 所示。

10.2　Polygon【多边形】对象的 Shape【形状】公共属性面板

Polygon【多边形】对象的 Shape【形状】公共属性是指 Polygon【多边形】对象形状共有的属性。属性面板卷展栏参数面板如图 10.9 所示。

图 10.8　Extra Attributes
【附加属性】卷展栏

图 10.9　Polygon【多边形】对象的 Shape【形状】
公共属性面板

在 Polygon【多边形】对象的 Shape【形状】公共属性面板中主要包括 Tessellation Attributes【细分属性】、Mesh Component Display【网络组件显示】、Mesh Controls【网络控制】、Tangent Space【切线空间】、Smooth【平滑网络】、Displacement Map【置换贴图】、Render Stats【渲染统计信息】、Object Display【对象显示】、Arnold、Node Behavior【节点行为】、【UUID】和 Extra Attributes【附加属性】12 个卷展栏参数。其中，Node Behavior【节点行为】、【UUID】和 Extra Attributes【附加属性】卷展栏参数在【Polygon 公共属性】面板中已经详细介绍，在此就不再赘述。其他各卷展栏参数介绍如下。

第 10 章 创建对象的属性

1. Tessellation Attributes【细分属性】卷展栏参数

Tessellation Attributes【细分属性】卷展栏参数，主要用来调节对象的最大三角形数、最大细分数、最大 UV、最小边长和最大边长等相关属性，如图 10.10 所示。

2. Mesh Component Display【网络组件显示】卷展栏参数

Mesh Component Display【网络组件显示】卷展栏参数，主要用来控制对象的网络组件的相关显示和隐藏，如图 10.11 所示。

图 10.10 Tessellation Attributes
【细分属性】卷展栏

图 10.11 Mesh Component Display
【网络组件显示】卷展栏

（1）Backface Culling【背面消隐】参数组：在该参数组中提供了 Off【禁用】、Wire【线】、Hard【硬边】和 Full【完全】4 种消隐方式，默认 Off【禁用】消隐方式。

（2）Display Edges【显示边】参数组：在该参数组中提供了 Standard【标准】、Soft/Hard【软/硬】、Hard（Color）【硬边（颜色）】和 Hard【硬边】4 种显示方式，默认 Standard【标准】显示方式。

（3）Display Color Channel【显示颜色通道】参数组：在该参数组中提供了 None【无】、Ambient【环境光】、Ambient+ Diffuse【环境光+漫反射】、Diffuse【漫反射】、Specular【镜面反射】和 Emission【发射】6 种显示颜色通道的方式。

（4）Material Blend【材质混合】参数组：在该参数组中提供了 Over Write【覆盖】、Add【相加】、Subtract【相减】、Multiply【相乘】、Divide【相除】、Average【平均】和 Modulate 2x【相乘再乘 2】7 种材质混合模式，默认 Overwrite【覆盖】混合类型。

（5）Normal Type【法线类型】参数组：在该参数组中提供了 Face【面】、Vertex【顶点】和 Vertex face【顶点面】3 种法线类型，默认 Face【面】法线类型。

3. Mesh Controls【网络控制】卷展栏参数

Mesh Controls【网络控制】卷展栏参数，主要用来调节顶点颜色源、四边形分割和顶点发现方法等相关参数，如图 10.12 所示。

（1）Vertex Color Source【顶点颜色源】参数组：该参数组中提供了 None【无】、Current Color Set【当前颜色集】和 Influence Colors【影响颜色】3 种顶点颜色源，默认 None【无】顶点颜色源。

（2）Quad Split【四边形分割】参数组：该参数组中提供了 Left【左侧】、Right【右侧】和 Best Shape【最佳形状】3 种顶点颜色源，默认 Best Shape【最佳形状】四边形分割方式。

（3）Vertex Normal Method【顶点法线方法】参数组：该参数组中提供了 Unweighted【未加权】、Angle Weighted【加权角度】、Area Weighted【加权区域】和 Angle And Area Weighted【加权角度和区域】4 种顶点颜色源，默认 Angle And Area Weighted【加权角度和区域】顶点法线方法。

4. Tangent Space【切线空间】卷展栏参数

Tangent Space【切线空间】卷展栏参数，主要用来调节坐标系、UV 缠绕顺序、切线平滑角度和切线法线阈值等相关参数，如图 10.13 所示。

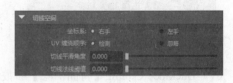

图 10.12　Mesh Controls【网络控制】卷展栏　　　图 10.13　Tangent Space【切线空间】卷展栏

5. Smooth【平滑网络】卷展栏参数

Smooth【平滑网络】卷展栏参数，主要用来调节对象的细分级别、Open Subdiv 控件和 Maya Catmull-Clark 控件等相关参数，如图 10.14 所示。

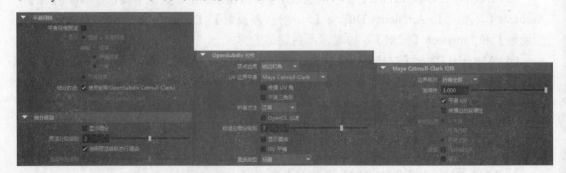

图 10.14　Smooth【平滑网络】卷展栏

（1）Vertex Boundary【顶点边界】参数组：该参数组中提供了 Sharp Edges and Corners【锐边和角】和 Sharp Edges【锐边】2 种顶点边界，默认 Sharp Edges and Corners【锐边和角】顶点边界方式。

（2）UV Boundary Smoothing【UV 边界平滑】参数组：该参数组中提供了 None【无】、Preserve Edges and Corners【保留边和角】、Preserve Edges【保留边】和【Maya Catmull-Clark】4 种 UV 边界平滑方式，默认【Maya Catmull-Clark】UV 边界平滑方式。

（3）Crease Method【折痕方法】参数组：该参数组中提供了 Normal【正常】和 Chaikin【细分】2 种折痕方法，默认 Normal【正常】折痕方法。

（4）Displacement Type【置换类型】参数组：该参数组中提供了 Scalar【标量】和 Vector（global space）【向量（全局空间）】2 种置换类型。默认 Scalar【标量】置换类型。

（5）Boundary Rules【边界规则】参数组：该参数组中提供了 Legacy【旧版】、Crease All【折痕全部】和 Crease Edges【折痕边】3 种边界规则，默认 Crease All【折痕全部】边界规则。

6. Displacement Map【置换贴图】卷展栏参数

Displacement Map【置换贴图】卷展栏参数，主要用来调节初始采样率、额外采样率、纹理阈值、法线阈值和边界框缩放等相关参数，如图 10.15 所示。

7. Render Stats【渲染统计信息】卷展栏

Render Stats【渲染统计信息】卷展栏参数，主要用来调节对象的阴影、可见性、显示方式和 3D 运动模糊等相关参数，如图 10.16 所示。

图 10.15 Displacement Map【置换贴图】卷展栏

图 10.16 Render Stats【渲染统计信息】卷展栏

8. Object Display【对象显示】卷展栏参数

Object Display【对象显示】卷展栏参数，主要用来调节对象显示、重影信息、边界框信息、绘制覆盖和选择等信息，如图 10.17 所示。

图 10.17 Object Display【对象显示】卷展栏

其中，Ghosting Control【重影控制】参数组提供了 Global Prefs【全局首选项】、Custom Frames【自定义帧】、Custom Frame Steps【自定义帧布数】、Custom Key Steps【自定义关键帧步数】和 Keyframes【关键帧】5 种控制重影的方式。默认 Global Prefs【全局首选项】重影控制方式。

9.【Arnold】卷展栏参数

【Arnold】卷展栏参数主要用来调节 Arnold【阿诺德】、Subdivision【细分】、Displacement Attributes【置换属性】和 Volume Attributes【体积属性】等参数，如图 10.18 所示。

图 10.18 【Arnold】卷展栏

（1）Arnold Translator【阿诺德翻译公司】参数组：在该参数组中提供了 Mesh _light【多边形网络光】、Poly mesh【多边形网络】和 Procedural【过程】3 种类型，默认 Poly mesh【多边形网络】类型。

（2）Motion Vector Unit【运动矢量单元】参数组：在该参数组中提供了 Per Frame【每帧】和 Per Second【每秒】2 种运动矢量单元。

（3）Type【类型】参数组：在该参数组中提供了 None【无】、Catclark【卡特莫尔克拉克】和 Linear【线性】3 种细分类型。

（4）Adaptive Metric【自适应度量】参数组：在该参数组中提供了 Auto【自动】、Edge_length【边缘长度】和 Flatness【平面度】3 种自适应度量方式。

（5）Adaptative Space【适应空间】参数组：在该参数组中提供了 raster【栅格】和 object【对象】2 种适应空间类型。

（6）UV Smoothing【UV 平滑】参数组：在该参数组中提供了 Pin_corners【大头针角】、Pin_borders【大头针边界】、Linear【线性】和 Smooth【平滑】。

10.3 Polygon Primitives【多边形基本体】对象属性面板

在 Maya 中主要有 Sphere【球体】、Cube【立方体】、Cylinder【圆柱体】、Cone【圆锥体】、Plane【平面】、Torus【圆环】、Prism【棱柱】、Pyramid【棱锥】、Pipe【管道】、Helix【螺旋线】、Soccer【足球】和 Platonic Solids【柏拉图多面体】12 个 Polygon Primitives【多边形基本体】对象。各个对象属性面板的具体功能介绍如下。

1. Sphere【球体】对象属性面板

Sphere【球体】对象属性面板主要用来调节球体的半径、轴向细分数、高度细分数和创建 UV 的方式。Sphere【球体】对象属性面板如图 10.19 所示。

在 Create UV【创建 UV】参数组中，提供了 None【无】、Pinched at Pole【已收缩到极点】和 Sawtooth at Pole【极点处的锯齿形】3 种创建 UV 的方式，默认 Sawtooth at Pole【极点处的锯齿形】方式。

2. Cube【立方体】对象属性面板

Cube【立方体】对象属性面板主要用来调节立方体的宽度、高度、深度、细分宽度、高度细分、深度细分和创建 UV 的方式。Cube【立方体】对象属性面板如图 10.20 所示。

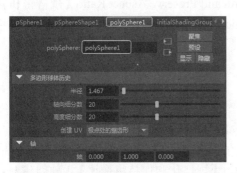

图 10.19　Sphere【球体】对象属性面板

图 10.20　Cube【立方体】对象属性面板

Create UV【创建 UV】参数组提供了 None【无】、Normalization Off【规格化关闭】、Normalize Each Face Separately【分别规格化每个面】、Normalize Collectively【整体规格化】和 Normalize Collectively and Preserve Aspect Ratio【整体规格化和保持纵横比】5 种创建 UV 的方式。默认 Normalize Collectively【整体规格化】方式。

3. Cylinder【圆柱体】对象属性面板

Cylinder【圆柱体】对象属性面板主要用来调节圆柱体的半径、高度、轴向细分数、高度细分数、端面细分数和创建 UV 方式。Cylinder【圆柱体】对象属性面板如图 10.21 所示。

在 Create UV【创建 UV】参数组中,提供了 None【无】、Normalization Off【规格化关闭】、Normalization【规格化】和 Normalization and Preserve Aspect Ratio【规格化和保持纵横比】4 种创建 UV 的方式。默认 Normalization and Preserve Aspect Ratio【规格化和保持纵横比】方式。

4. Cone【圆锥体】对象属性面板

Cone【圆锥体】对象属性面板,主要用来调节圆锥体的半径、高度、轴向细分数、高度细分数、端面细分数和创建 UV 方式。Cone【圆锥体】对象属性面板如图 10.22 所示。

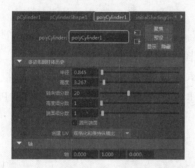   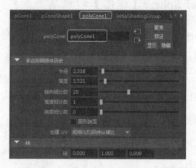

图 10.21　Cylinder【圆柱体】对象属性面板　　图 10.22　Cone【圆锥体】对象属性面板

在 Create UV【创建 UV】参数组中,提供了 None【无】、Normalization Off【规格化关闭】、Normalization【规格化】和 Normalization and Preserve Aspect Ratio【规格化和保持纵横比】4 种创建 UV 的方式。默认 Normalization and Preserve Aspect Ratio【规格化和保持纵横比】方式。

5. Plane【平面】对象属性面板

Plane【平面】对象属性面板,主要用来调节平面的宽度、高度、细分宽度、高度细分和创建 UV 方式。Plane【平面】对象属性面板如图 10.23 所示。

在 Create UV【创建 UV】参数组中,提供了 None【无】、Normalization Off【规格化关闭】和 Normalization and Preserve Aspect Ratio【规格化和保持纵横比】3 种创建 UV 的方式。默认 Normalization and Preserve Aspect Ratio【规格化和保持纵横比】方式。

6. Torus【圆环】对象属性面板

Torus【圆环】对象属性面板,主要用来调节圆环的半径、截面半径、扭曲、轴向细分数、高度细分数和是否创建 UV。Torus【圆环】对象属性面板如图 10.24 所示。

第 10 章 创建对象的属性

图 10.23　Plane【平面】对象属性面板

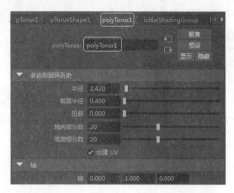

图 10.24　Torus【圆环】对象属性面板

7．Prism【棱柱】对象属性面板

Prism【棱柱】对象属性面板，主要用来调节棱柱的长度、侧面长度、边数、高度细分数、端面细分数和创建 UV 的方式。Prism【棱柱】对象属性面板如图 10.25 所示。

在 Create UV【创建 UV】参数组中，提供了 None【无】、Normalization Off【规格化关闭】、Normalization【规格化】和 Normalization and Preserve Aspect Ratio【规格化和保持纵横比】4 种创建 UV 的方式，默认 Normalization and Preserve Aspect Ratio【规格化和保持纵横比】方式。

8．Pyramid【棱锥】对象属性面板

Pyramid【棱锥】对象属性面板，主要用来调节棱锥的侧面长度、边数、高度细分数、端面细分和创建 UV 的方式。Pyramid【棱锥】对象属性面板如图 10.26 所示。

图 10.25　Prism【棱柱】对象属性面板

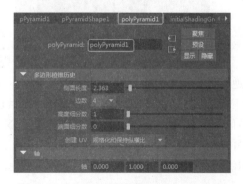

图 10.26　Pyramid【棱锥】对象属性面板

（1）在 Number Of Sides【边数】参数组中，提供了【3】、【4】和【5】3 种边的棱锥模式。

（2）在 Create UV【创建 UV】参数组中，提供了 None【无】、Normalization Off【规格化关闭】、Normalization【规格化】和 Normalization and Preserve Aspect Ratio【规格化和

保持纵横比】4 种创建 UV 的方式，默认 Normalization and Preserve Aspect Ratio【规格化和保持纵横比】方式。

9. Pipe【管道】对象属性面板

Pipe【管道】对象属性面板，主要用来调节管道的半径、高度、厚度、轴向细分数、高度细分数、端面细分数、圆形端面和创建 UV 方式。Pipe【管道】对象属性面板如图 10.27 所示。

10. Helix【螺旋线】对象属性面板

Helix【螺旋线】对象属性面板，主要用来调节螺旋线的圈数、高度、宽度、半径、轴向细分数、圈细分数、端面细分数、圆形端面、方向和创建 UV 方式。Helix【螺旋线】对象属性面板如图 10.28 所示。

图 10.27 Pipe【管道】对象属性面板

图 10.28 Helix【螺旋线】对象属性面板

（1）在 Direction【方向】参数组中，提供了 Clockwise【顺时针】e 和 Counterclockwise【逆时针】2 种方向创建方式。

（2）在 Create UV【创建 UV】参数组中，提供了 None【无】、Normalization Off【规格化关闭】、Normalization【规格化】和 Normalization and Preserve Aspect Ratio【规格化和保持纵横比】4 种创建 UV 的方式，默认 Normalization and Preserve Aspect Ratio【规格化和保持纵横比】方式。

11. Soccer【足球】对象属性面板

Soccer【足球】对象属性面板，主要用来调节足球的半径、侧面长度和创建 UV 的方式。Soccer【足球】对象属性面板，如图 10.29 所示。

在 Create UV【创建 UV】参数组中，提供了 None【无】、Normalization Off【规格化关闭】、Normalize Each Face Separately【分别规格化每个面】、Normalize Collectively【整体

规格化】和 Normalize Collectively and Preserve Aspect Ratio【整体规格化和保持纵横比】5
种创建 UV 的方式。默认 Normalize Collectively and Preserve Aspect Ratio【整体规格化和保
持纵横比】方式。

12. Platonic Solids【柏拉图多面体】属性面板

Platonic Solids【柏拉图多面体】属性面板，主要用来调节柏拉图多面体的半径、侧面
长度、实体类型和创建 UV 的方式。Platonic Solids【柏拉图多面体】属性面板，如图 10.30
所示。

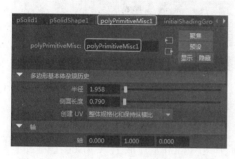

图 10.29　Soccer【足球】对象属性面板

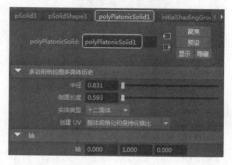

图 10.30　Platonic Solids【柏拉图多面体】属性面板

（1）在 Solid Type【实体类型】参数组中，提供了 Dodecahedron【十二面体】、Icosahedron
【二十面体】、Octahedron【八面体】和 Tetrahedron【四面体】4 种实体类型。

（2）在 Create UV【创建 UV】参数组中，提供了 None【无】、Normalization Off【规
格化关闭】、Normalize Each Face Separately【分别规格化每个面】、Normalize Collectively
【整体规格化】和 Normalize Collectively and Preserve Aspect Ratio【整体规格化和保持纵横
比】5 种创建 UV 的方式。默认 Normalize Collectively and Preserve Aspect Ratio【整体规格
化和保持纵横比】方式。

10.4　NURBS Primitives【NURBS 基本体】对象的 Shape【形状】公共属性面板

NURBS Primitives【NURBS 基本体】对象的 Shape【形状】公共属性是指 NURBS 对
象形状共有的属性。属性面板卷展栏参数面板如图 10.31 所示。

在 NURBS Primitives【NURBS 基本体】对象的 Shape【形状】公共属性面板中主要包
括了 NURBS Surface History【NURBS 曲面历史】、NURBS Surface Display【NURBS 曲面
显示】、Tessellation【细分】、Texture Map【纹理贴图】、Displacement Map【置换贴图】、
Render Stats【渲染统计信息】、Object Display【对象显示】、Arnold【阿诺德】、Behavior【节
点行为】、【UUID】和 Extra Attributes【附加属性】11 个卷展栏参数。其中 Arnold【阿诺德】、
Node Behavior【节点行为】、【UUID】和 Extra Attributes【附加属性】卷展栏参数请参考本

章"10.2 Polygon【多边形】对象的 Shape【形状】公共属性面板"对应的卷展栏参数,在此就不再赘述。其他各卷展栏参数介绍如下。

1. NURBS Surface History【NURBS 曲面历史】卷展栏参数

NURBS Surface History【NURBS 曲面历史】卷展栏参数,主要用来显示曲面的 U 向最小/最大范围、V 向最小/最大范围、UV 向跨度数、UV 向次数、形式 U 和形式 V 等参数,该卷展栏参数不能修改,如图 10.32 所示。

图 10.31　NURBS Primitives【NURBS 基本体】对象的 Shape【形状】公共属性面板

图 10.32　NURBS Surface History【NURBS 曲面历史】卷展栏参数面板

(1) Form U【形式 U】参数组:在该参数组提供了 Open【开放】、Closed【闭合】和 Periodic【周期】3 种形式。

(2) Form V【形式 V】参数组:在该参数组提供了 Open【开放】、Closed【闭合】和 Periodic【周期】3 种形式。

2. NURBS Surface Display【NURBS 曲面显示】卷展栏参数

NURBS Surface Display【NURBS 曲面显示】卷展栏参数,主要用来调节曲线精度、曲线精度着色、简化模式、U 向简化、V 向简化、U 向分段数、V 向简化、U 向分段数、V 向分段数和法线显示比例等参数,如图 10.33 所示。

3. Tessellation【细分】卷展栏参数

Tessellation【细分】卷展栏参数,主要用来调节简单细分选项、主细分属性、次细分属性和公用细分选项等参数,如图 10.34 所示。

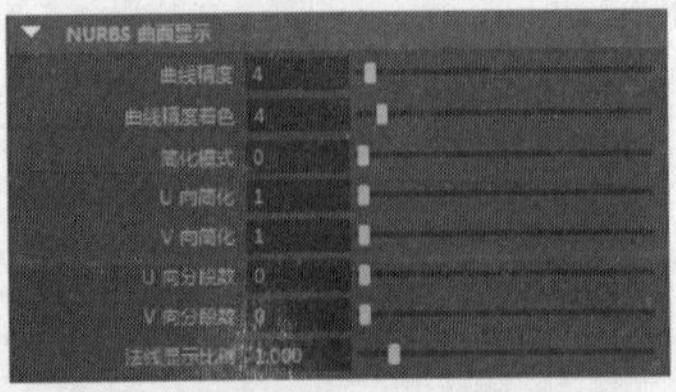

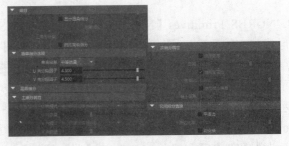

图 10.33　NURBS Surface Display【NURBS 曲面显示】卷展栏参数面板

图 10.34　Tessellation【细分】卷展栏参数面板

Curvature Tolerance【曲率容差】参数组：在该参数组提供了 Highest Quality【最高质量】、High Quality【高质量】、Medium Quality【中等质量】、Low Quality【低质量】和 No Curvature Check【不检查曲率】5 种曲率容差方式，默认 Medium Quality【中等质量】容差方式。

4. Texture Map【纹理贴图】卷展栏参数

Texture Map【纹理贴图】卷展栏参数主要用来修复纹理扭曲，卷展栏参数面板如图 10.35 所示。

5. Displacement Map【置换贴图】卷展栏参数

Displacement Map【置换贴图】卷展栏参数，主要用来调节特征置换、初始采样率、额外采样率、法线阈值和边界框缩放等参数，如图 10.36 所示。

6. Render Stats【渲染统计信息】卷展栏参数

Render Stats【渲染统计信息】卷展栏参数，主要用来调节是否启用投射阴影、接收阴影、透底、运动模糊、主可见性、平滑着色处理、在反射中可见、在折射中可见、双面、反射、反向、几何体抗锯齿覆盖、着色采样覆盖和 3D 运动模糊等参数，如图 10.37 所示。

图 10.35 Texture Map【纹理贴图】卷展栏参数面板

图 10.36 Displacement Map【置换贴图】卷展栏参数面板

图 10.37 Render Stats【渲染统计信息】卷展栏参数面板

7. Object Display【对象显示】卷展栏参数

Object Display【对象显示】卷展栏参数，主要用来调节对象显示、重影信息、边界框信息、绘制覆盖和选择等参数，如图 10.38 所示。

Ghosting Control【重影控制】参数组：在该参数组中提供了 Global Prefs【全局首选项】、Custom Frames【自定义帧】、Custom Frame Steps【自定义帧步数】、Custom Key Steps【自定义关键帧步数】和 Key frames【关键帧】5 种重影控制方式，默认 Global Prefs【全局首选项】控制方式。

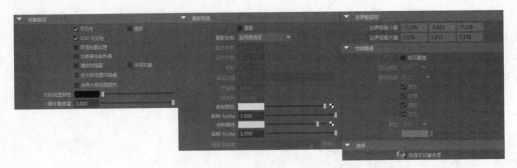

图 10.38　Object Display【对象显示】卷展栏参数面板

10.5　NURBS Curve【NURBS 曲线】对象的 Shape【形状】公共属性面板

NURBS Curve【NURBS 曲线】对象的 Shape【形状】公共属性是指 NURBS 曲线形状共有的属性。属性面板卷展栏参数面板如图 10.39 所示。

在 NURBS Curve【NURBS 曲线】对象的 Shape【形状】公共属性面板中主要包括了 NURBS Curve History【NURBS 曲线历史】、Component Display【组建显示】、Object Display【对象显示】、Arnold【阿诺德】、Node Behavior【节点行为】、【UUID】和 Extra Attributes【附加属性】7 卷展栏参数。其中 Object Display【对象显示】卷展栏参数请参考本章 "10.4 Object Display【对象显示】卷展栏参数" 面板，Arnold【阿诺德】、Node Behavior【节点行为】、【UUID】和 Extra Attributes【附加属性】4 个卷展栏参数请参考本章 "10.3 Polygon Primitives【多边形基本体】对象属性面板"，在此就不再赘述。

1. NURBS Curve History【NURBS 曲线历史】卷展栏参数

NURBS Curve History【NURBS 曲线历史】卷展栏参数，主要用来调节曲线的最小/最大值、跨度数、形式和次数等参数，如图 10.40 所示。

图 10.39　NURBS Curve【NURBS 曲线】　　　图 10.40　NURBS Curve History【NURBS
　对象的 Shape【形状】公共属性面板　　　　　　　曲线历史】卷展栏参数面板

Form【形式】参数组：在该参数组中提供了 Open【开放】、Closed【闭合】和 Periodic【周期】3 种形式，默认 Periodic【周期】形式。

第 10 章 创建对象的属性

2. Component Display【组建显示】卷展栏参数

Component Display【组建显示】卷展栏参数，主要用来控制是否启用显示 CV、显示 EP、显示壳线、显示几何体，如图 10.41 所示。

10.6　NURBS Primitives【NURBS 基本体】对象属性面板

在 Maya 中 NURBS 基本体对象主要包括 Sphere【球体】、Cube【立方体】、Cylinder【圆柱体】、Cone【圆锥体】、Plane【平面】、Torus【圆环】、Circle【圆形】和 Square【方形】8 个。各个对象属性面板的功能介绍如下。

1. Sphere【球体】对象属性面板

Sphere【球体】对象属性面板，主要用来调节开始扫描、结束扫描、半径、次数、容差、分段数、跨度数和位置属性等参数。Sphere【球体】对象属性面板如图 10.42 所示。

图 10.41　Component Display【组建显示】
卷展栏参数面板

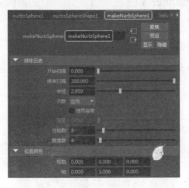

图 10.42　Sphere【球体】对象属性面板

Degree【次数】参数组：在该参数组中主要提供了 Linear【线性】和 Cubic【立方】2 种方式。

2. Cube【立方体】对象属性面板

Cube【立方体】对象属性面板，主要用来调节次数、U 向面片数、V 向面片数、宽度、长度比、高度比和位置属性等参数，如图 10.43 所示。

Degree【次数】参数组：在该参数组中主要提供了 Linear【线性】、Quadratic【二次方】、Cubic【立方】、Quintic【五次】和 Heptic【七次】5 种方式。

3. Cylinder【圆柱体】对象属性面板

Cylinder【圆柱体】对象属性面板，主要用来调节开始扫描、结束扫描、半径、次数、容差、分段数、跨度数、高度比和位置属性，如图 10.44 所示。

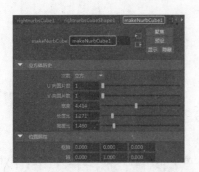 

图 10.43　Cube【立方体】对象属性面板　　　图 10.44　Cylinder【圆柱体】对象属性面板

Degree【次数】参数组：在该参数组中主要提供了 Linear【线性】和 Cubic【立方】2 种方式，默认 Cubic【立方】方式。

4. Cone【圆锥体】对象属性面板

Cylinder【圆柱体】对象属性面板，主要用来调节开始扫描、结束扫描、半径、容差、分段数、跨端数、高度比和位置属性，如图 10.45 所示。

Degree【次数】参数组：在该参数组中主要提供了 Linear【线性】和 Cubic【立方】2 种方式，默认 Cubic【立方】方式。

5. Plane【平面】对象属性面板

Plane【平面】对象属性面板，主要用来调节次数、U 向面片数、V 向面片数、宽度、长度比和位置属性，如图 10.46 所示。

图 10.45　Cylinder【圆柱体】对象属性面板　　　图 10.46　Plane【平面】对象属性面板

Degree【次数】参数组：在该参数组中主要提供了 Linear【线性】、Quadratic【二次方】、Cubic【立方】、Quintic【五次】和 Heptic【七次】5 种方式，默认 Cubic【立方】方式。

6. Torus【圆环】对象属性面板

Torus【圆环】对象属性面板，主要用来调节开始扫描、结束扫描、半径、次数、容差、

第 10 章 创建对象的属性

分段数、跨度数、高度比、次扫描和位置属性等，如图 10.47 所示。

Degree【次数】参数组：在该参数组中主要提供了 Linear【线性】和 Cubic【立方】2 种方式，默认 Cubic【立方】方式。

7．Circle【圆形】对象属性面板

Circle【圆形】对象属性面板，主要用来调节扫描、半径、次数、容差、分段数、法线、中心和第一点参数，如图 10.48 所示。

图 10.47　Torus【圆环】对象属性面板

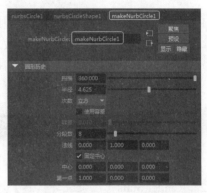

图 10.48　Circle【圆形】对象属性面板

8．Square【方形】对象属性面板

Square【方形】对象属性面板，主要用来调节侧面长度 1、侧面长度 2、每侧跨度数、次数、法线和中心等参数，如图 10.49 所示。

10.7　Volume Primitives【体积基本体】对象的 Shape 【形状】公共属性面板

Volume Primitives【体积基本体】对象的 Shape【形状】公共属性是指 Volume【体积】对象形状共有的属性，如图 10.50 所示。

图 10.49　Square【方形】对象属性面板

图 10.50　Volume Primitives【体积基本体】对象的 Shape【形状】公共属性面板

在 Volume Primitives【体积基本体】对象的 Shape【形状】公共属性面板中，主要包括 Render Sphere Attributes【渲染球体属性】、Render Stats【渲染统计信息】、Object Display【对象显示】、Node Behavior【节点行为】、【UUID】和 Extra Attributes【附加属性】6 个卷展栏参数，其中 Node Behavior【节点行为】、【UUID】和 Extra Attributes【附加属性】卷展栏参数请参考"10.2 Polygon【多边形】对象的 Shape【形状】公共属性面板"对应的卷展栏参数，在此就不再赘述。其他各卷展栏参数介绍如下。

1. Render Volume Attributes【渲染体积属性】卷展栏参数

Render Volume Attributes【渲染体积属性】卷展栏参数，主要用来调节体积对象的半径、大小、圆锥体角度和圆锥体端面等参数，如图 10.51 所示。

提示：在 Maya 2018 中，主要有 Sphere【球体】、Cube【长方体】和 Cone【圆锥体】3 种体积对象属性，不同体积对象的渲染对象属性也不同。

图 10.51　Render Sphere Attributes【渲染球体属性】卷展栏面板

2. Render Stats【渲染统计信息】卷展栏参数

Render Stats【渲染统计信息】卷展栏参数，主要用来控制是否启用投射阴影、接收阴影、主可见性、在反射中可见、在折射中可见、几何体抗锯齿覆盖、体积采样数覆盖和深度抖动等功能，如图 10.52 所示。

3. Object Display【对象显示】卷展栏参数

Object Display【对象显示】卷展栏参数，主要用来调节对象显示、重影信息、边界框信息、绘制覆盖和选择等参数，如图 10.53 所示。

图 10.52　Render Stats【渲染统计信息】卷展栏面板

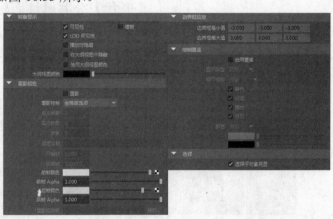

图 10.53　Object Display【对象显示】卷展栏面板

Ghosting Control【重影控制】参数组：在该参数组中提供了 Global Prefs【全局首选项】、Custom Frames【自定义帧】、Custom Frame Steps【自定义帧布数】、Custom Key Steps【自定义关键帧步数】和 Key frames【关键帧】5 种控制重影的方式。默认 Global Prefs【全局首选项】重影控制方式。

10.8 Volume Primitives【体积基本体】对象属性面板

在 Maya 2018 中，体积基本体主要包括 Sphere【球体】、Cube【长方体】和 Cone【圆锥体】3 个对象，各个对象属性面板的具体功能介绍如下。

1. Sphere【球体】对象属性面板

Sphere【球体】对象属性面板，主要用来调节体积雾属性、颜色渐变、衰减、蒙版不透明度和节点行为等属性，如图 10.54 所示。

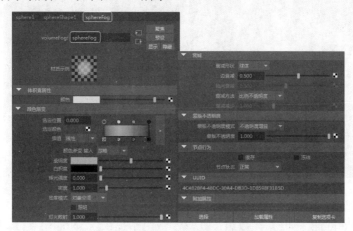

图 10.54　Sphere【球体】对象属性面板

（1）Interpolation【插值】参数组：在该参数组中提供了 None【无】、Linear【线性】、Smooth【平滑】和 Spline【样条线】4 种插值方式，默认 Linear【线性】插值方式。

（2）Color Ramp Input【颜色渐变输入】参数组：在该参数组中提供了 Ignore【忽略】、Transparency【透明度】、Concentric【同心】和 Y Gradient【Y 渐变】4 种输入方式，默认 Ignore【忽略】方式。

（3）Density Mode【密度模式】参数组：在该参数组中提供了 World Space【世界空间】和 Object Space【对象空间】两种密度模式，默认 Object Space【对象空间】密度模式。

（4）Drop off Shape【衰减形状】参数组：在该参数组中提供了 Off【禁用】、Sphere【球体】、Cube【立方体】、Cone【圆锥体】和 Light Cone【光锥】5 种衰减形状。默认 Sphere【球体】衰减形状。

（5）Drop off Method【衰减方法】参数组：在该参数组中提供了 Scale Opacity【比例

不透明度】和 Subtract Density【减少密度】2 种衰减方法。

（6）Matte Opacity Mode【蒙版不透明度模式】参数组：在该参数组中提供了 Black Hole【黑洞】、Solid Matte【匀值蒙版】和 Opacity Gain【不透明度增益】3 种蒙版，默认 Opacity Gain【不透明度增益】模式。

（7）Node State【节点状态】参数组：在该参数组中提供了 Normal【正常】、Has No Effect【无效果】、Blocking【阻塞】、Waiting-Normal【等待-正常】、Waiting-Has No Effect【等待-无效果】和 Waiting-Blocking【等待-阻塞】6 种节点状态，默认 Normal【正常】节点状态。

2. Cube【长方体】对象属性面板

Cube【长方体】对象属性面板，主要用来调节体积长方体的体积雾、颜色渐变、衰减、蒙版不透明度、节点行为、UUID 和附加属性，如图 10.55 所示。

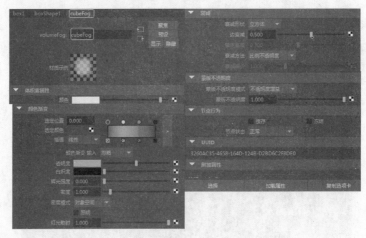

图 10.55　Cube【长方体】对象属性面板

（1）Interpolation【插值】参数组：在该参数组中提供了 None【无】、Linear【线性】、Smooth【平滑】和 Spline【样条线】4 种插值方式，默认 Linear【线性】插值方式。

（2）Color Ramp Input【颜色渐变输入】参数组：在该参数组中提供了 Ignore【忽略】、Transparency【透明度】、Concentric【同心】和 Y Gradient【Y 渐变】4 种输入方式，默认 Ignore【忽略】方式。

（3）Density Mode【密度模式】参数组：在该参数组中提供了 World Space【世界空间】和 Object Space【对象空间】两种密度模式，默认 Object Space【对象空间】密度模式。

（4）Drop off Shape【衰减形状】参数组：在该参数组中提供了 Off【禁用】、Sphere【球体】、Cube【立方体】、Cone【圆锥体】和 Light Cone【光锥】5 种衰减形状。默认 Cube【立方体】衰减形状。

（5）Drop off Method【衰减方法】参数组：在该参数组中提供了 Scale Opacity【比例不透明度】和 Subtract Density【减少密度】2 种衰减方法。

（6）Matte Opacity Mode【蒙版不透明度模式】参数组：在该参数组中提供了 Black Hole

【黑洞】、Solid Matte【匀值蒙版】和 Opacity Gain【不透明度增益】3 种蒙版,默认 Opacity Gain【不透明度增益】模式。

(7) Node State【节点状态】参数组:在该参数组中提供了 Normal【正常】、Has No Effect【无效果】、Blocking【阻塞】、Waiting-Normal【等待-正常】、Waiting-Has No Effect【等待-无效果】和 Waiting-Blocking【等待-阻塞】6 种节点状态,默认 Normal【正常】节点状态。

3. Cone【圆锥体】对象属性面板

Cone【圆锥体】对象属性面板,主要用来调节体积圆锥体的体积雾属性、颜色渐变、衰减、蒙版不透明度、节点行为、UUID 和附加属性,如图 10.56 所示。

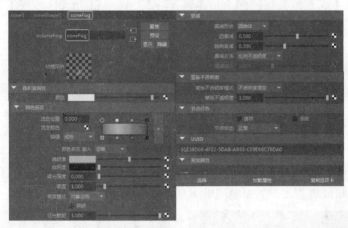

图 10.56　Cube【长方体】对象属性面板

(1) Interpolation【插值】参数组:在该参数组中提供了 None【无】、Linear【线性】、Smooth【平滑】和 Spline【样条线】4 种插值方式,默认 Linear【线性】插值方式。

(2) Color Ramp Input【颜色渐变输入】参数组:在该参数组中提供了 Ignore【忽略】、Transparency【透明度】、Concentric【同心】和 Y Gradient【Y 渐变】4 种输入方式,默认 Ignore【忽略】方式。

(3) Density Mode【密度模式】参数组:在该参数组中提供了 World Space【世界空间】和 Object Space【对象空间】两种密度模式,默认 Object Space【对象空间】密度模式。

(4) Drop off Shape【衰减形状】参数组:在该参数组中提供了 Off【禁用】、Sphere【球体】、Cube【立方体】、Cone【圆锥体】和 Light Cone【光锥】5 种衰减形状。默认 Cone【圆锥体】衰减形状。

(5) Drop off Method【衰减方法】参数组:在该参数组中提供了 Scale Opacity【比例不透明度】和 Subtract Density【减少密度】2 种衰减方法。

(6) Matte Opacity Mode【蒙版不透明度模式】参数组:在该参数组中提供了 Black Hole【黑洞】、Solid Matte【匀值蒙版】和 Opacity Gain【不透明度增益】3 种蒙版,默认 Opacity

Gain【不透明度增益】模式。

（7）Node State【节点状态】参数组：在该参数组中提供了 Normal【正常】、Has No Effect【无效果】、Blocking【阻塞】、Waiting-Normal【等待-正常】、Waiting-Has No Effect【等待-无效果】和 Waiting-Blocking【等待-阻塞】6 种节点状态，默认 Normal【正常】节点状态。

10.9 Lights【灯光】对象的 Shape【形状】属性面板

在 Maya 中，灯光对象类型主要包括 Ambient Light【环境光】、Directional Light【平行光】、Point Light【点光源】、Spot Light【聚光灯】、Area Light【区域光】和 Volume Light【体积光】6 种，各种灯光对象的形状属性面板介绍如下。

1. Ambient Light【环境光】对象的 Shape【形状】属性面板

Ambient Light【环境光】对象的 Shape【形状】属性面板，主要包括 Ambient Light Attributes【环境光属性】、Shadows【阴影】、Object Display【对象显示】、Node Behavior【节点行为】、【UUID】和 Extra Attributes【附加属性】6 个卷展栏参数，如图 10.57 所示。

1）Ambient Light Attributes【环境光属性】卷展栏参数

Ambient Light Attributes【环境光属性】卷展栏参数，主要用来调节灯光类型、颜色、强度和环境光明暗处理等参数，如图 10.58 所示。

图 10.57 Ambient Light【环境光】对象 Shape【形状】属性面板

图 10.58 Ambient Light Attributes【环境光属性】卷展栏参数面板

【类型】参数组：在该参数组中提供了 Ai Area Light【Ai 区域光】、Ai Barndoor【Ai 挡光板】、Ai Light Gobo【Ai 遮光布】、Ai Light Blocker【Ai 阳光阻断】、Ai Light Decay【Ai 灯光衰减】、Ai Photometric Light【Ai 光度学灯光】、Ai Sky Dome Light【Ai 穹顶天光】、Ambient Light【环境光】、Area Light【区域光】、Directional Light【平行光】、Point Light【点光源】、Spot Light【聚光灯】和 Volume Light【体积光】13 种灯光类型，默认 Ambient Light【环境光】灯光类型。

2）Shadows【阴影】卷展栏参数

Shadows【阴影】卷展栏参数，主要用来调节阴影颜色、阴影半径、阴影光线数和光线深度限制等参数，如图 10.59 所示。

3）Object Display【对象显示】卷展栏参数

Object Display【对象显示】卷展栏参数，主要用来调节对象显示、重影信息、边界框信息、绘制覆盖和选择等参数，如图10.60所示。

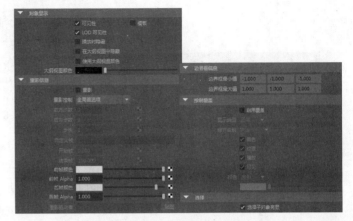

图10.59 Shadows【阴影】
卷展栏参数面板

图10.60 Object Display【对象显示】
卷展栏参数面板

Ghosting Control【重影控制】参数组：该参数组主要包括 Global Prefs【全局首选项】、Custom Frames【自定义帧】、Custom Frame Steps【自定义帧步数】、Custom Key Steps【自定义关键帧步数】和 Key Fraes【关键帧】5种重影控制方式，默认 Global Prefs【全局首选项】控制方式。

4）Node Behavior【节点行为】卷展栏参数

Node Behavior【节点行为】卷展栏参数，主要用来调节缓存、冻结和节点状态等参数，如图10.61所示。

Node State【节点状态】参数组：该参数组提供了 Normal【正常】、Has No Effect【无效果】、Blocking【阻塞】、Waiting-Normal【等待-正常】、Waiting-Has No Effect【等待-无效果】和 Waiting-Blocking【等待-阻塞】6种节点状态，默认 Normal【正常】节点状态。

5）【UUID】卷展栏参数

【UUID】卷展栏参数面板主要用来显示 UUID 的编号，如图10.62所示。

图10.61 Node Behavior【节点行为】卷展栏参数面板

图10.62 【UUID】卷展栏参数面板

6）Extra Attributes【附加属性】卷展栏参数

Extra Attributes【附加属性】卷展栏参数，主要用来调节定位器比例，如图10.61所示。

2. Directional Light【平行光】对象的 Shape【形状】属性面板

Directional Light【平行光】对象的 Shape【形状】属性面板，主要包括 Directional Light

Attributes【平行光属性】、Shadows【阴影】、Object Display【对象显示】、Arnold【阿诺德】、Node Behavior【节点行为】、UUID【UUID】和 Extra Attributes【附加属性】7个卷展栏参数，如图10.64所示。

图 10.63 Extra Attributes【附加属性】卷展栏参数面板

图 10.64 Directional Light【平行光】对象的 Shape【形状】属性面板

1）Directional Light Attributes【平行光属性】卷展栏参数

Directional Light Attributes【平行光属性】卷展栏参数，主要用来调节类型、颜色和强度参数以及是否启用默认照明、发射漫反射和发射镜面反射等参数。如图10.65所示。

Type【类型】参数组：在该参数组中提供了 Ai Area Light【Ai 区域光】、Ai Barndoor【Ai 挡光板】、Ai Light Gobo【Ai 遮光布】、Ai Light Blocker【Ai 阳光阻断】、Ai Light Decay【Ai 灯光衰减】、Ai Photometric Light【Ai 光度学灯光】、Ai Sky Dome Light【Ai 穹顶天光】、Ambient Light【环境光】、Area Light【区域光】、Directional Light【平行光】、Point Light【点光源】、Spot Light【聚光灯】和 Volume Light【体积光】13 种灯光类型，默认 Directional Light【平行光】灯光类型。

2）Shadows【阴影】卷展栏参数

Shadows【阴影】卷展栏参数，主要用来调节阴影颜色、深度贴图阴影属性和光线跟踪阴影属性等参数，如图10.66所示。

图 10.65 Directional Light Attributes【平行光属性】卷展栏参数面板

图 10.66 Shadows【阴影】卷展栏参数面板

3）Object Display【对象显示】卷展栏参数

Object Display【对象显示】卷展栏参数，主要用来调节对象显示、重影信息、边界框信息、绘制覆盖和选择等相关参数，如图 10.67 所示。

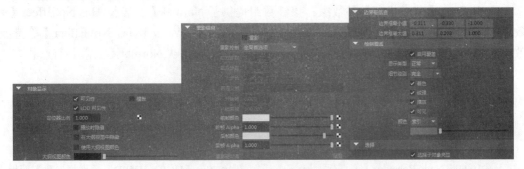

图 10.67　Object Display【对象显示】卷展栏参数面板

（1）Ghosting Control【重影控制】参数组：该参数组主要包括 Global Prefs【全局首选项】、Custom Frames【自定义帧】、Custom Frame Steps【自定义帧步数】、Custom Key Steps【自定义关键帧步数】和 Key fraes【关键帧】5 种重影控制方式。默认 Global Prefs【全局首选项】控制方式。

（2）Display Type【显示类型】参数组：在该参数组中主要提供了 Normal【正常】、Template【模板】和 Reference【引用】3 种显示类型，默认 Normal【正常】显示类型。

（3）Level of Detail【细节级别】参数组：在该参数组中主要提供了 Full【完全】和 Bounding Box【边界框】2 种细节级别，默认 Full【完全】细节级别。

（4）Color【颜色】参数组：在该参数组中主要提供了 Index【索引】和 RGB【RGB】2 种颜色绘制覆盖方式，默认 Index【索引】颜色绘制覆盖方式。

4）Arnold【阿诺德】卷展栏参数

Arnold【阿诺德】卷展栏参数，主要用来调节 Arnold【阿诺德】、Light Filters【灯光过滤】和 User Options【户用选项】等参数设置，如图 10.68 所示。

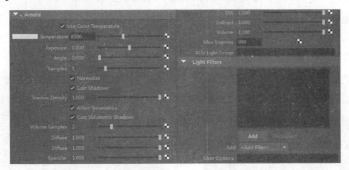

图 10.68　Arnold【阿诺德】卷展栏参数面板

Add【增加】参数组：该参数组提供了 Add Filter【增加过滤】和 Ai Light Blocker【ai 灯光挡板器】2 种增加方式。

5）Node Behavior【节点行为】卷展栏参数

Node Behavior【节点行为】卷展栏参数，主要用来调节缓存、冻结和节点状态等参数，如图10.69所示。

Node State【节点状态】参数组：该参数组提供了 Normal【正常】、Has No Effect【无效果】、Blocking【阻塞】、Waiting-Normal【等待-正常】、Waiting-Has No Effect【等待-无效果】和 Waiting-Blocking【等待-阻塞】6种节点状态。默认 Normal【正常】节点状态。

6）【UUID】卷展栏参数

【UUID】卷展栏参数，主要显示 UUID 的编号，如图10.70所示。

图 10.69　Node Behavior【节点行为】卷展栏参数面板

图 10.70　【UUID（UUID）卷展栏参数】面板

7）Extra Attributes【附加属性】卷展栏参数

Extra Attributes【附加属性】卷展栏参数，主要用来调节 Ai 颜色色温，如图10.71所示。

图 10.71　Extra Attributes【附加属性】卷展栏参数面板

3．Point Light【点光源】对象的 Shape【形状】属性面板

Point Light【点光源】对象的 Shape【形状】属性面板，主要包括 Point Light Attributes【点光源属性】、Light Effects【灯光效果】、Shadows【阴影】、Object Display【对象显示】、Arnold【阿诺德】、Node Behavior【节点行为】、【UUID】和 Extra Attributes【附加属性】8个卷展栏参数，如图10.72所示。关于 Arnold【阿诺德】、Node Behavior【节点行为】、【UUID】和 Extra Attributes【附加属性】4个卷展栏参数，请读者参考"Directional Light【平行光】对象的 Shape【形状】属性面板"中对应的卷展栏参数。

1）Point Light Attributes【点光源属性】卷展栏参数

Point Light Attributes【点光源属性】卷展栏参数，主要用来调节灯光类型、颜色、强度和衰退速率等参数，如图10.73所示。

图 10.72　Extra Attributes【附加属性】卷展栏参数面板

图 10.73　Point Light Attributes【点光源属性】卷展栏参数面板

第10章 创建对象的属性

（1）Type【类型】参数组：在该参数组中提供了 Ai Area Light【Ai 区域光】、Ai Barndoor【Ai 挡光板】、Ai Light Gobo【Ai 遮光布】、Ai Light Blocker【Ai 阳光阻断】、Ai Light Decay【Ai 灯光衰减】、Ai Photometric Light【Ai 光度学灯光】、Ai Sky Dome Light【Ai 穹顶天光】、Ambient Light【环境光】、Area Light【区域光】、Directional Light【平行光】、Point Light【点光源】、Spot Light【聚光灯】和 Volume Light【体积光】13 种灯光类型，默认 Point Light【点光源】灯光类型。

（2）Decay Rate【衰退速率】参数组：在该参数组中提供了 No Decay【无衰退】、Linear【线性】、Quadratic【二次方】和 Cubic【立方】4 种衰退方式，默认 No Decay【无衰退】衰退方式。

2）Light Effects【灯光效果】卷展栏参数

Light Effects【灯光效果】卷展栏参数，主要用来调节灯光雾、雾类型、雾半径、雾密度和灯光辉光灯参数，如图 10.74 所示。

图 10.74　Light Effects【灯光效果】卷展栏参数面板

Fog Type【雾类型】参数组：在该参数组中提供了 Normal【正常】、Linear【线性】和 Exponential【指数】3 种雾类型，默认 Normal【正常】雾类型。

3）Shadows【阴影】卷展栏参数

Shadows【阴影】卷展栏参数，主要用来调节阴影颜色、深度贴图阴影属性和光线跟踪阴影属性，如图 10.75 所示。

图 10.75　Shadows【阴影】卷展栏参数面板

Disk Based Dmap (s)【基于磁盘的深度贴图】参数组：在该参数组中提供了 Off【禁用】、Overwrite Existing Dmap (s)【覆盖现有深度贴图】和 Reuse Existing Dmap (s)【重用现有深度贴图】3 种方式，默认 Off【禁用】方式。

4）Object Display【对象显示】卷展栏参数

Object Display【对象显示】卷展栏参数，主要用来调节主要用来调节聚光的对象显示、重影信息、边界框信息、绘制覆盖和选择等参数，如图 10.76 所示。

图 10.76　Object Display【对象显示】卷展栏参数面板

Ghosting Control【重影控制】参数组：该参数组主要包括 Global Prefs【全局首选项】、Custom Frames【自定义帧】、Custom Frame Steps【自定义帧步数】、Custom Key Steps【自定义关键帧步数】和 Key Fraes【关键帧】5 种重影控制方式。默认 Global Prefs【全局首选项】控制方式。

4．Spot Light【聚光灯】对象的 Shape【形状】属性面板

Spot Light【聚光灯】对象的 Shape【形状】属性面板，主要包括 Spot Light Attributes【聚光灯属性】、Light Effects【灯光效果】、Shadows【阴影】、Object Display【对象显示】、Arnold【阿诺德】、Node Behavior【节点行为】、【UUID】和 Extra Attributes【附加属性】8 个卷展栏参数，如图 10.77 所示。关于 Arnold【阿诺德】、Node Behavior【节点行为】、UUID【UUID】和 Extra Attributes【附加属性】4 个卷展栏参数，请读者参考 "Directional Light【平行光】对象的 Shape【形状】属性面板" 中对应的卷展栏参数。

1）Spot Light Attributes【聚光灯属性】卷展栏参数

Spot Light Attributes【聚光灯属性】卷展栏参数，主要用来调节灯光类型、颜色、强度、衰退速率、圆锥角度、半影角度和衰减等参数，如图 10.78 所示。

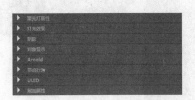　　　　　

图 10.77　Spot Light【聚光灯】　　　图 10.78　Spot Light Attributes【聚光灯属性】
　　　　对象的 Shape【形状】属性面板　　　　　　卷展栏参数面板

（1）Type【类型】参数组：在该参数组中提供了 Ai Area Light【Ai 区域光】、Ai Barndoor【Ai 挡光板】、Ai Light Gobo【Ai 遮光布】、Ai Light Blocker【Ai 阳光阻断】、Ai Light Decay【Ai 灯光衰减】、Ai Photometric Light【Ai 光度学灯光】、Ai Sky Dome Light【Ai 穹顶天光】、Ambient Light【环境光】、Area Light【区域光】、Directional Light【平行光】、Point Light【点光源】、Spot Light【聚光灯】和 Volume Light【体积光】13 种灯光类型，默认 Spot Light【聚光灯】灯光类型。

（2）Decay Rate【衰退速率】参数组：在该参数组中提供了 No Decay【无衰退】、Linear【线性】、Quadratic【二次方】和 Cubic【立方】4 种衰退方式，默认 No Decay【无衰退】衰退方式。

2）Light Effects【灯光效果】卷展栏参数

Light Effects【灯光效果】卷展栏参数，主要用来调节聚光灯的特效、挡光板和衰减区域等参数，如图 10.79 所示。

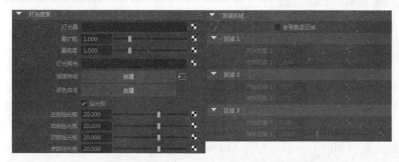

图 10.79　Light Effects【灯光效果】卷展栏参数面板

3）Shadows【阴影】卷展栏参数

Shadows【阴影】卷展栏参数，主要用来调节阴影颜色、深度贴图阴影属性和光线跟踪阴影属性，如图 10.80 所示。

图 10.80　Shadows【阴影】卷展栏参数面板

Disk Based Dmap(s)【基于磁盘的深度贴图】参数组：在该参数组中提供了 Off【禁用】、Overwrite Existing Dmap(s)【覆盖现有深度贴图】和 Reuse Existing Dmap(s)【重用现有深度贴图】3 种方式，默认 Off【禁用】方式。

4）Object Display【对象显示】卷展栏参数

Object Display【对象显示】参数卷展栏，主要用来调节聚光的对象显示、重影信息、边界框信息、绘制覆盖和选择等参数，如图 10.81 所示。

Ghosting Control【重影控制】参数组：在该参数组中主要包括 Global Prefs【全局首选项】、Custom Frames【自定义帧】、Custom Frame Steps【自定义帧步数】、Custom Key Steps【自定义关键帧步数】和 Key fraes【关键帧】5 种重影控制方式。默认 Global Prefs【全局首选项】控制方式。

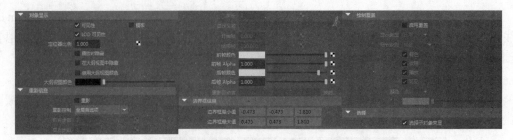

图 10.81 Object Display【对象显示】卷展栏参数面板

5. Area Light【区域光】对象的 Shape【形状】属性面板

Area Light【区域光】对象的 Shape【形状】属性面板，主要包括 Area Light Attributes【区域光属性】、Light Effects【灯光效果】、Shadows【阴影】、Object Display【对象显示】、Arnold【阿诺德】、Node Behavior【节点行为】、【UUID】和 Extra Attributes【附加属性】8 个卷展栏参数，如图 10.82 所示。关于 Arnold【阿诺德】、Node Behavior【节点行为】、UUID【UUID】和 Extra Attributes【附加属性】4 个卷展栏参数，请读者参考 "Directional Light【平行光】对象的 Shape【形状】属性面板" 中对应的卷展栏参数。

1）Area Light Attributes【区域光属性】卷展栏参数

Area Light Attributes【区域光属性】卷展栏参数，主要用来调节灯光类型、颜色、强度和衰退速率等参数，如图 10.83 所示。

图 10.82　Area Light【区域光】　　　　图 10.83　Area Light Attributes【区域光属性】
对象的 Shape【形状】属性面板　　　　　　　卷展栏参数面板

（1）Type【类型】参数组：在该参数组中提供了 Ai Area Light【Ai 区域光】、Ai Barndoor【Ai 挡光板】、Ai Light Gobo【Ai 遮光布】、Ai Light Blocker【Ai 阳光阻断】、Ai Light Decay【Ai 灯光衰减】、Ai Photometric Light【Ai 光度学灯光】、Ai Sky Dome Light【Ai 穹顶天光】、Ambient Light【环境光】、Area Light【区域光】、Directional Light【平行光】、Point Light【点光源】、Spot Light【聚光灯】和 Volume Light【体积光】13 种灯光类型，默认 Area Light【区域光】灯光类型。

（2）Decay Rate【衰退速率】参数组：在该参数组中提供了 No Decay【无衰退】、Linear【线性】、Quadratic【二次方】和 Cubic【立方】4 种衰退方式，默认 No Decay【无衰退】衰退方式。

2）Light Effects【灯光效果】卷展栏参数

Light Effects【灯光效果】卷展栏参数，主要用来调节灯光的辉光，如图 10.84 所示。

第10章 创建对象的属性

图10.84 Light Effects【灯光效果】卷展栏参数面板

3）Shadows【阴影】卷展栏参数

Shadows【阴影】卷展栏参数，主要用来调节阴影颜色、深度贴图阴影属性和光线跟踪阴影属性，如图10.85所示。

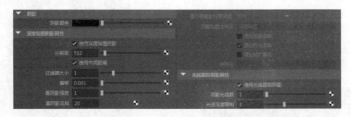

图10.85 Shadows【阴影】卷展栏参数面板

Disk Based Dmap (s)【基于磁盘的深度贴图】参数组：在该参数组中提供了 Off【禁用】、Overwrite Existing Dmap (s)【覆盖现有深度贴图】和 Reuse Existing Dmap (s)【重用现有深度贴图】3种方式，默认 Off【禁用】方式。

4）Object Display【对象显示】卷展栏参数

Object Display【对象显示】卷展栏参数，主要用来调节主要用来调节聚光的对象显示、重影信息、边界框信息、绘制覆盖和选择等参数，如图10.86所示。

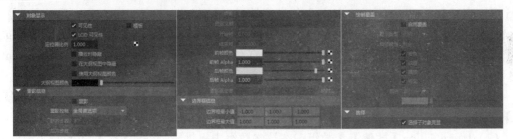

图10.86 Object Display【对象显示】卷展栏参数面板

Ghosting Control【重影控制】参数组：在该参数组中主要包括 Global Prefs【全局首选项】、Custom Frames【自定义帧】、Custom Frame Steps【自定义帧步数】、Custom Key Steps【自定义关键帧步数】和 Keyfraes【关键帧】5种重影控制方式。默认 Global Prefs【全局首选项】控制方式。

6. Volume Light【区域光】对象的 Shape【形状】属性面板

Volume Light【区域光】对象的 Shape【形状】属性面板，主要包括 Volume Light Attributes【体积光属性】、Light Effects【灯光效果】、Shadows【阴影】、Object Display【对象显示】、Node Behavior【节点行为】、【UUID】和 Extra Attributes【附加属性】7个卷展栏参数，如图10.87所示。关于 Node Behavior【节点行为】、【UUID】和 Extra Attributes【附加属性】

3个卷展栏参数，请读者参考"Directional Light【平行光】对象的 Shape【形状】属性面板"中对应的卷展栏参数。

1）Volume Light Attributes【体积光属性】卷展栏参数

Volume Light Attributes【体积光属性】卷展栏参数，主要用来调节灯光类型、颜色、强度、灯光形状、颜色范围和半影等参数，如图 10.88 所示。

图 10.87　Volume Light【区域光】对象的 Shape【形状】属性面板

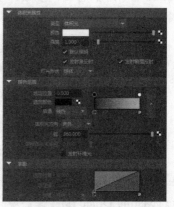

图 10.88　Volume Light Attributes【体积光属性】卷展栏参数面板

（1）Type【类型】参数组：在该参数组中提供了 Ai Area Light【Ai 区域光】、Ai Barndoor【Ai 挡光板】、Ai Light Gobo【Ai 遮光布】、Ai Light Blocker【Ai 阳光阻断】、Ai Light Decay【Ai 灯光衰减】、Ai Photometric Light【Ai 光度学灯光】、Ai Sky Dome Light【Ai 穹顶天光】、Ambient Light【环境光】、Area Light【区域光】、Directional Light【平行光】、Point Light【点光源】、Spot Light【聚光灯】和 Volume Light【体积光】13 种灯光类型，默认 Volume Light【体积光】灯光类型。

（2）Light Shape【灯光形状】参数组：在该参数组中提供了 Box【长方体】、Sphere【球体】、Cylinder【圆柱体】和 Cone【圆锥体】4 种灯光形状，默认 Sphere【球体】灯光形状。

（3）Interpolation【插值】参数组：在该参数组中提供了 None【无】、Linear【线性】、Smooth【平滑】和 Spline【样条线】4 种插值方式，默认 None【无】插值方式。

（4）【体积光方向】参数组：在该参数组中提供了 Outward【向外】、Inward【向内】和 Down Axis【向下轴】3 种体积光方向，默认 Outward【向外】体积光方向。

2）Light Effects【灯光效果】卷展栏参数

Light Effects【灯光效果】卷展栏参数，主要用来灯光雾、雾密度和灯光辉光参数，如图 10.89 所示。

图 10.89　Light Effects【灯光效果】卷展栏参数面板

3）Shadows【阴影】卷展栏参数

Shadows【阴影】卷展栏参数，主要用来调节阴影颜色、深度贴图阴影属性和光线跟踪阴影属性，如图 10.90 所示。

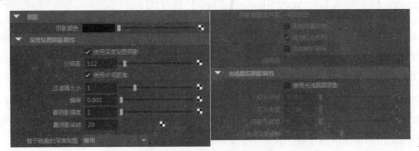

图 10.90　Shadows【阴影】卷展栏参数面板

Disk Based Dmap (s)【基于磁盘的深度贴图】参数组：在该参数组中提供了 Off【禁用】、Overwrite Existing Dmap (s)【覆盖现有深度贴图】和 Reuse Existing Dmap (s)【重用现有深度贴图】3 种方式，默认 Off【禁用】方式。

4）Object Display【对象显示】卷展栏参数

Object Display【对象显示】卷展栏参数，主要用来调节主要用来调节聚光的对象显示、重影信息、边界框信息、绘制覆盖和选择等参数，如图 10.91 所示。

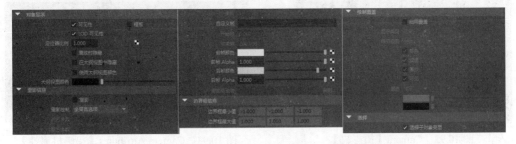

图 10.91　Object Display【对象显示】卷展栏参数面板

Ghosting Control【重影控制】参数组：在该参数组中主要包括 Global Prefs【全局首选项】、Custom Frames【自定义帧】、Custom Frame Steps【自定义帧步数】、Custom Key Steps【自定义关键帧步数】和 Key fraes【关键帧】5 种重影控制方式。默认 Global Prefs【全局首选项】控制方式。

10.10　Cameras【摄影机】对象属性面板

在 Maya 中，摄影机类型主要包括 Camera【摄影机】、Camera and Aim【摄影机和目标】、Camera，Aim and Up【摄影机、目标和上方向】、Stereo Camera【立体摄影机】和 Multi Stereo Rig【多重摄影机装配】5 种。其中，Stereo Camera【立体摄影机】和 Multi Stereo Rig【多

重摄影机装配】2种摄影机的属性与前面两种基本相同，在此就不再详细介绍。摄影机对象的属性面板功能介绍如下。

1. Camera【摄影机】对象的 Shape【形状】属性面板

Camera【摄影机】对象的 Shape【形状】属性面板，主要包括 Camera Attributes【摄影机属性】、Frustum Display Controls【视锥显示控件】、Film Back【胶片背】、Depth of Field【景深】、Output Settings【输出设置】、Environment【环境】、Special Effects【特殊效果】、Display Options【显示选项】、Movement Options【移动选项】、Orthographic Views【正交视图】、Object Display【对象显示】、Arnold【阿诺德】、Node Behavior【节点行为】、UUID【UUID】和 Extra Attributes【附加属性】15个卷展栏参数，如图10.92所示。

1）Camera Attributes【摄影机属性】卷展栏参数

Camera Attributes【摄影机属性】卷展栏参数，主要用来调节控制、视角、焦距、摄影机比例、自动渲染剪切平面、剪切平面和远剪平面等参数，如图10.93所示。

图 10.92　Camera【摄影机】
对象的 Shape【形状】属性面板

图 10.93　Camera Attributes【摄影机属性】
卷展栏参数面板

Controls【控制】参数组：该参数组提供了 Camera【摄影机】、Camera and Aim【摄影机和目标】和 Camera，Aim and Up【摄影机、目标和上方向】3种摄影机切换方式，默认 Camera【摄影机】。

2）Frustum Display Controls【视锥显示控件】卷展栏参数

Frustum Display Controls【视锥显示控件】卷展栏参数，主要用来控制是否启用显示近剪裁平面、显示远剪裁平面和显示视锥，如图10.94所示。

3）Film Back【胶片背】卷展栏参数

Film Back【胶片背】卷展栏参数，主要用来调节胶片门、摄影机光圈（单位：英寸）、摄影机光圈（单位：mm）、胶片纵横比、镜头挤压比、适配分辨门、胶片适配偏移、胶片偏移、、前所放、胶片平移、胶片滚动枢轴、胶片滚转顺序和后缩放等参数，如图10.95所示。

第 10 章 创建对象的属性

图 10.94　Frustum Display Controls
【视锥显示控件】卷展栏参数面板

图 10.95　Film Back【胶片背】
卷展栏参数面板

（1）Film Gate【胶片门】参数组：在该参数组中提供了 User【用户】、16mm Theatrical【16mm 剧院】、Super 16mm【超 16mm】、35mm Academy【35mm 学院】、35mm TV Projection【35mmTV 投影】、35mm Full Aperture【35mm 全光圈】、35mm 1.85 Projection【35mm 1.85 投影】、35mm Anamorphic【35mm 变形】、70mm Projection【70mm 投影】、Vista Vision【景深电影】和 Imax【巨幕电影】11 种胶片门方式，默认 User【用户】胶片门方式。

（2）【适配分辨门】参数组：在该参数组中提供了 Fill【填充】、Horizontal【水平】、Vertical【垂直】和 Over scan【过扫描】4 种适配方式，默认 Fill【填充】方式。

（3）Film Roll Order【胶片滚转顺序】参数组：在该参数组中提供了 Rotate-Translate【旋转平移】和 Translate-Rotate【平移旋转】2 种滚转顺序，默认【旋转平移】滚转顺序。

4）Depth of Field【景深】卷展栏参数

Depth of Field【景深】卷展栏参数，主要用来调节景深、聚焦距离、F 制光圈和聚焦区域比例参数，如图 10.96 所示。

5）Output Settings【输出设置】卷展栏参数

Output Settings【输出设置】卷展栏参数，主要用来调节可渲染、图像、遮罩、深度、深度类型、基于透明度的深度和预合成模板等参数，如图 10.97 所示。

图 10.96　Depth of Field【景深】
卷展栏参数面板

图 10.97　Output Settings【输出设置】
卷展栏参数面板

Depth【深度类型】参数组：该参数组中提供了 Closest Visible Depth【最近可见深度】和 Furthest Visible Depth【最远可见深度】2 种深度类型，默认 Furthest Visible Depth【最远可见深度】深度类型。

6）Environment【环境】卷展栏参数

Environment【环境】卷展栏参数，主要用来调节背景色和创建图像平面，如图 10.98 所示。

7）Special Effects【特殊效果】卷展栏参数

Special Effects【特殊效果】卷展栏参数，主要用来调节快门角度，如图 10.99 所示。

图 10.98　Environment【环境】
卷展栏参数面板

图 10.99　Special Effects【特殊效果】
卷展栏参数面板

8）Display Options【显示选项】卷展栏参数

Display Options【显示选项】卷展栏参数，主要用来调节显示相关参数、二维平移和缩放等参数，如图 10.100 所示。

9）Movement Options【移动选项】卷展栏参数

Movement Options【移动选项】卷展栏参数，主要用来调节兴趣中心和翻转枢轴等参数，如图 10.101 所示。

图 10.100　Display Options【显示选项】
卷展栏参数面板

图 10.101　Movement Options【移动选项】
卷展栏参数面板

10）Orthographic Views【正交视图】卷展栏参数

Orthographic Views【正交视图】卷展栏参数，主要用来控制是否启用正交和调节正交宽度参数，如图 10.102 所示。

图 10.102　Orthographic Views【正交视图】卷展栏参数面板

11）Object Display【对象显示】卷展栏参数

Object Display【对象显示】卷展栏参数，主要用来调节对象显示、重影信息、边界框信息、绘制覆盖和选择等参数，如图 10.103 所示。

第 10 章 创建对象的属性

图 10.103　Object Display【对象显示】卷展栏参数面板

（1）Ghosting Control【重影控制】参数组：在该参数组中主要包括 Global Prefs【全局首选项】、Custom Frames【自定义帧】、Custom Frame Steps【自定义帧步数】、Custom Key Steps【自定义关键帧步数】和 Key Frames【关键帧】5 种重影控制方式，默认 Global Prefs【全局首选项】控制方式。

（2）Display Detail【显示类型】参数组：在该参数组中主要包括 Normal【正常】、Template【模板】和 Reference【引用】3 种类型，默认 Normal【正常】类型。

（3）Level of Detail【细节级别】参数组：在该参数组中主要包括 Full【完全】和 Bounding Box【边界框】2 种类型，默认 Full【完全】类型。

12）Arnold【阿诺德】卷展栏参数

Arnold【阿诺德】卷展栏参数，主要用来调节摄影机的类型、曝光、文件贴图、旋转快门、持续旋转快门、启用景深、聚焦距离、控件大小、光圈片、光圈片曲率、孔径旋转、光圈纵横比、UV 重映射、摄影机运动模糊、快门开始、快门结束、快门类型、快门曲线、数值、位置和用户选项等参数，如图 10.104 所示。

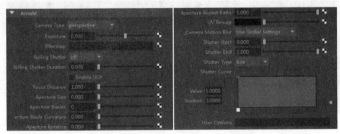

图 10.104　Arnold【阿诺德】卷展栏参数面板

（1）Camera Type【摄影机类型】参数组：在该参数组中提供了 Cylindrical【圆柱体】、Fisheye【鱼眼】、mtoa_shaders【Mtoa_着色器】、Orthographic【正交】、Perspective【透视图】和 Spherical【球形】6 种摄影机类型，默认 Perspective【透视图】摄影机类型。

（2）Rolling Shutter【旋转快门】参数组：在该参数组中提供了 Off【关闭】、Top【顶视图】、Bottom【底视图】、【左视图】和【右视图】5 种旋转快门类型，默认【Off】摄影机类型。

（3）Camera Motion Blur【摄影机运动模糊】参数组：在该参数组中提供了 Use Global

Settings【使用全局设置】、开启【On】和关闭【Off】3种运动模糊方式。

（4）Shutter Type【快门类型】参数组：在该参数组中提供了 Box【立方体】、Triangle【三角形】和 Curve【曲线】3 种快门类型。

13）Node Behavior【节点行为】卷展栏参数

Node Behavior【节点行为】参数卷展栏，主要用来调节缓存、冻结和节点状态等参数，如图 10.105 所示。

Node State【节点状态】参数组：该参数组提供了 Normal【正常】、Has No Effect【无效果】、Blocking【阻塞】、Waiting-Normal【等待-正常】、Waiting-Has No Effect【等待-无效果】和 Waiting-Blocking【等待-阻塞】6 种节点状态。默认 Normal【正常】节点状态。

14）【UUID】卷展栏参数

【UUID】卷展栏参数面板只显示 UUID 序列号，如图 10.106 所示。

图 10.105　Node Behavior【节点行为】卷展栏参数面板

图 10.106　【UUID】卷展栏参数面板

15）Extra Attributes【附加属性】卷展栏参数

Extra Attributes【附加属性】卷展栏参数，主要用来调节 Ai 多边形网络、Ai 偏移值、Ai 网络大小、Ai U 向偏移值、Ai V 向偏移值、Ai 位置、Ai 看、Ai 矩阵、Ai 附近剪辑、Ai 远剪辑、Ai 偏手性、Ai 时间采样、Ai 屏幕窗口最小值和 Ai 屏幕窗口最大值参数，如图 10.107 所示。

图 10.107　Extra Attributes【附加属性】卷展栏参数面板

Ai Handedness【Ai 偏手性】参数组：在该参数组中提供了 Right【右】和 Left【左】两种方式。

2. Camera and Aim【摄影机和目标】对象的 Shape【形状】属性面板

Camera and Aim【摄影机和目标】对象的 Shape【形状】属性面板，主要包括 Camera Attributes【摄影机属性】、Frustum Display Controls【视锥显示控件】、Film Back【胶片背】、Depth of Field【景深】、Output Settings【输出设置】、Environment【环境】、Special Effects【特殊效果】、Display Options【显示选项】、Movement Options【移动选项】、Orthographic Views

第 10 章 创建对象的属性

【正交视图】、Object Display【对象显示】、Arnold【阿诺德】、Node Behavior【节点行为】、【UUID】和 Extra Attributes【附加属性】15 个卷展栏参数，如图 10.108 所示。

图 10.108　Camera and Aim【摄影机和目标】对象的 Shape【形状】属性面板

提示：在"Camera and Aim【摄影机和目标】对象的 Shape【形状】属性面板"中，各卷展栏参数与"10.10.1 Camera【摄影机】对象的 Shape【形状】属性面板"的各卷展栏参数完全相同，在此就不再赘述。

3. Camera and Aim【摄影机和目标】对象的 Group【群组】属性面板

Camera and Aim【摄影机和目标】对象的 Group【群组】属性面板，主要包括 Up Vector Properties【上方向向量特性】、Constraint Properties【约束特性】、Transform Properties【变换特性】和 Extra Attributes【附加属性】4 个卷展栏参数，如图 10.109 所示。

1）Up Vector Properties【上方向向量特性】卷展栏参数

Up Vector Properties【上方向向量特性】卷展栏参数，主要用来调节世界上方向类型和世界上方向向量和扭曲参数，如图 10.110 所示。

图 10.109　Camera and Aim【摄影机和目标】　　图 10.110　Up Vector Properties
　　　　对象的 Group【群组】属性面板　　　　　　　【上方向向量特性】卷展栏参数面板

World Up Type【世界上方向类型】参数组：在该参数组中提供了 Scene Up【场景上方向】、Object Up【对象上方向】、Object Rotation Up【对象旋转上方向】、Vector【向量】和 None【无】5 种世界上方向类型。

2）Constraint Properties【约束特性】卷展栏参数

Constraint Properties【约束特性】卷展栏参数，主要用来调节间距、目标向量和上方向向量和显示连接器参数，如图 10.111 所示。

3）Transform Properties【变换特性】卷展栏参数

Transform Properties【变换特性】卷展栏参数，主要用来调节 Group【群组】的变换属性、枢轴、限制信息、显示、节点行为和 UUID 参数，如图 10.112 所示。

图 10.111　Constraint Properties【约束特性】
卷展栏参数面板

图 10.112　Transform Properties【变换特性】
卷展栏参数面板

提示：在"Transform Properties【变换特性】卷展栏参数"中的二级卷展栏参数与"10.1 Polygon【多边形】对象公共属性面板"中的卷展栏参数完全相同，在此就不再赘述。

4）Extra Attributes【附加属性】卷展栏参数

Extra Attributes【附加属性】卷展栏参数，主要用来调节启用静止位置、锁定输出、比例补偿、方向缩放、偏移、静止旋转和使用旧偏移计算参数，如图 10.113 所示。

4．Camera and Aim【摄影机和目标】对象的 Aim Shape【目标形态】属性面板

Camera and Aim【摄影机和目标】对象的 Aim Shape【目标形态】属性面板，主要包括 Locator Attributes【定位器属性】、Object Display【对象显示】、Node Behavior【节点行为】、【UUID】和 Extra Attributes【附加属性】5 个卷展栏参数，如图 10.114 所示。

图 10.113　Extra Attributes【附加属性】
卷展栏参数】面板

图 10.114　Camera and Aim【摄影机和目标】
对象的 Aim Shape【目标形态】属性面板

1）Locator Attributes【定位器属性】卷展栏参数

Locator Attributes【定位器属性】卷展栏参数，主要用来调节局部位置和局部比例参数，如图 10.115 所示。

图 10.115　Locator Attributes【定位器属性】卷展栏参数面板

2）Object Display【对象显示】卷展栏参数

Object Display【对象显示】卷展栏参数，主要用来调节对象显示、重影信息、边界框信息、绘制覆盖和选择等参数，如图 10.116 所示。

（1）Ghosting Control【重影控制】参数组：在该参数组中主要包括 Global Prefs【全局首选项】、Custom Frames【自定义帧】、Custom Frame Steps【自定义帧步数】、Custom Key Steps【自定义关键帧步数】和 Key Frames【关键帧】5 种重影控制方式。默认 Global Prefs【全局首选项】控制方式。

第 10 章 创建对象的属性

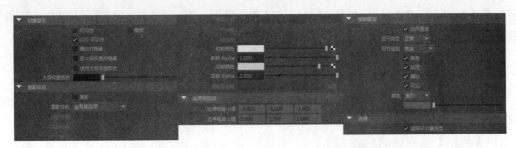

图 10.116　Object Display【对象显示】卷展栏参数面板

（2）Display Detail【显示类型】参数组：在该参数组中主要包括 Normal【正常】、Template【模板】和 Reference【引用】3 种类型，默认 Normal【正常】类型。

（3）Level of Detail【细节级别】参数组：在该参数组中主要包括 Full【完全】和 Bounding Box【边界框】2 种类型，默认 Full【完全】类型。

3）Node Behavior【节点行为】卷展栏参数

Node Behavior【节点行为】卷展栏参数，主要用来调节缓存、冻结和节点状态参数，如图 10.117 所示。

Node State【节点状态】参数组：该参数组提供了 Normal【正常】、Has No Effect【无效果】、Blocking【阻塞】、Waiting-Normal【等待-正常】、Waiting-Has No Effect【等待-无效果】和 Waiting-Blocking【等待-阻塞】6 种节点状态，默认 Normal【正常】节点状态。

4）【UUID】卷展栏参数

【UUID】卷展栏参数面板只显示 UUID 序列号，如图 10.118 所示。

图 10.117　Node Behavior【节点行为】
　　　　　卷展栏参数面板

图 10.118　【UUID】卷展栏参数面板

5）Extra Attributes【附加属性】卷展栏参数

关于 Extra Attributes【附加属性】卷展栏参数，在默认情况下没有任何参数，如图 10.119 所示。

图 10.119　Extra Attributes【附加属性】卷展栏参数面板

10.11　Type【类型】对象属性面板

在菜单栏中单击【创建】→【类型】命令，在视图中自动创建一个 Type【类型】文字，按键盘上的"Ctrl+A"组合键，显示【类型】的对象属性面板，如图 10.120 所示。

在该属性面板中主要包括 Type 1【类型 1】、Vertor Adjust 1【矢量调节】、Poly Remesh 1

【多边形重齿合 1】、Poly Auto Proj 1【多边形自动项目 1】、Tweak 2【扭曲 2】、Shell Deformer 1【壳变形器 1】、Type Extrude 1【挤出类型 1】和 Tweak 1【扭曲 1】8 个选项。

1. Type1【类型 1】属性面板

在 Type1【类型 1】属性面板中，主要用来调节 Type【类型】、Text【文本】、Geometry【几何体】、Texturing【纹理】、Animation【动画】和 Font Size【字体大小】等属性。

1）【文本】选项参数

Text【文本】选项参数，主要用来调节文本的对齐方式、字体大小、跟踪、字距微调比例、前导比例和空间宽度比例等参数，如图 10.121 所示。

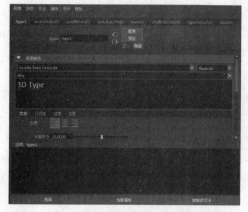
图 10.120　Type【类型】对象属性面板

图 10.121　Text【文本】选项参数面板

2）Geometry【几何体】选项参数

Geometry【几何体】选项参数，主要用来调节网络设置、可变形类型、挤出、倒角、外部倒角和内部倒角等参数，如图 10.122 所示。

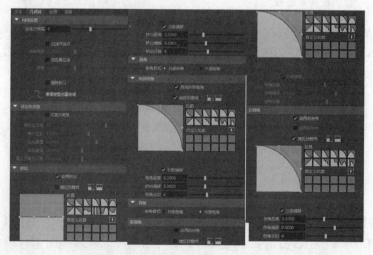
图 10.122　【几何体】选项参数面板

第 10 章 创建对象的属性

3）Texturing【纹理】选项参数

Texturing【纹理】选项参数，主要用来调节默认着色器、类型着色器、倒角着色器和挤出着色器等参数，如图 10.123 所示。

Default Shader【默认着色器】参数组：在该参数组主要提供了 Blinn【布林】、Lambert【兰伯特】、Phong【冯】、Phong E【冯 E】和 RampShader【渐变节点】5 种着色类型。

4）Animation【动画】选项参数

Animation【动画】选项参数，主要用来调节动画模式、动画枢轴、延迟等参数，如图 10.124 所示。

图 10.123　Texturing【纹理】选项参数面板

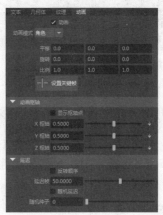

图 10.124　Animation【动画】选项参数面板

Animation Mode【动画模式】参数组：在该参数组中提供 Character【角色】、Word【单词】和 Line【线】3 种模式。默认【角色】动画模式。

2. Vertor Adjust 1【矢量调节 1】属性面板

Vertor Adjust 1【矢量调节 1】属性面板，主要用来修改矢量调节的名称和顶点组 ID 参数，如图 10.125 所示。

3. Poly Remesh 1【多边形重齿合】属性面板

Poly Remesh 1【多边形重齿合】属性面板，主要用来调节节点行为和 UUID 卷展栏参数，如图 10.126 所示。

（1）Node State【节点状态】参数组：该参数组提供了 Normal【正常】、Has No Effect【无效果】、Blocking【阻塞】、Waiting-Normal【等待-正常】、Waiting-Has No Effect【等待-无效果】和 Waiting-Blocking【等待-阻塞】6 种节点状态，默认 Has No Effect【无效果】节点状态。

（2）【插值类型】参数组：该参数组提供了【线性】、【立方】和【等待】3 种插值类型。默认【线性】插值类型。

图 10.125　Vertor Adjust 1【矢量调节 1】属性面板

图 10.126　Poly Remesh 1【多边形重齿合 1】属性面板

4. Poly AutoProj 1【多边形自动项目 1】属性面板

Poly AutoProj 1【多边形自动项目 1】属性面板，主要用来调节变换属性、多边形自动投影、节点行为、UUID 和附加属性等相关属性，如图 10.127 所示。

（1）Optimize【优化】参数组：该参数组主要提供了 Less Distortion【较少的扭曲】和 Fewer Pieces【较少的片数】2 种优化方式，默认 Fewer Pieces【较少的片数】优化方式。

（2）Layout【排布】参数组：该参数组主要提供了 Overlap【重叠】、Along U【沿 U 方向】、Into Square【置于方形】和 Tile【平铺】4 种排布方式，默认 Into Square【置于方形】排布方式。

（3）Layout Method【布局方法】参数组：该参数组主要提供了 Block Stacking【块堆叠】和 Shape Stacking【形状堆叠】2 种布局方法，默认 Block Stacking【块堆叠】布局方法。

（4）Scale Mode【缩放模式】参数组：该参数组主要提供 Uniform【一致】None【无】和 Stretch to Square【拉伸至方形】3 种缩放模式，默认 Uniform【一致】缩放模式。

（5）Node State【节点状态】参数组：该参数组提供了 Normal【正常】、Has No Effect【无效果】、Blocking【阻塞】、Waiting-Normal【等待-正常】、Waiting-Has No Effect【等待-无效果】和 Waiting-Blocking【等待-阻塞】6 种节点状态。默认 Has No Effect【无效果】节点状态。

5. Tweak【扭曲】属性面板

在 Type【类型】对象中，一个类型一般包括 2 个 Tweak【扭曲】属性面板，但这两个属性面板的参数完全相同，Tweak【扭曲】属性面板主要用来调节调整属性、变形器属性、节点行为、UUID 和附加属性等参数，如图 10.128 所示。

图 10.127　Poly Auto Proj 1
【多边形自动项目 1】属性面板

图 10.128　Tweak【扭曲】属性面板

6. Shell Deformer 1【壳变形器 1】属性面板

Shell Deformer 1【壳变形器 1】属性面板，主要用来调节变换、步长、枢轴和附加属性等参数，如图 10.129 所示。

7. Type Extrude 1【挤出类型 1】属性面板

Type Extrude 1【挤出类型 1】属性面板，主要用来调节挤出和倒角等参数，如图 10.130 所示。

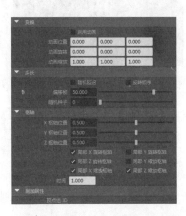
图 10.129　Shell Deformer 1
【壳变形器 1】属性面板

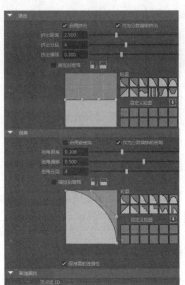
图 10.130　Type Extrude 1
【挤出类型 1】属性面板

第 11 章 Rigging【装备】和 FX 中的对象属性

本章主要介绍 Maya 2018 中的 Rigging【装备】和 FX 中的对象属性，主要包括 Joint【关节】、Particle【粒子】、Emitter【发射器】、3D Container【3D 容器】和 2D Container【2D 容器】、Ocean【海洋】、Pond【池塘】、Air【空气】、Drag【阻力】、Gravity【重力】、Newton【牛顿】、Radial【径向】、Turbulence【湍流】、Uniform【统一】和 Vortex【漩涡】15 个对象。

11.1 Joint【关节】对象属性面板

Joint【关节】对象属性面板，主要包括 Transform Attributes【变换属性】、Joint【关节】、Limit Information【限制信息】、Display【显示】、Node Behavior【节点行为】、【UUID】和 Extra Attributes【附加属性】7 个卷展栏参数，如图 11.1 所示。

1. Transform Attributes【变换属性】卷展栏

Transform Attributes【变换属性】卷展栏参数主要用来调节对象的位置、旋转、斜切和旋转轴，详细参数如图 11.2 所示。

图 11.1 Joint【关节】对象属性面板

图 11.2 Transform Attributes【变换属性】卷展栏

在 Rotate Order【旋转顺序】参数组中，有【zyz】、【yzx】、【zxy】、【xzy】、【yxz】和【zyx】6 种旋转顺序供选择。

2. Joint【关节】卷展栏

Joint【关节】卷展栏参数，主要用来调节关节、关节标签设置和关节旋转限制阻尼等参数，如图 11.3 所示。

（1）Draw Style【绘制样式】参数组：在该参数组中主要提供了 Bone【骨骼】、Multi-child as Box【作为外框的多子对象】和 None【无】3 种绘制模式，默认 Bone【骨骼】绘制模式。

第 11 章　Rigging【装备】和 FX 中的对象属性

（2）Side【侧】参数组：在该参数组中主要提供了 Center【中心】、Left【左侧】、Right【右侧】和 None【无】4 种侧模式，默认 Center【中心】侧模式。

（3）【类型】参数组：在该参数组中主要提供了 None【无】、Root【根】、Hip【髋部】、Knee【膝盖】、Foot【脚】、Tone【脚趾】、Spine【脊椎】、Neck【颈部】、Head【头】、Collar【锁骨】、Shoulder【肩部】、Elbow【肘部】、Hand【手】、Finger【手指】、Thumb【大趾】、Prop A【道具 A】、Prop B【道具 B】、Prop C【道具 C】、Other【其他】、Index Finger【食指】、Middle Finger【中指】、Ring Finger【无名指】、Pinky Finger【小指】、Extra Finger【附加手指】、Big Toe【大趾】、Index Toe【二趾】、Middle Toe【中趾】、Four Toe【四趾】、Little Toe【小趾】和 Foot Big Toe【脚大趾】30 种类型，默认 None【无】类型。

3. Limit Information【限制信息】卷展栏

Limit Information【限制信息】卷展栏参数主要用来调节平移、旋转和缩放的最大/当前/最小值范围，详细参数如图 11.4 所示。

图 11.3　Joint【关节】卷展栏

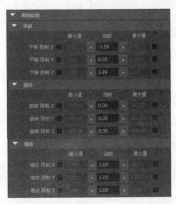
图 11.4　Limit Information【限制信息】卷展栏

4. Display【显示】卷展栏

Display【显示】卷展栏参数主要用来调节选择控制柄、超纵器、重影信息、边界框信息、绘制覆盖和选择等相关参数，详细参数如图 11.5 所示。

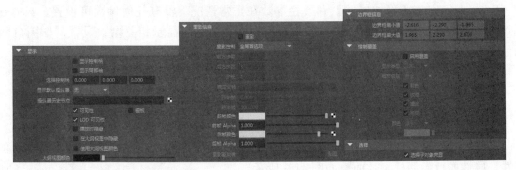
图 11.5　Display【显示】卷展栏

（1）Show Manip Default【显示默认操纵柄】参数组：在该参数组中提供了 None【无】、Translate【转换】、Rotate【旋转】、Scale【比例】、Transform【变换】、Global Default【全局默认】、Smart【智能】和 Specified【以指定】8 种操纵柄方式。

（2）Ghosting Control【重影控制】参数组：在该参数组中提供了 Global Prefs【全局首选项】、Global Frames【自定义帧】、Custom Frame Steps【自定义帧步数】、Custom Key Steps【自定义关键帧步数】和 Key frames【关键帧】5 种控制重影的方式。

（3）Display Type【显示类型】参数组：在该参数组中主要提供了 Normal【正常】、Template【模板】和 Reference【引用】3 种显示类型，默认 Normal【正常】显示类型。

（4）Level of Detail【细节级别】参数组：在该参数组中主要提供了 Full【完全】和 Bounding Box【边界框】2 种显示类型，默认 Full【完全】显示类型。

（5）Color【颜色】参数组：在该参数组中主要提供了 Index【索引】和【RGB】2 种显示类型，默认 Index【索引】类型。

5. Node Behavior【节点行为】卷展栏参数

Node Behavior【节点行为】卷展栏参数主要用来调节缓存、冻结和节点状态等参数，详细参数如图 11.6 所示。

在 Node State【节点状态】参数组中提供 Normal【正常】、Has No Effect【无效果】、Blocking【阻塞】、Waiting-Normal【等待-正常】、Waiting-Has No Effect【等待-无效果】和 Waiting-Blocking【等待-阻塞】6 种状态供选择。

6.【UUID】参数卷展栏参数

【UUID】参数卷展栏中没有相关参数，只显示了【UUID】的序列号，如图 11.7 所示。

图 11.6 Node Behavior【节点行为】卷展栏　　　　图 11.7 【UUID】卷展栏

7. Extra Attributes【附加属性】参数卷展栏参数

Extra Attributes【附加属性】卷展栏参数在正常情况下没有任何参数可设置，只有在通过 Load Attributes【加载属性】按钮加载属性之后才有显示相关参数设置，如图 11.8 所示。

11.2 Particle【粒子】对象的 Shape【形状】属性面板

Particle【粒子】对象的 Shape【形状】属性面板是指 Particle【粒子】对象形状共有的属性。属性面板卷展栏参数面板如图 11.9 所示。

第 11 章 Rigging【装备】和 FX 中的对象属性

图 11.8 Extra Attributes【附加属性】卷展栏

图 10.9 Particle【粒子】对象的 Shape【形状】公共属性面板

在【Polygon（多边形）对象的 Shape（形状）公共属性】面板中主要包括 General Control Attributes【常规控制属性】、Emission Attributes（See also emitter tabs）【发射属性（另请参见发射器选项卡）】、Lifespan Attributes（See also per-particle tab）【寿命属性（另请参见每粒子选项卡）】、Time Attributes【时间属性】、Collision Attributes【碰撞属性】、Soft Body Attributes【柔体属性】、Goal Weights and Objects【目标权重和对象】、Instancer（Geometry Replacement）【实例化器（几何体替换）】、Emission Random Stream Seeds【发射随机流种子】、Render Attributes【渲染属性】、Render Stats【渲染统计信息】、Per Particle（Array）Attributes【每粒子（数组）属性】、Add Dynamic Attributes【添加动态属性】、Clip Effects Attributes【片段效果属性】、Sprite Attributes【精灵属性】、Object Display【对象显示】、Arnold【阿诺德】、Node Behavior【节点行为】、【UUID】和 Extra Attributes【附加属性】20 个卷展栏参数。

1. General Control Attributes【常规控制属性】卷展栏参数

General Control Attributes【常规控制属性】卷展栏参数，主要用来调节动力学权重、计数和事件总数等相关属性，如图 11.10 所示。

2. Emission Attributes（See also emitter tabs）【发射属性（另请参见发射器选项卡）】卷展栏参数

Emission Attributes（See also emitter tabs）【发射属性（另请参见发射器选项卡）】卷展栏参数，主要用来调节最大计数、细节级别和继承因子等参数，如图 11.11 所示。

图 11.10 General Control Attributes【常规控制属性】卷展栏

图 11.11 Emission Attributes（See also emitter tabs）【发射属性（另请参见发射器选项卡）】卷展栏

3. Lifespan Attributes（See also per-particle tab）【寿命属性（另请参见每粒子选项卡）】卷展栏参数

Lifespan Attributes（See also per-particle tab）【寿命属性（另请参见每粒子选项卡）】卷展栏参数，主要用来调节粒子的寿命模式、寿命、寿命随机和常规种子等参数，如图11.12所示。

Lifespan Mode【寿命模式】参数组：主要提供了Live forever【永生】、Constant【恒定】、Random range【随机范围】和lifespan PP【每粒子】only【仅寿命】4种寿命模式。默认Live forever【永生】模式。

4. me Attributes【时间属性】卷展栏参数

Time Attributes【时间属性】卷展栏参数，主要用来开始帧和当前时间参数，如图11.13所示。

图11.12　Lifespan Attributes（See also per-particle tab）【寿命属性（另请参见每粒子选项卡）】卷展栏

图11.13　Time Attributes【时间属性】卷展栏

5. Collision Attributes【碰撞属性】卷展栏参数

Collision Attributes【碰撞属性】卷展栏参数，主要用来调节碰撞的跟踪深度大小，如图11.14所示。

6. Soft Body Attributes【柔体属性】卷展栏参数

Soft Body Attributes【柔体属性】卷展栏参数，主要用来调节输入几何体空间和目标几何体空间的方式，如图11.15所示。

图11.14　Collision Attributes【碰撞属性】卷展栏

图11.15　Soft Body Attributes【柔体属性】卷展栏

7. Goal Weights and Objects【目标权重和对象】卷展栏参数

Goal Weights and Objects【目标权重和对象】卷展栏参数，主要用来调节目标平滑度的大小，如图11.16所示。

图11.16　Goal Weights and Objects【目标权重和对象】卷展栏

第 11 章　Rigging【装备】和 FX 中的对象属性

8. Instancer（Geometry Replacement）【实例化器（几何体替换）】卷展栏参数

Instancer（Geometry Replacement）【实例化器（几何体替换）】卷展栏参数，主要用来调节实例化器节点、允许所有数据类型、常规选项、旋转选项和循环选项，如图 11.17 所示。

图 11.17　Instancer（Geometry Replacement）【实例化器（几何体替换）】卷展栏

（1）Position【位置】参数组：在该参数组中，提供了 Acceleration【加速】、Force【力】、Position【位置】、Ramp Acceleration【渐变加速】、Ramp Position【渐变位置】、Ramp Velocity【渐变速度】、Velocity【速度】、World Position【世界位置】和 World Velocity【世界速度】9 种位置方式，默认 World Position【世界位置】方式。

（2）Scale【比例】参数组：在该参数组中，提供了 None【无】、Acceleration【加速】、【力】、Position【位置】、Ramp Acceleration【渐变加速】、Ramp Position【渐变位置】、Ramp Velocity【渐变速度】、Velocity【速度】、World Position【世界位置】和 World Velocity【世界速度】10 种位置方式，默认 None【无】方式。

（3）【斜切】参数组：在该参数组中，提供了 None【无】、Acceleration【加速】、Force【力】、Position【位置】、Ramp Velocity【渐变加速】、Ramp Position【渐变位置】、Ramp Velocity【渐变速度】、Velocity【速度】、World Position【世界位置】和 World Velocity【世界速度】10 种位置方式，默认 None【无】方式。

（4）Visibility【可见性】、Object Index【对象索引】和 Rotation【旋转类型】3 个参数组提供的选项完全相同，主要提供了 None【无】、Age【年龄】、Birth Time【出生时间】、Mass【质量】、Particle ID【粒子 ID】和 lifespan PP【寿命 PP】6 种位置方式，默认 None【无】方式。

（5）Rotation【旋转】、Aim Direction【目标方向】、Aim Position【目标位置】、Aim Axis【目标轴】、Aim Up Axis【目标上方向轴】和 Aim World Up【目标世界上方向】7 个参数组提供的选项完全相同，主要提供了 None【无】、Acceleration【加速】、Force【力】、Position【位置】、Ramp Velocity【渐变加速】、Ramp Position【渐变位置】、Ramp Velocity【渐变速

度】、Velocity【速度】、World Position【世界位置】和 World Velocity【世界速度】10 种位置方式，默认 None【无】方式。

（6）Cycle Start Object【循环开始对象】和 Age【年龄】2 个参数组提供的选项完全相同，主要提供了 None【无】、Age【年龄】、Birth Time【出生时间】、Mass【质量】、Particle ID【粒子 ID】和 lifespan PP【寿命 PP】6 种位置方式。Cycle Start Object【循环开始对象】参数组默认【无】方式，Age【年龄】参数组默认 Age【年龄】。

9. Emission Random Stream Seeds【发射随机流种子】卷展栏参数

Emission Random Stream Seeds【发射随机流种子】卷展栏参数，主要用来调节发射器参数，如图 11.18 所示。

图 11.18　Emission Random Stream Seeds【发射随机流种子】卷展栏

10. Render Attributes【渲染属性】卷展栏参数

Render Attributes【渲染属性】卷展栏参数，主要用来调节粒子渲染类型和添加属性等参数，如图 11.19 所示。

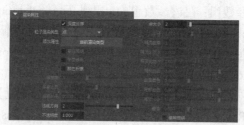

图 11.19　Render Attributes【渲染属性】卷展栏

Particle Render Type【粒子渲染类型】参数组：在该参数组中，主要提供了 MultPoint【多点】、MultStreak【多条纹】、Numeric【数值】、Points【点】、Spheres【球体】、Sprites【精灵】、Streak【条纹】、Blobby Surface（s/w）【滴状曲面（s/w）】、Cloud（s/w）【云（s/w）】和 Tube（s/w）【管状体（s/w）】10 种粒子渲染类型，默认 Points【点】粒子渲染类型。

11. Render Stats【渲染统计信息】卷展栏参数

Render Stats【渲染统计信息】卷展栏参数，主要用来在发射中可见、在折射中可见、投射阴影、接收阴影、运动模糊和主可见性等参数，如图 11.20 所示。

12. Per Particle（Array）Attributes【每粒子（数组）属性】卷展栏参数

Per Particle（Array）Attributes【每粒子（数组）属性】卷展栏参数，主要用来调节位置、渐变位置、速度、渐变速度、加速、渐变加速、质量、寿命 PP（每粒子）和世界速度等参数，如图 11.21 所示。

第 11 章　Rigging【装备】和 FX 中的对象属性

图 11.20　Render Stats
【渲染统计信息】卷展栏

图 11.21　Per Particle（Array）Attributes
【每粒子（数组）属性】卷展栏

13. Add Dynamic Attributes【添加动态属性】卷展栏参数

Add Dynamic Attributes【添加动态属性】卷展栏参数，主要用来添加常规、不透明度和颜色属性，如图 11.22 所示。

14. Clip Effects Attributes【片段效果属性】卷展栏参数

关于 Clip Effects Attributes【片段效果属性】卷展栏参数，在默认情况下没有任何参数可调，如图 11.23 所示。

图 11.22　Add Dynamic Attributes
【添加动态属性】卷展栏

图 11.23　Clip Effects Attributes
【片段效果属性】卷展栏

15. Sprite Attributes【精灵属性】卷展栏参数

关于 Sprite Attributes【精灵属性】卷展栏参数，在默认情况下没有任何参数可调，如图 11.24 所示。

图 11.24【Sprite Attributes（精灵属性）】卷展栏

16. Object Display【对象显示】卷展栏参数

Object Display【对象显示】卷展栏参数主要用来调节对象显示、边界框信息、绘制覆盖和选择等参数，如图 11.25 所示。

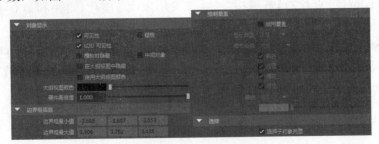

图 11.25　Object Display【对象显示】卷展栏

355

（1）Display Type【显示类型】参数组：在该参数组中主要提供了 Normal【正常】、Template【模板】和 Reference【引用】3 种显示类型，默认 Normal【正常】显示类型。

（2）Level of Detail【细节级别】参数组：在该参数组中主要提供了 Full【完全】和 Bounding Box【边界框】2 种显示类型，默认 Full【完全】显示类型。

（3）Color【颜色】参数组：在该参数组中主要提供了 Index【索引】和 GRB【RGB】2 种显示类型，默认 Index【索引】类型。

17. Arnold【阿诺德】卷展栏参数

Arnold【阿诺德】卷展栏参数，主要用来调节 Self Shadows【自投影】、Opaque【不透明体】、Visible in Diffuse【在传播中可见】、Visible in Glossy【在光纸上可见】、Matte【蒙版】、Trace Sets【足迹设置】、Render Points AS【渲染点】、Min Particle Radius【最小粒子半径】、Radius Multiplier【半径增值】、Max Particle Radius【最大粒子半径】、Min Pixel Width【最小粒子宽度】、Export Particle ID【输出粒子 ID】、Export Attributes【输出属性】、Delete Dead Particles【删除死亡粒子】、Volume Step Size【体积帧布大小】和 User Options【用户选项】等参数，如图 11.26 所示。

图 11.26【Arnold（阿诺德）】卷展栏

Render Points AS【渲染点 AS】参数组：该参数组主要包括 Points【点】、Spheres【球体】和 Quads【四倍】3 种渲染点方式，默认 Points【点】渲染方式。

18. Node Behavior【节点行为】卷展栏参数

Node Behavior【节点行为】卷展栏参数，主要用来调节缓存、冻结和节点状态等参数，如图 11.27 所示。

在 Node State【节点状态】参数组中，提供 Normal【正常】、Has No Effect【无效果】、Blocking【阻塞】、Waiting-ormal【等待-正常】、Waiting-as No Effect【等待-无效果】和 Waiting-locking【等待-阻塞】6 种状态。

19.【UUID】卷展栏参数

【UUID】卷展栏参数主要用来显示 UUID 的序列号，如图 11.28 所示。

第 11 章　Rigging【装备】和 FX 中的对象属性

图 11.27　Node Behavior【节点行为】卷展栏　　　图 11.28　【UUID】卷展栏

20. Extra Attributes【附加属性】卷展栏参数

Extra Attributes【附加属性】卷展栏参数，主要用来调节 Ai sss Set Name【Ai sss 集名称】、Ai Interpolate Blur【Ai 插值模糊】、Ai Evaluate Every【Ai 每个数值】和 Goal UV Set Name【目标 UV 集名称】等参数，如图 11.29 所示。

11.3　Emitter【发射器】属性面板

在 Emitter【发射器】属性面板中，主要包括 Transform Attributes【变换属性】、Basic Emitter Attributes【基本发射器属性】、Distance/Direction Attributes【距离/方向属性】、Basic Emission Speed Attributes【基础发射速率属性】、Volume Emitter Attributes【体积发射器属性】、Volume Speed Attributes【体积速率属性】、Texture Emission Attributes（UNRBS/Poly Surfaces only）【纹理发射器属性（仅 UNRBS/多边形曲面）】和 Extra Attributes【附加属性】8 个卷展栏参数，如图 11.30 所示。

图 11.29　Extra Attributes【附加属性】卷展栏　　　图 11.30　Emitter【发射器】属性面板

1. Transform Attributes【变换属性】卷展栏参数

Transform Attributes【变换属性】卷展栏参数，主要用来调节平移、旋转、缩放、斜切、旋转顺序、旋转轴和继承变换等参数，如图 11.31 所示。

2. Basic Emitter Attributes【基本发射器属性】卷展栏参数

Basic Emitter Attributes【基本发射器属性】卷展栏参数，主要用来调节发射器类型、速率（粒子/秒）、循环发射和循环间隔等参数，如图 11.32 所示。

图 11.31　Transform Attributes　　　图 11.32　Basic Emitter Attributes
　　　【变换属性】卷展栏　　　　　　　　【基本发射器属性】卷展栏

（1）Emitter Type【发射器类型】参数组：在该参数组中主要提供了 Directional【方向】、Omni【泛向】、Surface【表面】、Curve【曲线】和 Volume【体积】5 种发射器类型，默认 Omni【泛向】发射器类型。

（2）Cycle Emission【循环发射】参数组：在该参数组中主要提供了 None（Time Random off）【无（禁用 Time Random）】和 Frame（Time Random on）【帧（启用）Time Random】5 种发射器类型，默认 None（Time Random off）【无（禁用 Time Random）】发射器类型。

3. Distance/Direction Attributes【距离/方向属性】卷展栏参数

Distance/Direction Attributes【距离/方向属性】卷展栏参数，主要用来调节最小距离、最大距离、方向和扩散等参数，如图 11.33 所示。

4. Basic Emission Speed Attributes【基础发射速率属性】卷展栏参数

Basic Emission Speed Attributes【基础发射速率属性】卷展栏参数，主要用来调节速率、速率随机、切线速率和法线速率等参数，如图 11.34 所示。

图 11.33　Distance/Direction Attributes
【距离/方向属性】卷展栏

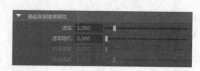

图 11.34　Basic Emission Speed Attributes
【基础发射速率属性】卷展栏

5. Volume Emitter Attributes【体积发射器属性】卷展栏参数

Volume Emitter Attributes【体积发射器属性】卷展栏参数，主要用来调节体积形状、体积偏移、体积扫描和截面半径等参数，如图 11.35 所示。

Volume Shape【体积形状】参数组：在该参数组中提供了 Cube【立方体】、Sphere【球体】、Cylinder【圆柱体】、Cone【圆锥体】和 Torus【圆环】5 种体积形状，默认 Cube【立方体】体积形状。

6. Volume Speed Attributes【体积速率属性】卷展栏参数

Volume Speed Attributes【体积速率属性】卷展栏参数，主要用来调节远离中心、远离轴、沿轴、绕轴、随机方向和平行光速率等参数，如图 11.36 所示。

图 11.35　Volume Emitter Attributes
【体积发射器属性】卷展栏

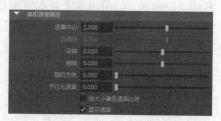

图 11.36　Volume Speed Attributes
【体积速率属性】卷展栏

第 11 章 Rigging【装备】和 FX 中的对象属性

7. Texture Emission Attributes（UNRBS/Poly Surfaces only）【纹理发射器属性（仅 UNRBS/多边形曲面）】卷展栏参数

Texture Emission Attributes（UNRBS/Poly Surfaces only）【纹理发射器属性（仅 UNRBS/多边形曲面）】卷展栏参数，主要用来调节粒子颜色和纹理速率等参数，如图 11.37 所示。

8. Extra Attributes【附加属性】卷展栏参数

Extra Attributes【附加属性】卷展栏参数，主要用来调节大纲视图颜色和线框颜色 RGB 等参数，如图 11.38 所示。

图 11.37　Texture Emission Attributes（UNRBS/Poly Surfaces only）【纹理发射器属性（仅 UNRBS/多边形曲面）】卷展栏

图 11.38　Extra Attributes【附加属性】卷展栏

11.4　3D Container【3D 容器】的 Shape【形状】属性面板

在 3D Container【3D 容器】的 Shape【形状】属性面板中，主要包括 Container Properties【容器特性】、Contents Method【内容方法】、Display【显示】、Dynamic Simulation【动力学模拟】、Liquids【液体】、Auto Resize【自动调整大小】、Self Attraction and Repulsion【自吸引和排斥】、Contents Details【内容详细信息】、Grids Cache【栅格缓存】、Surface【表面】、Output Mesh【输出网络】、Shading【着色】、Shading Quality【着色质量】、Textures【纹理】、Lighting【照明】、Render Stats【渲染统计信息】、Object Display【对象显示】、Arnold【阿诺德】、Node Behavior【节点行为】、【UUID】和 Extra Attributes【附加属性】21 个卷展栏参数，如图 11.39 所示。

1. Container Properties【容器特性】卷展栏参数

Container Properties【容器特性】卷展栏参数，主要用来调节基本分辨率、分辨率、大小和边界等参数，如图 11.40 所示。

图 11.39　3D Container【3D 容器】的 Shape【形状】属性面板　　　图 11.40　Container Properties【容器特性】卷展栏

Boundary X/Boundary Y/Boundary Z【边界X/边界Y/边界Z】参数组：在该参数组中提供了None【无】、Both Sides【两侧】、-X Side【-X侧】、X Side【X侧】和Wrapping【折回】5种边界方式，默认Both Sides【两侧】边界方式。

2. Contents Method【内容方法】卷展栏参数

Contents Method【内容方法】卷展栏参数，主要用来调节密度、速度、温度、燃料、颜色方法和衰减方法等参数，如图11.41所示。

（1）Density【密度】参数组：在该参数组中提供了Off（zero）【禁用（零）】、Static Grid【静态栅格】、Dynamic Grid【动态栅格】和Gradient【渐变】4种密度方式，默认Dynamic Grid【动态栅格】密度方式。

（2）Velocity【速度】参数组：在该参数组中提供了Off（zero）【禁用（零）】、Static Grid【静态栅格】、Dynamic Grid【动态栅格】和Gradient【渐变】4种速度方式，默认Dynamic Grid【动态栅格】速度方式。

（3）Temperature【温度】参数组：在该参数组中提供了Off（zero）【禁用（零）】、Static Grid【静态栅格】、Dynamic Grid【动态栅格】和Gradient【渐变】4种温度方式，默认Off（zero）【禁用（零）】温度方式。

（4）Fuel【燃料】参数组：在该参数组中提供了Off（zero）【禁用（零）】、Dynamic Grid【动态栅格】和Gradient【渐变】3种燃料方式，默认Off（zero）【禁用（零）】燃料方式。

（5）Color Method【颜色方法】参数组：在该参数组中提供了Use Shading Color【使用着色颜色】、Static Grid【静态栅格】和Dynamic Grid【动态栅格】3种颜色方法，默认Use Shading Color【使用着色颜色】颜色方法。

（6）Fall off Method【衰减方法】参数组：在该参数组中提供了Off（zero）【禁用（零）】和Static Grid【静态栅格】2种衰减方法，默认Off（zero）【禁用（零）】衰减方法。

（7）Density Gradient【密度渐变】、Velocity Gradient【速度渐变】、Temperature Gradient【温度渐变】和Fuel Gradient【燃料渐变】参数组：这4个参数组提供的选择完全相同，主要提供了Constant【恒定】、X Constant【X渐变】、Y Constant【Y渐变】、Z Constant【Z渐变】、-X Constant【-X渐变】、-Y Constant【-Y渐变】、-Z Constant【-Z渐变】和Center Gradient【中心渐变】8种渐变方式，默认的都是Constant【恒定】渐变方式。

3. Display【显示】卷展栏参数

Display【显示】卷展栏参数，主要用来调节着色显示、不透明度预览增益、每个体素的切片数、体素质量、边界绘制、数值显示、线框显示、速度绘制跳过和绘制长度等参数，如图11.42所示。

（1）Shaded Display【着色显示】参数组：在该参数组中，主要提供了Off【禁用】、As Render【作为渲染】、Density【密度】、Temperature【温度】、Fuel【燃料】、Collision【碰撞】、Density And Color【密度和颜色】、Density And Temp【密度和温度】、Density And Fuel【密度和燃料】、Density And Collision【密度和碰撞】和Fall off【衰减】11种着色显示，默认As Render【作为渲染】着色显示方式。

第 11 章　Rigging【装备】和 FX 中的对象属性

图 11.41　Contents Method【内容方法】卷展栏

图 11.42　Display【显示】卷展栏

（2）Voxel Quality【体素质量】参数组：在该参数组中，主要提供了 Faster【更快】和 Better【更好】2 种体素质量，默认 Faster【更快】方式。

（3）Boundary Draw【边界绘制】参数组：在该参数组中，主要提供了 Bottom【底】、Reduced【精简】、Outline【轮廓】、Full【完全】、Bounding Box【边界框】和 None【无】6 种边界绘制方式，默认 Bottom【底】边界绘制模式。

（4）Numeric Display【数值显示】参数组：在该参数组中，主要提供了 Off【禁用】、Density【密度】、Temperature【温度】和 Fuel【燃料】4 种数值显示方式，默认 Off【禁用】数值显示方式。

（5）Wireframe Display【线框显示】参数组：在该参数组中，主要提供了 Off【禁用】、Rectangles【矩形】和 Particles【粒子】3 种线框显示，默认 Particles【粒子】线框显示。

4．Dynamic Simulation【动力学模拟】卷展栏参数

Dynamic Simulation【动力学模拟】卷展栏参数，主要用来调节重力、黏度、摩擦力、阻尼、解算器、高细节解算、子步、解算器质量、栅格插值器、开始帧和模拟速度比例等参数，如图 11.43 所示。

（1）Solver【解算器】参数组：在该参数组中，主要提供了 None【无】、Navier-Stokes【方程】和 Spring Mesh【弹簧网络】3 种解算器模式。默认 Navier-Stokes【方程】解算器模式。

（2）High Detail Solve【高细节解算】参数组：在该参数组中，主要提供了 Off【禁用】、All Grids Except Velocity【除速度之外的所有栅格】、Velocity only【仅速度】和 All Grids【所有栅格】4 种高细节解算器模式，默认 Off【禁用】高细节解算模式。

（3）Grid Interpolator【栅格插值器】参数组：在该参数组中，主要提供了 Linear【线性】和 Hermite【艾尔米特】2 种栅格插值器，默认 Linear【线性】栅格插值器方式。

5．Liquids【液体】卷展栏参数

Liquids【液体】卷展栏参数，主要用来调节液体方法、液体最小密度、液体喷雾、质量范围、密度张力、密度压力和密度压力阈值等参数，如图 11.44 所示。

图 11.43 Dynamic Simulation【动力学模拟】卷展栏

图 11.44 Liquids【液体】卷展栏

Liquid Method【液体方法】参数组：该参数组主要提供了 Liquid and Air【液体和空气】和 Density Based Mass【基于密度的质量】2 种密度压力阈值，默认 Liquid and Air【液体和空气】密度压力阈值。

6．Auto Resize【自动调整大小】卷展栏参数

Auto Resize【自动调整大小】卷展栏参数，主要用来调节最大分辨率、动态偏移、自动调整阈值大小和自动调整边界大小等参数，如图 11.45 所示。

7．Self Attraction and Repulsion【自吸引和排斥】卷展栏参数

Self Attraction and Repulsion【自吸引和排斥】卷展栏参数，主要用来调节自作用力、自吸引、自排斥、平衡值和自作用力距离等参数，如图 11.46 所示。

图 11.45 Auto Resize
【自动调整大小】卷展栏

图 11.46 Self Attraction and Repulsion
【自吸引和排斥】卷展栏

Self Force【自作用力】参数组：该参数组主要提供了 Off（zero）【禁用（零）】、Density【密度】和 Temperature【温度】3 种自作用力，默认 Off（zero）【禁用（零）】自作用力方式。

8．Contents Details【内容详细信息】卷展栏参数

Contents Details【内容详细信息】卷展栏参数，主要用来调节密度、速度、湍流、温度、燃料和颜色等参数，如图 11.47 所示。

第 11 章 Rigging【装备】和 FX 中的对象属性

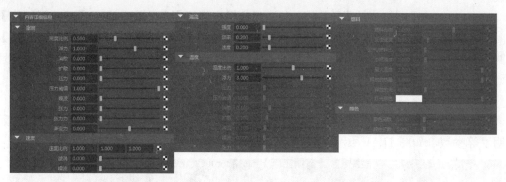

图 11.47 Contents Details【内容详细信息】卷展栏

9. Grids Cache【栅格缓存】卷展栏参数

Grids Cache【栅格缓存】卷展栏参数，主要用来确定是否启用读取密度、读取速度、读取温度、读取燃料、读取颜色、读取纹理坐标和读取衰减等功能，如图 11.48 所示。

10. Surface【表面】卷展栏参数

Surface【表面】卷展栏参数，主要用来调节表面阈值、表面容差、镜面反射颜色、余弦、选定位置、选定颜色、插值和折射率等参数，如图 11.49 所示。

图 11.48 Grids Cache【栅格缓存】卷展栏

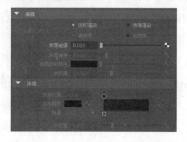

图 11.49 Surface【表面】卷展栏

Interpolation【插值】参数组：在该参数组中，主要提供了 None【无】、Linear【线性】、Smooth【平滑】和 Spline【样条线】4 种插值方式，默认 Linear【线性】插值方式。

11. Output Mesh【输出网络】卷展栏参数

Output Mesh【输出网络】卷展栏参数，主要用来调节网络方法、网络分辨率和网络平滑迭代次数等参数，如图 11.50 所示。

图 11.50 Output Mesh【输出网络】卷展栏

Mesh Method【网络方法】参数组：在该参数组中提供了 Triangle Mesh【三角形网络】、Quad Mesh【四边形网络】、Tetrahedron【四面体】和 Acute Tetrahedron【锐角四面体】4 种网络方法，默认 Triangle Mesh【三角形网络】网络方法。

12. Shading【着色】卷展栏参数

Shading【着色】卷展栏参数，主要用来调节着色、颜色、白炽度、不透明度和蒙版不透明度等参数，如图 11.51 所示。

图 11.51　Shading【着色】卷展栏

（1）Drop off Shape【衰减形状】参数组：在该参数组中主要提供了 Off【禁用】、Sphere【球体】、Cube【立方体】、Cone【圆锥体】、Double Cone【双锥体】、X Gradient【X 渐变】、Y Gradient【Y 渐变】、Z Gradient【Z 渐变】、-X Gradient【-X 渐变】、- Gradient【-Y 渐变-】、-Z Gradient【-Z 渐变】和 Use Falloff Grid【使用衰减栅格】12 种衰减形式，默认 Cube【立方体】衰减形式。

（2）Interpolation【插值】参数组：在该参数组中主要提供了 None【无】、Linear【线性】、Smooth【平滑】和 Spline【样条线】4 种插值方式，默认 Linear【线性】插值方式。

（3）【颜色输入】参数组：在该参数组中主要提供了 Constant【恒定】、X Gradient【X 渐变】、Y Gradient【Y 渐变】、Z Gradient【Z 渐变】、Center Gradient【中心渐变】、Density【密度】、Temperature【温度】、Fuel【燃料】、Pressure【压力】、Speed【速度】和 Density And Fuel【密度和颜料】11 种颜色输入方式，默认【恒定】颜色输入方式。

提示：在【白炽度】和【不透明度】卷展栏中，Interpolation【插值】参数组和【颜色输入】参数组提供的选项，与【颜色】卷展栏中的 Interpolation【插值】参数组和【颜色输入】参数组提供的选项完全相同，在此就不再赘述。

（4）Matte Opacity Mode【蒙版不透明度模式】参数组：在该参数组中主要提供了 Black Hole【黑洞】、Solid Matte【匀值蒙版】和 Opacity Gain【不透明度增益】3 种蒙版不透明度模式，默认 Opacity Gain【不透明度增益】蒙版不透明度模式。

13. Shading Quality【着色质量】卷展栏参数

Shading Quality【着色质量】卷展栏参数，主要用来调节质量、对比度容差、采样方法和渲染插值器等参数，如图 11.52 所示。

第 11 章 Rigging【装备】和 FX 中的对象属性

图 11.52 Shading Quality【着色质量】卷展栏

（1）Sample Method【采样方法】参数组：在该参数组中主要提供了 Uniform【一致】、Jittered【抖动】、Adaptive【自适应】和 Adaptive Jittered【自适应抖动】4 种采样方法，默认 Adaptive Jittered【自适应抖动】采样方法。

（2）Render Interpolator【渲染插值器】参数组：在该参数组中主要提供了 Linear【线性】和 Smooth【平滑】两种渲染插值器，默认 Linear【线性】渲染插值器。

14．Textures【纹理】卷展栏参数

Textures【纹理】卷展栏参数主要用来调节纹理类型、坐标类型、坐标方法、坐标速度、颜色纹理增益、白炽纹理增益、不透明度纹理增益、阈值、振幅、比率、频率比、最大深度、纹理时间、缩放因子、纹理原点、纹理比例、纹理旋转、内爆和内爆中心等参数，如图 11.53 所示。

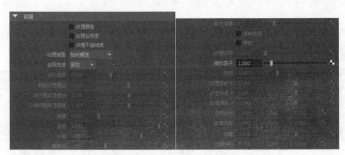

图 11.53 Textures【纹理】卷展栏

（1）Texture【纹理类型】参数组：在该参数组中主要提供了 Perlin Noise【柏林噪波】、Billow【翻转】、Volume Wave【体积波浪】、Wispy【束状】、SpaceTime【空间时间】和 Mandelbrot【满德尔布罗特】6 种纹理类型，默认 Perlin Noise【柏林噪波】类型。

（2）Coordinate Method【坐标方法】参数组：在该参数组中主要提供了 Fixed【固定】和 Grid【栅格】两种坐标方法，默认 Fixed【固定】坐标方法。

15．Lighting【照明】卷展栏参数

Lighting【照明】卷展栏参数，主要用来调节阴影不透明度、阴影扩散、灯光类型、灯光亮度、灯光颜色、环境光亮度、环境光扩散、环境色、平行光、点光源和点光源衰退等参数，如图 11.54 所示。

（1）Light Brightness【灯光类型】参数组：在该参数组中主要提供了 Diagonal【对角】、Directional【平行】和 Point【点】3 种灯光类型，默认 Directional【平行】灯光类型。

（2）Point Light Decay【点光源衰退】参数组：在该参数组中主要提供了 No Decay【无衰退】、Linear【线性】、Quadratic【二次方】和 Cubic【立方】4 种点光源衰退模式。

16. Render Stats【渲染统计信息】卷展栏参数

Render Stats【渲染统计信息】卷展栏参数，主要用来调节着色采样数、最大着色采样数和体积采样数等参数，如图 11.55 所示。

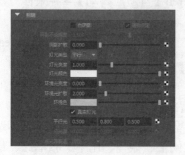

图 11.54　Lighting【照明】卷展栏

图 11.55　Render Stats【渲染统计信息】卷展栏

17. Object Display【对象显示】卷展栏参数

Object Display【对象显示】卷展栏参数，主要用来调节对象显示、重影信息、边界框信息、绘制覆盖和选择等参数，如图 11.56 所示。

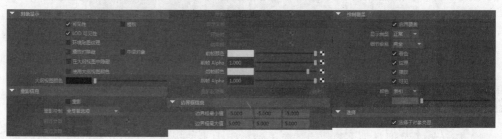

图 11.56　Object Display【对象显示】卷展栏

（1）Ghosting Control【重影控制】参数组：在该参数组中主要提供了 Global Prefs【全局首选项】、Custom Frames【自定义帧】、Custom Frame Steps【自定义帧步数】、Custom Key Steps【自定义关键帧步数】和 Key Frames【关键帧】5 种重影控制方式。默认 Global Prefs【全局首选项】重影控制方式。

（2）Display Type【显示类型】参数组：在该参数组中主要提供了 Normal【正常】、Template【模板】和 Reference【引用】3 种显示类型，默认 Normal【正常】显示类型。

（3）Level of Detail【细节级别】参数组：在该参数组中主要提供了 Full【完全】和 Bounding Box【边界框】3 种细节级别，默认 Full【完全】细节级别。

（4）Color【颜色】参数组：在该参数组中主要提供了 Index【索引】和 RGB【RGB】2 种颜色方式，默认 Index【索引】颜色方式。

18. Arnold【阿诺德】卷展栏参数

Arnold【阿诺德】卷展栏参数，主要用来调节 Step Size【帧步尺寸】、Enable Deformation Blur【使变形模糊】、Motion Vector Scale【运动矢量比例】、Filter Type【过滤器类型】、Function Anisotropy【功能异向性】、Visible In Diffuse【在传播中可见】、Visible In Glossy【在光纸上可见】、Custom Texture【自定义纹理】、Override Fluid Texture【超控液体纹理】、Texture Color【纹理颜色】、Texture Incandescence【纹理白热化】、Texture Opacity【纹理不透明性】、Coordinate Method【整合法】、Texture【纹理】和 User Options【用户选项】等参数，如图 11.57 所示。

图 11.57　Arnold【阿诺德】卷展栏

19. Node Behavior【节点行为】卷展栏参数

Node Behavior【节点行为】卷展栏参数，主要用来调节缓存、冻结和节点状态等参数，如图 11.58 所示。

在 Node State【节点状态】参数组中，主要有 Normal【正常】、Has No Effect【无效果】、Blocking【阻塞】、Waiting-Normal【等待-正常】、Waiting-Has No Effect【等待-无效果】和 Waiting-Blocking【等待-阻塞】6 种状态供选择。

20.【UUID】卷展栏参数

【UUID】卷展栏参数面板中没有参数可调，只显示 UUID 序列号，如图 11.59 所示。

图 11.58　Node Behavior【节点行为】卷展栏　　　图 11.59　【UUID】卷展栏

21. Extra Attributes【附加属性】卷展栏参数

Extra Attributes【附加属性】卷展栏参数，主要用来调节透底、纹理原点和从缓存播放等参数，如图 11.60 所示。

11.5　2D Container【2D 容器】的 Shape【形状】属性面板

在 2D Container【2D 容器】的 Shape【形状】属性面板中，主要包括 Container Properties【容器特性】、Contents Method【内容方法】、Display【显示】、Dynamic Simulation【动力学模拟】、Liquids【液体】、Auto Resize【自动调整大小】、Self Attraction and Repulsion【自吸引和排斥】、Contents Details【内容详细信息】、Grids Cache【栅格缓存】、Surface【表面】、Output Mesh【输出网络】、Shading【着色】、Shading Quality【着色质量】、Textures【纹理】、Lighting【照明】、Render Stats【渲染统计信息】、Object Display【对象显示】、Arnold【阿诺德】、Node Behavior【节点行为】、【UUID】和 Extra Attributes【附加属性】21 个卷展栏参数，如图 11.61 所示。

图 11.60　Extra Attributes
【附加属性】卷展栏

图 11.61　2D Container【2D 容器】
的 Shape【形状】属性面板

提示：2D Container【2D 容器】的 Shape【形状】属性面板中的卷展栏参数，与 3D Container【3D 容器】的 Shape【形状】属性面板中的卷展栏参数基本相同，在此就不再赘述。请读者参考"11.4 3D Container【3D 容器】的 Shape【形状】属性面板"的卷展栏参数。

11.6　Ocean【海洋】对象属性面板

Ocean【海洋】对象属性面板包括 transform【变换】、Ocean Preview Plane【海洋预览平面】、Ocean Shader【海洋着色】、Time【时间】和 Default Render Utility List【默认渲染实用程序列表】5 个属性选项面板，如图 11.62 所示。

图 11.62　Ocean【海洋】对象属性面板

1. Ocean【海洋】对象属性的 Transform【变换】属性面板

Ocean【海洋】对象属性的 Transform【变换】属性面板，主要包括 Transform Attributes【变换属性】、Pivots【枢轴】、Limit Information【限制信息】、Display【显示】、Node Behavior

【节点行为】、【UUID】和 Extra Attributes【附加属性】7 个卷展栏参数，如图 11.63 所示。

提示：Ocean【海洋】对象属性的 Transform【变换】属性面板的参数，与 Polygon【多边形】对象公共属性面板的参数完全相同，在此就不再详细讲解。请读者参考"10.1 Polygon【多边形】对象公共属性面板"。

2. Ocean【海洋】对象属性的 Ocean Preview Plane【海洋预览平面】属性面板

Ocean【海洋】对象属性的 Ocean Preview Plane【海洋预览平面】属性面板，主要包括 Object Display【对象显示】、Node Behavior【行为节点】、【UUID】和 Extra Attributes【附加属性】4 个卷展栏参数，如图 11.64 所示。

图 11.63　Ocean【海洋】对象属性的 transform【变换】属性面板

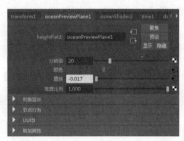
图 11.64　Ocean【海洋】对象属性的 Ocean Preview Plane【海洋预览平面】属性面板

1）Object Display【对象显示】卷展栏参数

Object Display【对象显示】卷展栏参数，主要用来调节对象显示、重影信息、边界框信息、绘制覆盖和选择等参数，如图 11.65 所示。

图 11.65　Object Display【对象显示】卷展栏

（1）Ghosting Control【重影控制】参数组：在该参数组中主要提供了 Global Prefs【全局首选项】、Custom Frames【自定义帧】、Custom Frame Steps【自定义帧步数】、Custom Key Steps【自定义关键帧步数】和 Key frames【关键帧】5 种重影控制方式，默认 Global Prefs【全局首选项】重影控制方式。

（2）Display Type【显示类型】参数组：在该参数组中主要提供了 Normal【正常】、Template【模板】和 Reference【引用】3 种显示类型，默认 Normal【正常】显示类型。

（3）Level of Detail【细节级别】参数组：在该参数组中主要提供了 Full【完全】和 Bounding Box【边界框】3 种细节级别，默认 Full【完全】细节级别。

（4）Color【颜色】参数组：在该参数组中主要提供了 Index【索引】和【RGB】2 种颜色方式，默认 Index【索引】颜色方式。

2）Node Behavior【行为节点】卷展栏参数

Node Behavior【行为节点】卷展栏参数，主要用来调节缓存、冻结和节点状态等参数，如图 11.66 所示。

在 Node State【节点状态】参数组中有 Normal【正常】、Has NO Effect【无效果】、Blocking【阻塞】、Waiting—Normal【等待-正常】、Waiting—Has NO Effect【等待-无效果】和 Waiting—Blocking【等待-阻塞】6 种状态供选择。

3）【UUID】卷展栏参数

【UUID】卷展栏中没有参数可调，只显示 UUID 序列号，如图 11.67 所示。

图 11.66 Node Behavior【行为节点】卷展栏　　　　图 11.67　【UUID】卷展栏

4）Extra Attributes【附加属性】卷展栏参数

Extra Attributes【附加属性】卷展栏参数，主要用来调节 Color Set【颜色集】、Collision Offset Velocity Increment【碰撞偏移速度增量】、Collision Depth Velocity Increment【碰撞深度速度增量】、Collision Offset Velocity Multiplier【碰撞偏移速度倍增】和 Collision Depth Velocity Multiplier【碰撞深度速度倍增】等参数，如图 11.68 所示。

3. Ocean【海洋】对象属性的 Ocean Shader【海洋着色】属性面板

Ocean【海洋】对象属性的 Ocean Shader【海洋着色】属性面板，主要包括 Ocean Attributes【海洋属性】、Common Material Attributes【公用材质属性】、Specular Shading【镜面反射着色】、Glow【辉光】、Matte Opacity【蒙版不透明度】、Ray trace Options【光线跟踪选项】、Hardware Texturing【硬件纹理】、Node Behavior【节点行为】、【UUID】和 Extra Attributes【附加属性】10 个卷展栏参数，如图 11.69 所示。

图 11.68　Extra Attributes【附加属性】卷展栏　　　　图 11.69　Ocean【海洋】对象属性的 Ocean Shader【海洋着色】属性面板

第 11 章 Rigging【装备】和 FX 中的对象属性

在【类型】参数组中,主要提供了 Shader FX Shader【着色器特效渲染】、Stingray PBS【Stingray 基于物理的渲染】、Ai Ambient Occlusion【Ai 环境光遮蔽】、Ai Hair【Ai 头发】、Ai Ray Switch【Ai 射线开关】、Ai Shadow Catcher【Ai 阴影捕捉】、Ai Skin【Ai 蒙皮】、Ai Standard【Ai 标准】、Ai Utility【Ai 实用程序】、Ai Wire Frame【Ai 线框】、Anisotropic【各向异性】、Bifrost Aero Material【Bifrost Aero 材质】、Bifrost Foam Material【Bifrost 泡沫材质】、Bifrost Liquid Material【Bifrost 液体材质】、Blinn【布林】、Hair Tube Shader【头发管着色】、Lambert【兰伯特】、Layered Shader【分层着色】、Ocean Shader【海洋着色】、Phong【冯】、Phong E【冯 E】、Ramp Shader【渐变着色】、Shading Map【着色贴图】、Surface Shader【表面着色】和 Use Background【使用背景】26 种材质类型。

1)Ocean Attributes【海洋属性】卷展栏参数

Ocean Attributes【海洋属性】卷展栏参数,主要用来调节海洋属性、波高度、波湍流和波峰等参数,如图 11.70 所示。

Interpolation【插值】参数组:在该参数组中主要提供了 None【无】、Linear【线性】、Smooth【平滑】和 Spline【样条线】4 种插值方式,默认 Linear【线性】插值方式。

2)Common Material Attributes【公用材质属性】卷展栏参数

Common Material Attributes【公用材质属性】卷展栏参数,主要用来调节水颜色、泡沫颜色、透明度、折射率、白炽度、环境色、漫反射、波谷阴影、半透明、半透明聚焦和半透明深度等参数,如图 11.71 所示。

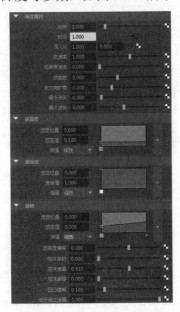
图 11.70 Ocean Attributes 【海洋属性】卷展栏

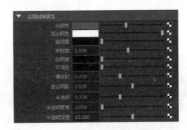
图 11.71 Common Material Attributes 【公用材质属性】卷展栏

3)Specular Shading【镜面反射着色】卷展栏参数

Specular Shading【镜面反射着色】卷展栏参数,主要用来调节镜面反射着色和环境等参数，如图 11.72 所示。

Interpolation【插值】参数组：在该参数组中主要提供了 None【无】、Linear【线性】、Smooth【平滑】和 Spline【样条线】4 种插值方式，默认 Smooth【平滑】插值方式。

4）Glow【辉光】卷展栏参数

Glow【辉光】卷展栏参数，主要用来调节辉光强度和镜面反射辉光参数，如图 11.73 所示。

图 11.72　Specular Shading【镜面反射着色】卷展栏

图 11.73　Glow【辉光】卷展栏

5）Matte Opacity【蒙版不透明度】卷展栏参数

Matte Opacity【蒙版不透明度】卷展栏参数，主要用来调节蒙版不透明度模式和蒙版不透明参数，如图 11.74 所示。

Matte Opacity Mode【蒙版不透明度模式】参数组：在该参数组中提供了 Black Hole【黑洞】、Solid Matte【匀值蒙版】和 Opacity Gain【不透明度增益】3 种蒙版不透度模式。默认 Opacity Gain【不透明度增益】蒙版不透明度模式。

6）Ray trace Options【光线跟踪选项】卷展栏参数

Ray trace Options【光线跟踪选项】卷展栏参数，主要用来调节折射限制、阴影衰减、反射限制和镜面反射度等参数，如图 11.75 所示。

图 11.74　Matte Opacity
【蒙版不透明度】卷展栏

图 11.75　Ray trace Options
【光线跟踪选项】卷展栏

7）Hardware Texturing【硬件纹理】卷展栏参数

Hardware Texturing【硬件纹理】卷展栏参数，主要用来调节带纹理通道、纹理分辨率和纹理过滤器等参数，如图 11.76 所示。

（1）Textured channel【带纹理通道】参数组：在该参数组中主要提供了 None【无】、Water Color【水颜色】、Foam Color【泡沫颜色】、Transparency【透明度】、Ambient Color【环境色】、Incandescence【白炽度】、Bump Map【凹凸贴图】、Diffuse【漫反射】、Translucence【半透明】、Eccentricity【偏心率】、Specular Color【镜面反射颜色】、Reflectivity【反射率】、Reflectivity Color【反射的颜色】和 Combined Textures【组合纹理】14 种带纹理通道的方式，默认 None【无】带纹理通道方式。

（2）Texture resolution【纹理分辨率】参数组：在该参数组中，主要提供了 Default【默认】、Low（32×32）【低（32×32）】、Medium（64×64）【中（64×64）】、High（128×128）【高（128×128）】和 Highest（256×256）【最高（256×256）】5 种纹理分辨率，默认 Default【默认】纹理分辨率。

（3）Texture Filter【纹理过滤器】参数组：在该参数组中主要提供了 Use Global Settings【使用全局设置】、Nearest（Unfiltered）【最近（未过滤）】、Bilinear【双线性】、Mipmap Nearest【Mipmap 最近】、Mipmap Linear【Mipmap 线性】、Mipmap Bilinear【Mipmap 双线性】和 Mipmap【Mipmap Trilinear 三线性】7 种纹理过滤器，默认 Use Global Settings【使用全局设置】纹理过滤器。

8）Node Behavior【行为节点】卷展栏参数

Node Behavior【行为节点】卷展栏参数，主要用来调节缓存、冻结和节点状态等参数，如图 11.77 所示。

图 11.76 Hardware Texturing
【硬件纹理】卷展栏

图 11.77 Node Behavior
【行为节点】卷展栏

在 Node State【节点状态】参数组中提供 Normal【正常】、Has No Effect【无效果】、Blocking【阻塞】、Waiting-Normal【等待-正常】、Waiting-Has No Effect【等待-无效果】和 Waiting-Blocking【等待-阻塞】6 种状态供选择。

9）【UUID】卷展栏参数

【UUID】卷展栏参数中没有参数可调，只显示 UUID 序列号，如图 11.78 所示。

10）Extra Attributes【附加属性】卷展栏参数

Extra Attributes【附加属性】卷展栏参数，在默认情况下没有任何参数，如图 11.79 所示。

图 11.78 【UUID】卷展栏

图 11.79 Extra Attributes【附加属性】卷展栏

4. Ocean【海洋】对象属性的 Time【时间】属性面板

Ocean【海洋】对象属性的 Time【时间】属性面板，主要包括 Time Attributes【时间属性】、Node Behavior【节点行为】、【UUID】和 Extra Attributes【附加属性】4 个属性卷展栏参数。

在 Node State【节点状态】参数组中，有 Normal【正常】、Has No Effect【无效果】、Blocking【阻塞】、Waiting-Normal【等待-正常】、Waiting-Has No Effect【等待-无效果】和 Waiting-Blocking【等待-阻塞】6 种状态供选择。

5. default Render Utility List【默认渲染实用程序列表】属性面板

default Render Utility List【默认渲染实用程序列表】属性面板，主要包括行为节点、UUID 和附加属性 3 个参数选项。

图 11.80　Ocean【海洋】对象属性的
Time【时间】属性面板

图 11.81　default Render Utility List
【默认渲染实用程序列表】属性面板

在 Node State【节点状态】参数组中有 Normal【正常】、Has No Effect【无效果】、Blocking【阻塞】、Waiting-Normal【等待-正常】、Waiting-Has No Effect【等待-无效果】和 Waiting-Blocking【等待-阻塞】6 种状态供选择。

11.7　Pond【池塘】对象属性面板

Pond【池塘】对象属性面板，主要包括 Pond【池塘】、Pond Shape【池塘形状】和 Time【时间】3 个属性选项面板，如图 11.82 所示。

图 11.82　Pond【池塘】对象属性面板

1. Pond【池塘】对象属性面板

Pond【池塘】对象属性面板，主要包括 Transform Attributes【变换属性】、Pivots【枢轴】、Limit Information【限制信息】、Display【显示】、Node Behavior【节点行为】、【UUID】和 Extra Attributes【附加属性】7 个卷展栏参数，如图 11.83 所示。

提示：Pond【池塘】对象属性面板的参数与 Polygon【多边形】对象公共属性面板的参数完全相同，在此就不再详细讲解，请读者参考"10.1 Polygon【多边形】对象公共属性面板"。

2. Pond【池塘】的 Pond Shape【池塘形状】属性面板

Pond【池塘】的 Pond Shape【池塘形状】属性面板，主要包括 Container Properties【容器特性】、Contents Method【内容方法】、Display【显示】、Dynamic Simulation【动力学模拟】、Liquids【液体】、Auto Resize【自动调整大小】、Self Attraction and Repulsion【自吸引和排斥】、Contents Details【内容详细信息】、Grids Cache【栅格缓存】、Surface【表面】、

Output【输出网络】、Shading【着色】、Shading Quality【着色质量】、Textures【纹理】、Lighting【照明】、Render Stats【渲染统计信息】、Object Display【对象显示】、Arnold【阿诺德】、Node Behavior【节点行为】、【UUID】和 Extra Attributes【附加属性】21 个卷展栏参数，如图 11.84 所示。

图 11.83　Pond【池塘】对象属性面板

图 11.84　Pond【池塘】的 Pond Shape【池塘形状】属性面板

1）Container Properties【容器特性】卷展栏参数

Container Properties【容器特性】卷展栏参数，主要用来调节 Base Resolution【基本分辨率】、Size【大小】和 Boundary【边界】参数，如图 11.85 所示。

（1）Boundary X【边界 X】参数组：在该参数组中，主要提供了 None【无】、Both Sides【两侧】、-X Side【-X 侧】、X Side【X 侧】和 Wrapping【折回】5 种边界方式，默认 Both Sides【两侧】边界方式。

（2）Boundary Y【边界 Y】参数组：在该参数组中主要提供了 None【无】、Both Sides【两侧】、-Y Side【-Y 侧】、Y Side【Y 侧】和 Wrapping【折回】5 种边界方式，默认 Both Sides【两侧】边界方式。

2）Contents Method【内容方法】卷展栏参数

Contents Method【内容方法】卷展栏参数，主要用来调节密度、密度渐变、速度、速度渐变、温度、温度渐变、燃料、燃料渐变、颜色方法和衰减方法等参数，如图 11.86 所示。

图 11.85　Container Properties【容器特性】属性面板

图 11.86　Contents Method【内容方法】属性面板

（1）Density【密度】、Velocity【速度】和 Temperature【温度】参数组：这三个参数提供的选项完全相同，主要提供了 Off（zero）【禁用（零）】、Static Grid【静态栅格】、Dynamic

Grid【动态栅格】和 Gradient【渐变】4 种方式，默认 Dynamic Grid【动态栅格】方式。

（2）Density Gradient【密度渐变】、Velocity Gradient【速度渐变】、Temperature Gradient【温度渐变】和 Fuel Gradient【颜料渐变】参数组：这 4 个参数组提供的选项完全相同，主要提供了 Constant【恒定】、X Gradient【X 渐变】、Y Gradient【Y 渐变】、Z Gradient【Z】、-X Gradient【-X 渐变】、-Y Gradient【-Y 渐变】、-Z Gradient【-Z 渐变】和 Center Gradient【中心渐变】8 种渐变方式，默认 Constant【恒定】渐变方式。

（3）Fule【燃料】参数组：在该参数组中提供了 Off（zero）【禁用（零）】、Dynamic Grid【动态栅格】和 Gradient【渐变】3 种方式。默认 Off（zero）【禁用（零）】方式。

（4）Color Method【颜色方法】参数组：在该参数组中提供了 Use Shading Color【使用着色颜色】、Static Grid【静态栅格】和 Dynamic Grid【动态栅格】3 种颜色方法。

（5）Falloff Method【衰减方法】参数组：在该参数组中提供了 Off（zero）【禁用（零）】和 Static Grid【静态栅格】2 种衰减方法，默认 Off（zero）【禁用（零）】衰减方法。

3）Display【显示】卷展栏参数

Display【显示】卷展栏参数，主要用来调节着色显示、不透明度预览增益、体素质量、边界绘制、数值显示、线框显示、速度和长度等参数，如图 11.87 所示。

（1）Shaded Display【着色显示】参数组：在该参数组中提供了 Off【禁用】、As Render【作为渲染】、Density【密度】、Temperature【温度】、Fuel【燃料】、Collision【碰撞】、Density and Color【密度和颜色】、Density and Temp【密度和温度】、Density and Fuel【密度和燃料】、Density and Collision【密度和碰撞】和 Falloff【衰减】11 种着色显示方式，默认 As Render【作为渲染】着色显示方式。

（2）Voxel Quality【体素质量】参数组：在该参数组中提供了 Faster【更快】和 Better【更好】2 个参数组，

（3）Boundary Draw【边界绘制】参数组：在该参数组中提供了 Outline【轮廓】、Full【完全】、Bounding【边界框】和 None【无】4 种数值显示方式，默认 Bounding【边界框】数值显示方式。

（4）Numeric Display【数值显示】参数组：在该参数组中提供了 Off【禁用】、Density【密度】、Temperature【温度】和 Fuel【燃料】4 种数值显示方式，默认 Off【禁用】数值显示方式。

（5）Wire frame Display【线框显示】参数组：在该参数组中提供了 Off【禁用】、Rectangles【矩形】和 Particles【粒子】3 种数值显示方式，默认 Particles【粒子】数值显示方式。

4）Dynamic Simulation【动力学模拟】卷展栏参数

Dynamic Simulation【动力学模拟】卷展栏参数，主要用来调节重力、黏度、摩擦力、阻尼、解算器、高精细节解算、子步、解算器质量、栅格插值器、开始帧、模拟速度比例等参数，如图 11.88 所示。

第 11 章 Rigging【装备】和 FX 中的对象属性

图 11.87 Display【显示】卷展栏

图 11.88 Dynamic Simulation【动力学模拟】卷展栏

（1）Solver【解算器】参数组：在该参数组中提供了 None【无】、Navier-Stokes【纳维叶-斯托克斯】和 Spring Mesh【弹簧网络】3 种解算器，默认 Spring Mesh【弹簧网络】解算器。

（2）High Detail Solver【高精细节解算】参数组：在该参数组中提供了 Off【禁用】、All Grids Except Velocity【除速度之外的所有栅格】、Velocity only【仅速度】和 All Grids【所有栅格】4 种高精细节解算，默认 Off【禁用】高精细节解算。

（3）Grid Interpolator【栅格插值器】参数组：在该参数组中提供了 Linear【线性】和 hermite【埃尔米特】2 种解算器，默认 Linear【线性】栅格插值器。

5）Liquids【液体】卷展栏参数

Liquids【液体】卷展栏参数，主要用来调节液体方法、液体最小密度、液体喷雾、质量范围、密度张力、张力、密度压力和密度压力阈值等参数，如图 11.89 所示。

Liquid Method【液体方法】参数组：在该参数组中主要提供了 Liquid and Air【液体和空气】和 Density Based Mass【基于密度的质量】2 种液体方法，默认 Liquid and Air【液体和空气】液体方法。

6）Auto Resize【自动调整大小】卷展栏参数

Auto Resize【自动调整大小】卷展栏参数，主要用来调节最大分辨率、动态偏移、自动调整阈值大小和自动调整边界大小等参数，如图 11.90 所示。

图 11.89 Liquids【液体】卷展栏

图 11.90 Auto Resize【自动调整大小】卷展栏

7）Self Attraction and Repulsion【自吸引和排斥】卷展栏参数

Self Attraction and Repulsion【自吸引和排斥】卷展栏参数，主要用来调节自作用力、自吸引、自排斥、平衡值和自作用力距离等参数，如图11.91所示。

图11.91　Self Attraction and Repulsion【自吸引和排斥】卷展栏

Self Force【自作用力】参数组：在该参数组中提供了Off（zero）【禁用（零）】、Density【密度】和Temperature【温度】3种自作用力方式，默认Off（zero）【禁用（零）】自作用力方式。

8）Contents Details【内容详细信息】卷展栏参数

Contents Details【内容详细信息】卷展栏参数，主要用来调节密度、速度、湍流、温度、燃料和颜色等参数，如图11.92所示。

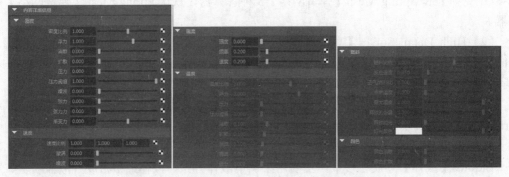

图11.92　Contents Details【内容详细信息】卷展栏

9）Grids Cache【栅格缓存】卷展栏参数

Grids Cache【栅格缓存】卷展栏参数，主要用来是否启用读取密度、读取速度、读取温度、读取燃料、读取颜色、读取纹理坐标和读取衰减等功能，如图11.93所示。

10）Surface【表面】卷展栏参数

Surface【表面】卷展栏参数，主要用来调节表面和环境等相关参数，如图11.94所示。

 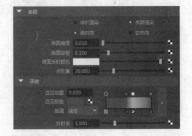

图11.93　Grids Cache【栅格缓存】卷展栏　　图11.94　Surface【表面】卷展栏

Interpolation【插值】参数组：在该参数组中提供了None【无】、Linear【线性】、Smooth【平滑】和Spline【样条线】4种插值方式，默认Linear【线性】插值方式。

第 11 章 Rigging【装备】和 FX 中的对象属性

11）Output Mesh【输出网络】卷展栏参数

Output Mesh【输出网络】卷展栏参数，主要用来调节网络方法、网络分辨率和网络平滑迭代次数等参数，如图 11.95 所示。

图 11.95　Output Mesh【输出网络】卷展栏

Mesh Resolution【网络方法】参数组：在该参数组中提供了 Triangle Mesh【三角形网络】、Quad Mesh【四边形网络】、Tetrahedron【四面体】和 Acute Tetrahedron【锐角四面体】4 种网络方法，默认 Acute Tetrahedron【三角形网络】网络方法。

12）Shading【着色】卷展栏参数

Shading【着色】卷展栏参数，主要用来调节着色、颜色、白炽度、不透明度和蒙版不透明度等参数，如图 11.96 所示。

图 11.96　Shading【着色】卷展栏

（1）Drop off Shape【衰减形式】参数组：在该参数组中提供了 Off【禁用】、Sphere【球体】、Cub【立方体】、Cone【圆锥体】、Double Cone【双锥体】、X Gradient【X 渐变】、Y Gradient【Y 渐变】、Z Gradient【Z 渐变】、-X Gradient【-X 渐变】、-Y Gradient【-Y 渐变】、-Z Gradient【-Z 渐变】和 Use Falloff Grid【使用衰减栅格】12 种衰减形式，默认 Cub【立方体】衰减形式。

（2）Interpolation【插值】参数组：在该参数组中提供了 None【无】、Linear【线性】、Smooth【平滑】和 Spline【样条线】4 种插值方式，默认 Linear【线性】插值方式。

（3）Color Input【颜色输入】、Incandescence Input【白炽度输入】和 Opacity Input【不透明度输入】3 个参数组提供的参数完全相同，在这 3 个参数组中提供了 Constant【恒定】、X Gradient【X 渐变】、Y Gradient【Y 渐变】、Z Gradient【Z 渐变】、Center Gradient【中心渐变】、Density【密度】、Temperature【温度】、Fuel【燃料】、Pressure【压力】、Speed【速度】和 Density and Fuel【密度和燃料】11 种颜色输入模式，默认 Temperature【温度】颜色输入模式。

（4）Matte Opacity Mode【蒙版不透明度模式】参数组：在该参数组中提供了 Black Hole【黑洞】、Solid Matte【匀值蒙版】和 Opacity Gain【不透明度增益】3 种蒙版不透明度模式，默认 Opacity Gain【不透明度增益】蒙版不透明度模式。

13）Shading Quality【着色质量】卷展栏参数

Shading Quality【着色质量】卷展栏参数，主要用来调节质量、对比度容差、采样方法和渲染插值器等参数，如图 11.97 所示。

（1）Sample Method【采样方法】参数组：在该参数组中提供了 Uniform【一致】、Jittered【抖动】、Adaptive Jittered【自适应】和 Adaptive Jittered【自适应抖动】4 种采样方法，默认 Adaptive Jittered【自适应抖动】采样方法。

（2）Render Interpolator【渲染插值器】参数组：在该参数组中提供了 Linear【线性】和 Smooth【平滑】2 种渲染插值器，默认 Linear【线性】渲染插值器。

14）Textures【纹理】卷展栏参数

Textures【纹理】卷展栏参数，主要用来调节纹理类型、坐标方法、坐标速度、颜色纹理增益、白炽度纹理增益、不透明度纹理增益、阈值、振幅、比率、频率比、最大深度、纹理时间、缩放因子、频率、纹理原点、纹理比例、纹理旋转、内爆和内爆中心等参数，如图 11.98 所示。

图 11.97　Shading Quality
【着色质量】卷展栏

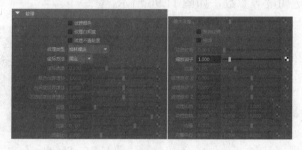
图 11.98　Textures
【纹理】卷展栏

（1）Texture Type【纹理类型】参数组：在该参数组中提供了 Perlin Noise【柏林噪波】、Billow【翻滚】、Volume Wave【体积波浪】、Wispy【束状】、SpaceTime【空间时间】和 Mandelbrot【曼德尔布罗特】6 种纹理类型，默认 Perlin Noise【柏林噪波】纹理类型。

（2）Coordinate Method【坐标方法】参数组：在该参数组中提供了 Fixed【固定】和 Grid【栅格】2 种坐标方法。

15）Lighting【照明】卷展栏参数

Lighting【照明】卷展栏参数，主要用来调节阴影不透明度、阴影扩散、灯光类型、灯光亮度、灯光颜色、环境光亮度、环境光扩散、环境色、平行光、点光源和点光源衰退等参数，如图 11.99 所示。

Light Type【灯光类型】参数组：在该参数组中提供了 Diagonal【对角】、Directional【平行】和 Point【点】3 种灯光类型，默认 Directional【平行】灯光类型。

16）Render Stats【渲染统计信息】卷展栏参数

Render Stats【渲染统计信息】卷展栏参数，主要用来调节着色采样数、最大着色采样数和体积采样数等参数，如图 11.100 所示。

第 11 章 Rigging【装备】和 FX 中的对象属性

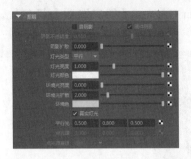

图 11.99 Lighting【照明】卷展栏

图 11.100 Render Stats【渲染统计信息】卷展栏

17) Object Display【对象显示】卷展栏参数

Object Display【对象显示】卷展栏参数，主要用来调节对象显示、重影信息、边界框信息、绘制覆盖和选择等参数，如图 11.101 所示。

图 11.101 Object Display【对象显示】卷展栏

（1）Ghosting Control【重影控制】参数组：在该参数组中主要提供了 Global Prefs【全局首选项】、Custom Frames【自定义帧】、Custom Frame Steps【自定义帧步数】、Custom Key Steps【自定义关键帧步数】和 Key Frames【关键帧】5 种重影控制方式，默认 Global Prefs【全局首选项】重影控制方式。

（2）Display Type【显示类型】参数组：在该参数组中主要提供了 Normal【正常】、Template【模板】和 Reference【引用】3 种显示类型，默认 Normal【正常】显示类型。

（3）Level of Detail【细节级别】参数组：在该参数组中主要提供了 Full【完全】和 Bounding Box【边界框】3 种细节级别，默认 Full【完全】细节级别。

（4）Color【颜色】参数组：在该参数组中主要提供了 Index【索引】和【RGB】2 种颜色方式，默认 Index【索引】颜色方式。

18) Arnold【阿诺德】卷展栏参数

Arnold【阿诺德】卷展栏参数，主要用来调节 Step Size【帧步尺寸】、Enable Deformation Blur【使变形模糊】、Motion Vector Scale【运动矢量比例】、Filter Type【过滤器类型】、Function Anisotropy【功能异向性】、Visible In Diffuse【在传播中可见】、Visible in Glossy【在光纸上可见】、Custom Texture【自定义纹理】、Override Fluid Texture【超控液体纹理】、Texture Color【纹理颜色】、Texture Incandescence【纹理白热化】、Texture Opacity【纹理不透明性】、

Coordinate Method【整合法】、Texture【纹理】和 User Options【用户选项】等参数，如图 11.102 所示。

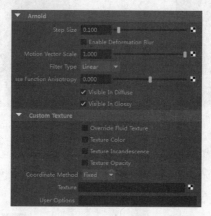

图 11.102　Arnold【阿诺德】卷展栏

19）Node Behavior【节点行为】卷展栏参数

Node Behavior【节点行为】卷展栏参数，主要用来调节缓存、冻结和节点状态等参数，如图 11.103 所示。

在 Node State【节点状态】参数组中，有 Normal【正常】、Has No Effect【无效果】、Blocking【阻塞】、Waiting-Normal【等待-正常】、Waiting-Has No Effect【等待-无效果】和 Waiting-Blocking【等待-阻塞】6 种状态供选择。

20）【UUID】卷展栏参数

【UUID】卷展栏参数中没有参数可调，只显示 UUID 序列号，如图 11.104 所示。

图 11.103　Node Behavior【节点行为】卷展栏　　图 11.104　【UUID】卷展栏

21）Extra Attributes【附加属性】卷展栏参数

Extra Attributes【附加属性】卷展栏参数，主要用来调节透底、纹理原点和从缓存播放等参数，如图 11.105 所示。

3. Pond【池塘】的 Time【时间】属性面板

Pond【池塘】的 Time【时间】属性面板主要包括 Time Attributes【时间属性】、Node Behavior【节点行为】、【UUID】和 Extra Attributes【附加属性】4 个卷展栏参数，如图 11.106 所示。

在 Node State【节点状态】参数组中，有 Normal【正常】、Has No Effect【无效果】、Blocking【阻塞】、Waiting-Normal【等待-正常】、Waiting-Has No Effect【等待-无效果】和 Waiting-Blocking【等待-阻塞】6 种状态供选择。

第 11 章 Rigging【装备】和 FX 中的对象属性

图 11.105 Extra Attributes【附加属性】卷展栏

图 11.106 Extra Attributes【附加属性】卷展栏

11.8 Air【空气】属性面板

Air Field【空气场】属性面板，主要包括 Transform【变换属性】、Predefined Settings【预定义设置】、Air Field Attributes【空气场属性】、Spread【扩散】、Distance【距离】、Volume Control Attributes【体积控制属性】、Special Effects【特效效果】和 Extra Attributes【附加属性】8 个卷展栏参数，如图 11.107 所示。

1. Transform【变换属性】卷展栏参数

Transform【变换属性】卷展栏参数，主要用来调节平移、旋转、缩放、倾斜、旋转顺序和旋转轴等参数，如图 11.108 所示。

图 11.107 Air【空气】属性面板

图 11.108 Transform【变换属性】卷展栏

Rotate Order【旋转顺序】参数组：在该参数组中提供了【xyz】、【yzx】、【zxy】、【xzy】、【yxz】和【zyx】6 种旋转顺序，默认【xyz】旋转顺序。

2. Predefined Settings【预定义设置】卷展栏参数

Predefined Settings【预定义设置】卷展栏参数，主要提供了 Wind【风】、Wake【尾迹】和 Fan【扇】3 种预定义设置，如图 11.109 所示。

3. Air Field Attributes【空气场属性】卷展栏参数

Air Field Attributes【空气场属性】卷展栏参数，主要用来调节幅值、衰减、方向、速度和继承速度等参数，如图 11.110 所示。

图 11.109　Predefined Settings
【预定义设置】卷展栏

图 11.110　Air Field Attributes
【空气场属性】卷展栏

4. Spread【扩散】卷展栏参数

Spread【扩散】卷展栏参数，主要用来调节扩散参数，如图 11.111 所示。

5. Distance【距离】卷展栏参数

Distance【距离】卷展栏参数，主要用来调节最大距离、选定位置、选定值和插值等参数，如图 11.112 所示。

图 11.111　Spread【扩散】卷展栏

图 11.112　Distance【距离】卷展栏

Interpolation【插值】参数组：在该参数组中提供了 None【无】、Linear【线性】、Smooth【平滑】和 Spline【样条线】4 种插值方式，默认 Linear【线性】插值方式。

6. Volume Control Attributes【体积控制属性】卷展栏参数

Volume Control Attributes【体积控制属性】卷展栏参数，主要用来调节体积形状、轴向幅值和曲线半径等参数，如图 11.113 所示。

图 11.113　Volume Control Attributes【体积控制属性】卷展栏

（1）Volume Shape【体积形状】参数组：在该参数组中提供了 None【无】、Cube【立方体】、Sphere【球体】、Cylinder【圆柱体】、Cone【圆锥体】、Torus【圆环】和 Curve【曲线】7 种体积形状，默认 None【无】体积形状。

（2）Interpolation【插值】参数组：在该参数组中提供了 None【无】、Linear【线性】、Smooth【平滑】和 Spline【样条线】4 种插值方式，默认 Linear【线性】插值方式。

7. Special Effects【特效效果】卷展栏参数

Special Effects【特效效果】卷展栏参数，主要用来控制是否启用逐顶点应用功能，如图 11.114 所示。

8. Extra Attributes【附加属性】卷展栏参数

Extra Attributes【附加属性】卷展栏参数，在默认情况下没有任何参数，如图 11.115 所示。

图 11.114 Special Effects 【特效效果】卷展栏

图 11.115 Extra Attributes 【附加属性】卷展栏

11.9 Drag【阻力】属性面板

Drag【阻力】属性面板，主要包括 Transform Attributes【变换属性】、Drag Field Attributes【阻力场属性】、Distance【距离】、Volume Control Attributes【体积控制属性】、Special Effects【特殊效果】和 Extra Attributes【附加属性】6 个卷展栏参数，如图 11.116 所示。

1. Transform Attributes【变换属性】卷展栏参数

Transform Attributes【变换属性】卷展栏参数，主要主要用来调节平移、旋转、缩放、倾斜、旋转顺序和旋转轴等参数，如图 11.117 所示。

图 11.116 Drag【阻力】属性面板

图 11.117 Transform Attributes【变换属性】卷展栏

Rotate Order【旋转顺序】参数组：在该参数组中提供了【xyz】、【yzx】、【zxy】、【xzy】、【yxz】和【zyx】6 种旋转顺序，默认【xyz】旋转顺序。

2. Drag Field Attributes【阻力场属性】卷展栏参数

Drag Field Attributes【阻力场属性】卷展栏参数，主要用来调节幅值、衰减、速度衰减、方向、继承速度和运动衰减等参数，如图 11.118 所示。

3. Distance【距离】卷展栏参数

Distance【距离】卷展栏参数，主要用来调节最大距离、选定位置、选定值和插值等参数，如图 11.119 所示。

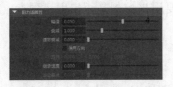

图 11.118　Drag Field Attributes【阻力场属性】卷展栏　　　图 11.119　Distance【距离】卷展栏

Interpolation【插值】参数组：在该参数组中提供了 None【无】、Linear【线性】、Smooth【平滑】和 Spline【样条线】4 种插值方式，默认 Linear【线性】插值方式。

4. Volume Control Attributes【体积控制属性】卷展栏参数

Volume Control Attributes【体积控制属性】卷展栏参数，主要用来调节体积控制属性、轴向幅值和曲线半径等参数，如图 11.120 所示。

（1）Volume Shape【体积形状】参数组：在该参数组中提供了 None【无】、Cube【立方体】、Sphere【球体】、Cylinder【圆柱体】、Cone【圆锥体】、Torus【圆环】和 Curve【曲线】7 种体积形状，默认 None【无】体积形状。

（2）Interpolation【插值】参数组：在该参数组中提供了 None【无】、Linear【线性】、Smooth【平滑】和 Spline【样条线】4 种插值方式，默认 Linear【线性】插值方式。

5. Special Effects【特殊效果】卷展栏参数

Special Effects【特殊效果】卷展栏参数，主要用来控制是否启用逐顶点应用功能，如图 11.121 所示。

图 11.120　Volume Control Attributes　　　　图 11.121　Special Effects
　　　【体积控制属性】卷展栏　　　　　　　　　　【特殊效果】卷展栏

第 11 章 Rigging【装备】和 FX 中的对象属性

6. Extra Attributes【附加属性】卷展栏参数

Extra Attributes【附加属性】卷展栏参数，主要用来调节大纲视图颜色和线框颜色 RGB 等参数，如图 11.122 所示。

11.10 Gravity【重力】属性面板

Gravity【重力】属性面板，主要包括 Transform Attributes【变换属性】、Gravity Field Attributes【重力场属性】、Distance【距离】、Volume Control Attributes【体积控制属性】、Special Effects【特殊效果】和 Extra Attributes【附加属性】6 个卷展栏参数，如图 11.123 所示。

图 11.122 Extra Attributes【附加属性】卷展栏

图 11.123 Gravity【重力】属性面板

1. Transform Attributes【变换属性】卷展栏参数

Transform Attributes【变换属性】卷展栏参数，主要用来调节主要主要用来调节平移、旋转、缩放、倾斜、旋转顺序和旋转轴等参数，如图 11.124 所示。

Rotate Order【旋转顺序】参数组：在该参数组中提供了【xyz】、【yzx】、【zxy】、【xzy】、【yxz】和【zyx】6 种旋转顺序，默认【xyz】旋转顺序。

2. Gravity Field Attributes【重力场属性】卷展栏参数

Gravity Field Attributes【重力场属性】卷展栏参数，主要用来调节重力场的幅值、衰减和方向等参数，如图 11.125 所示。

图 11.124 Transform Attributes
【变换属性】卷展栏

图 11.125 Gravity Field Attributes
【重力场属性】卷展栏

3. Distance【距离】卷展栏参数

Distance【距离】卷展栏参数，主要用来调节最大距离、选定位置和插值等参数，如

图 11.126 所示。

Interpolation【插值】参数组：在该参数组中提供了 None【无】、Linear【线性】、Smooth【平滑】和 Spline【样条线】4 种插值方式，默认 Linear【线性】插值方式。

4. Volume Control Attributes【体积控制属性】卷展栏参数

Volume Control Attributes【体积控制属性】卷展栏参数，主要用来调节体积控制属性、轴向幅值和曲线半径等参数，如图 11.127 所示。

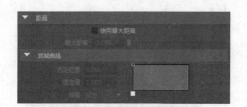

图 11.126 Distance【距离】卷展栏

图 11.127 Volume Control Attributes【体积控制属性】卷展栏

（1）Volume Shape【体积形状】参数组：在该参数组中提供了 None【无】、Cube【立方体】、Sphere【球体】、Cylinder【圆柱体】、Cone【圆锥体】、Torus【圆环】和 Curve【曲线】7 种体积形状，默认 None【无】体积形状。

（2）Interpolation【插值】参数组：在该参数组中提供了 None【无】、Linear【线性】、Smooth【平滑】和 Spline【样条线】4 种插值方式，默认 Linear【线性】插值方式。

5. Special Effects【特殊效果】卷展栏参数

Special Effects【特殊效果】卷展栏参数，主要用来控制是否启用逐顶点应用功能，如图 11.128 所示。

6. Extra Attributes【附加属性】卷展栏参数

Extra Attributes【附加属性】卷展栏参数，主要用来调节大纲视图颜色和线框颜色 RGB 等参数，如图 11.129 所示。

图 11.128 Special Effects【特殊效果】卷展栏

图 11.129 Extra Attributes【附加属性】卷展栏

11.11 Newton【牛顿】属性面板

Newton【牛顿】属性面板,主要包括 Transform Attributes【变换属性】、Newton Field Attributes【牛顿场属性】、Distance【距离】、Volume Control Attributes【体积控制属性】、Special Effects【特殊效果】和 Extra Attributes【附加属性】6个卷展栏参数,如图 11.130 所示。

1. Transform Attributes【变换属性】卷展栏

Transform Attributes【变换属性】卷展栏参数,主要用来调节主要主要用来调节平移、旋转、缩放、倾斜、旋转顺序和旋转轴等参数,如图 11.131 所示。

图 11.130　Newton【牛顿】属性面板　　　图 11.131　Transform Attributes【变换属性】卷展栏

Rotate Order【旋转顺序】参数组:在该参数组中提供了【xyz】、【yzx】、【zxy】、【xzy】、【yxz】和【zyx】6种旋转顺序,默认【xyz】旋转顺序。

2. Newton Field Attributes【牛顿场属性】卷展栏参数

Newton Field Attributes【牛顿场属性】卷展栏参数,主要用来调节牛顿场的幅值、衰减和最小距离等参数,如图 11.132 所示。

3. Distance【距离】卷展栏参数

Distance【距离】卷展栏参数,主要用来调节最大距离、选定位置和插值等参数,如图 11.133 所示。

图 11.132　Newton Field Attributes【牛顿场属性】卷展栏　　　图 11.133　Distance【距离】卷展栏

Interpolation【插值】参数组:在该参数组中提供了 None【无】、Linear【线性】、Smooth【平滑】和 Spline【样条线】4种插值方式,默认 Linear【线性】插值方式。

4. Volume Control Attributes【体积控制属性】卷展栏参数

Volume Control Attributes【体积控制属性】卷展栏参数，主要用来调节体积控制属性、轴向幅值和曲线半径等参数，如图 11.134 所示。

（1）Volume Shape【体积形状】参数组：在该参数组中提供了 None【无】、Cube【立方体】、Sphere【球体】、Cylinder【圆柱体】、Cone【圆锥体】、Torus【圆环】和 Curve【曲线】7 种体积形状，默认 None【无】体积形状。

（2）Interpolation【插值】参数组，在该参数组中提供了 None【无】、Linear【线性】、Smooth【平滑】和 Spline【样条线】4 种插值方式，默认 Linear【线性】插值方式。

5. Special Effects【特殊效果】卷展栏参数

Special Effects【特殊效果】卷展栏参数，主要用来控制是否启用逐顶点应用功能，如图 11.135 所示。

图 11.134　Volume Control Attributes
【体积控制属性】卷展栏

图 11.135　Special Effects
【特殊效果】卷展栏

6. Extra Attributes【附加属性】卷展栏参数

Extra Attributes【附加属性】卷展栏参数，主要用来调节大纲视图颜色和线框颜色 RGB 等参数，如图 11.136 所示。

11.12　Radial【径向】属性面板

Radial【径向】属性面板，主要包括 Transform Attributes【变换属性】、Radial Field Attributes【径向场属性】、Distance【距离】、Volume Control Attributes【体积控制属性】、Special Effects【特殊效果】和 Extra Attributes【附加属性】6 个卷展栏参数，如图 11.137 所示。

第 11 章 Rigging【装备】和 FX 中的对象属性

图 11.136 Extra Attributes
【附加属性】卷展栏

图 11.137 Radial
【径向】卷展栏

1. Transform Attributes【变换属性】卷展栏

Transform Attributes【变换属性】卷展栏参数，主要用来调节主要主要用来调节平移、旋转、缩放、倾斜、旋转顺序和旋转轴等参数，如图 11.138 所示。

Rotate Order【旋转顺序】参数组：在该参数组中提供了【xyz】、【yzx】、【zxy】、【xzy】、【yxz】和【zyx】6 种旋转顺序，默认【xyz】旋转顺序。

2. Radial Field Attributes【径向场属性】卷展栏参数

Radial Field Attributes【径向场属性】卷展栏参数，主要用来调节径向场的幅值、衰减和径向类型等参数，如图 11.139 所示。

图 11.138 Transform Attributes
【变换属性】卷展栏

图 11.139 Radial Field Attributes
【牛顿场属性】卷展栏

3. Distance【距离】卷展栏参数

Distance【距离】卷展栏参数，主要用来调节最大距离、选定位置和插值等参数，如图 11.140 所示。

Interpolation【插值】参数组：在该参数组中提供了 None【无】、Linear【线性】、Smooth【平滑】和 Spline【样条线】4 种插值方式，默认 Linear【线性】插值方式。

4. Volume Control Attributes【体积控制属性】卷展栏参数

Volume Control Attributes【体积控制属性】卷展栏参数，主要用来调节体积控制属性、轴向幅值和曲线半径等参数，如图 11.141 所示。

（1）Volume Shape【体积形状】参数组：在该参数组中提供了 None【无】、Cube【立方体】、Sphere【球体】、Cylinder【圆柱体】、Cone【圆锥体】、Torus【圆环】和 Curve【曲线】7 种体积形状，默认 None【无】体积形状。

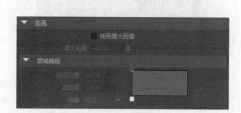

图 11.140 Distance
【距离】卷展栏

图 11.141 Volume Control Attributes
【体积控制属性】卷展栏

（2）Interpolation【插值】参数组：在该参数组中提供了 None【无】、Linear【线性】、Smooth【平滑】和 Spline【样条线】4 种插值方式，默认 Linear【线性】插值方式。

5. Special Effects【特殊效果】卷展栏参数

Special Effects【特殊效果】卷展栏参数，主要用来控制是否启用逐顶点应用功能，如图 11.142 所示。

6. Extra Attributes【附加属性】卷展栏参数

Extra Attributes【附加属性】卷展栏参数，主要用来调节大纲视图颜色和线框颜色 RGB 等参数，如图 11.143 所示。

图 11.142 Special Effects
【特殊效果】卷展栏

图 11.143 Extra Attributes
【附加属性】卷展栏

11.13 Turbulence【湍流】属性面板

Turbulence【湍流】属性面板，主要包括 Transform Attributes【变换属性】、Turbulence Field Attributes【湍流场属性】、Distance【距离】、Volume Control Attributes【体积控制属性】、Special Effects【特殊效果】和 Extra Attributes【附加属性】6 个卷展栏参数，如图 11.144 所示。

1. Transform Attributes【变换属性】卷展栏参数

Transform Attributes【变换属性】卷展栏参数，主要用来调节平移、旋转、缩放、倾斜、旋转顺序和旋转轴等参数，如图 11.145 所示。

图 11.144 Turbulence
【湍流】卷展栏

图 11.145 Transform Attributes
【变换属性】卷展栏

Rotate Order【旋转顺序】参数组：在该参数组中提供了【xyz】、【yzx】、【zxy】、【xzy】、【yxz】和【zyx】6 种旋转顺序，默认【xyz】旋转顺序。

2. Turbulence Field Attributes【湍流场属性】卷展栏参数

Turbulence Field Attributes【湍流场属性】卷展栏参数，主要用来调节湍流场的幅值、衰减、频率、相位、插值类型、噪波级别和噪波比等参数，如图 11.146 所示。

Interpolation【插值类型】参数组：在该参数组中提供了 Linear【线性】和 Quadratic【二次方】两种插值类型，默认 Linear【线性】插值类型。

3. Distance【距离】卷展栏参数

Distance【距离】卷展栏参数，主要用来调节最大距离、选定位置和插值等参数，如图 11.147 所示。

图 11.146 Turbulence Field Attributes
【湍流场属性】卷展栏

图 11.147 Distance
【距离】卷展栏

Interpolation【插值】参数组：在该参数组中提供了 None【无】、Linear【线性】、Smooth【平滑】和 Spline【样条线】4 种插值方式，默认 Linear【线性】插值方式。

4. Volume Control Attributes【体积控制属性】卷展栏参数

Volume Control Attributes【体积控制属性】卷展栏参数，主要用来调节体积控制属性、轴向幅值和曲线半径等参数，如图 11.148 所示。

（1）Volume Shape【体积形状】参数组：在该参数组中提供了 None【无】、Cube【立方体】、Sphere【球体】、Cylinder【圆柱体】、Cone【圆锥体】、Torus【圆环】和 Curve【曲线】7 种体积形状，默认 None【无】体积形状。

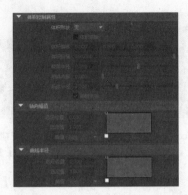

图 11.148　Volume Control Attributes【体积控制属性】卷展栏

（2）Interpolation【插值】参数组：在该参数组中提供了 None【无】、Linear【线性】、Smooth【平滑】和 Spline【样条线】4 种插值方式，默认 Linear【线性】插值方式。

5．Special Effects【特殊效果】卷展栏参数

Special Effects【特殊效果】卷展栏参数，主要用来控制是否启用逐顶点应用功能，如图 11.149 所示。

6．Extra Attributes【附加属性】卷展栏参数

Extra Attributes【附加属性】卷展栏参数，主要用来调节大纲视图颜色和线框颜色 RGB 等参数，如图 11.150 所示。

图 11.149　Special Effects【特殊效果】卷展栏　　　图 11.150　Extra Attributes【附加属性】卷展栏

11.14　Uniform【统一】属性面板

Uniform【统一】属性面板，主要包括 Transform Attributes【变换属性】、Uniform Field Attributes【统一场属性】、Distance【距离】、Volume Control Attributes【体积控制属性】、Special Effects【特殊效果】和 Extra Attributes【附加属性】6 个卷展栏参数，如图 11.151 所示。

1．Transform Attributes【变换属性】卷展栏参数

Transform Attributes【变换属性】卷展栏参数，主要用来调节平移、旋转、缩放、倾斜、旋转顺序和旋转轴等参数，如图 11.152 所示。

图 11.151　Uniform
【统一】卷展栏

图 11.152　Transform Attributes
【变换属性】卷展栏

Rotate Order【旋转顺序】参数组：在该参数组中提供了【xyz】、【yzx】、【zxy】、【xzy】、【yxz】和【zyx】6 种旋转顺序，默认【xyz】旋转顺序。

2. Turbulence Field Attributes【湍流场属性】卷展栏参数

Turbulence Field Attributes【湍流场属性】卷展栏参数，主要用来调节统一场的幅值、衰减和方向等参数，如图 11.153 所示。

Interpolation【插值类型】参数组：在该参数组中提供了 Linear【线性】和 Quadratic【二次方】两种插值类型，默认 Linear【线性】插值类型。

3. Distance【距离】卷展栏参数

Distance【距离】卷展栏参数，主要用来调节最大距离、选定位置和插值等参数，如图 11.154 所示。

图 11.153　Turbulence Field Attributes
【湍流场属性】卷展栏

图 11.154　Distance
【距离】卷展栏

Interpolation【插值】参数组：在该参数组中提供了 None【无】、Linear【线性】、Smooth【平滑】和 Spline【样条线】4 种插值方式，默认 Linear【线性】插值方式。

4. Volume Control Attributes【体积控制属性】卷展栏参数

Volume Control Attributes【体积控制属性】卷展栏参数，主要用来调节体积控制属性、轴向幅值和曲线半径等参数，如图 11.155 所示。

（1）Volume Shape【体积形状】参数组：在该参数组中提供了 None【无】、Cube【立方体】、Sphere【球体】、Cylinder【圆柱体】、Cone【圆锥体】、Torus【圆环】和 Curve【曲线】7 种体积形状，默认 None【无】体积形状。

（2）Interpolation【插值】参数组：在该参数组中提供了 None【无】、Linear【线性】、

Smooth【平滑】和 Spline【样条线】4 种插值方式，默认 Linear【线性】插值方式。

5. Special Effects【特殊效果】卷展栏参数

Special Effects【特殊效果】卷展栏参数，主要用来控制是否启用逐顶点应用功能，如图 11.156 所示。

图 11.155　Volume Control Attributes
【体积控制属性】卷展栏

图 11.156　Special Effects
【特殊效果】卷展栏

6. Extra Attributes【附加属性】卷展栏参数

Extra Attributes【附加属性】卷展栏参数，主要用来调节大纲视图颜色和线框颜色 RGB 等参数，如图 11.157 所示。

11.15　Vortex【漩涡】属性面板

Vortex【漩涡】属性面板，主要包括 Transform Attributes【变换属性】、Vortex Field Attributes【漩涡场属性】、Distance【距离】、Volume Control Attributes【体积控制属性】、Special Effects【特殊效果】和 Extra Attributes【附加属性】6 个卷展栏参数，如图 11.158 所示。

图 11.157　Extra Attributes【附加属性】卷展栏

图 11.158　Uniform【统一】卷展栏

1. Transform Attributes【变换属性】卷展栏参数

Transform Attributes【变换属性】卷展栏参数，主要用来调节平移、旋转、缩放、倾斜、

第 11 章 Rigging【装备】和 FX 中的对象属性

旋转顺序和旋转轴等参数，如图 11.159 所示。

Rotate Order【旋转顺序】参数组：在该参数组中提供了【xyz】、【yzx】、【zxy】、【xzy】、【yxz】和【zyx】6 种旋转顺序，默认【xyz】旋转顺序。

2. Turbulence Field Attributes【湍流场属性】卷展栏参数

Turbulence Field Attributes【湍流场属性】卷展栏参数，主要用来调节漩涡场的幅值、衰减和轴等参数，如图 11.160 所示。

图 11.159　Transform Attributes
【变换属性】卷展栏

图 11.160　Turbulence Field Attributes
【湍流场属性】卷展栏

Interpolation【插值类型】参数组：在该参数组中提供了 Linear【线性】和 Quadratic【二次方】2 种插值类型，默认 Linear【线性】插值类型。

3. Distance【距离】卷展栏参数

Distance【距离】卷展栏参数，主要用来调节最大距离、选定位置和插值等参数，如图 11.161 所示。

Interpolation【插值】参数组：在该参数组中提供了 None【无】、Linear【线性】、Smooth【平滑】和 Spline【样条线】4 种插值方式，默认 Linear【线性】插值方式。

4. Volume Control Attributes【体积控制属性】卷展栏参数

Volume Control Attributes【体积控制属性】卷展栏参数，主要用来调节体积控制属性、轴向幅值和曲线半径等参数，如图 11.162 所示。

图 11.161　Distance
【距离】卷展栏

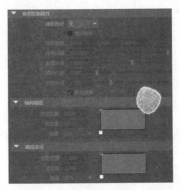

图 11.162　Volume Control Attributes
【体积控制属性】卷展栏

（1）Volume Shape【体积形状】参数组：在该参数组中提供了 None【无】、Cube【立方体】、Sphere【球体】、Cylinder【圆柱体】、Cone【圆锥体】、Torus【圆环】和 Curve【曲线】7 种体积形状，默认 None【无】体积形状。

（2）Interpolation【插值】参数组：在该参数组中提供了 None【无】、Linear【线性】、Smooth【平滑】和 Spline【样条线】4 种插值方式，默认 Linear【线性】插值方式。

5. Special Effects【特殊效果】卷展栏参数

Special Effects【特殊效果】卷展栏参数，主要用来控制是否启用逐顶点应用功能，如图 11.163 所示。

6. Extra Attributes【附加属性】卷展栏参数

Extra Attributes【附加属性】卷展栏参数，主要用来调节大纲视图颜色和线框颜色 RGB 等参数，如图 11.164 所示。

图 11.163　Special Effects
【特殊效果】卷展栏

图 11.164　Extra Attributes
【附加属性】卷展栏

第 12 章　Render Settings【渲染设置】面板

本章主要介绍 Maya Software（软件）渲染器、Maya Hardware（硬件）渲染器、Maya Hardware 2.0（硬件 2.0）渲染器和 Maya Vector（向量）渲染器的相关属性。

12.1　Render Settings【渲染设置】面板中的 Common【公用】属性面板

Maya Software（软件）渲染器、Maya Hardware（硬件）渲染器、Maya Hardware2.0（硬件 2.0）渲染器、Maya Vector（向量）渲染器的 Common【公用】属性面板的参数基本相同，主要包括 Color Management【颜色管理】、File Output【文件输出】、Frame Range【帧范围】、Render able Cameras【可渲染摄影机】、Image Size【图像大小】、Scene Assembly【场景集合】和 Render Options【渲染选项】7 个卷展栏参数，如图 12.1 所示。

1. Color Management【颜色管理】卷展栏参数

Color Management【颜色管理】卷展栏参数，主要用来调节【将输出变换应用于渲染器】和【输出变换】参数，如图 12.2 所示。

图 12.1　Joint【关节】对象属性面板面板　　图 12.2　Color Management【颜色管理】卷展栏

【输出变换】参数组：该参数组主要提供了 ACES RRT v0.7、ACES RRT v1.0、Log、1.8 gamma【1.8 伽玛】、2.2 gamma【2.2 伽玛】、Rec 709 gamma【Rec 709 伽玛】、sRGB gamma【sRGB 伽玛】、Raw、【Stingray tone-map】、Use View Transform【使用视图变换】和 Add New【添加新的】11 种输出变换，默认 Use View Transform【使用视图变换】输出变换。

2. File Output【文件输出】卷展栏参数

File Output【文件输出】卷展栏参数，主要用来调节文件名前缀、图像格式、编码、质量、帧/动画扩展名、帧填充、帧缓冲区命名、自定义命名字符串和扩展名等参数，如图 12.3 所示。

（1）Image format【图像格式】参数组：该参数组提供了【Alias PIX（als）】、【AVI（avi）】、【Cineon（cin）】、【DDS（dds）】、【EPS（eps）】、【GIF（gif）】、【JPEG（jpg）】、【Maya IFF（iff）】、【Maya 16 IFF（iff）】、【PSD（psd）】、【分层 PSD（psd）】、【PNG（png）】、【Quantel（yuv）】、【Quicktime 影片（mov）】、【RLA（rla）】、【SGI（sgi）】、【SGI16（sgi）】、【SoftImage（pic）】、【Targa（tga）】、【Tiff（tif）】、【Tiff16（tif）】、【Windows 位图（bmp）】、【GIF（gif）】、【Sony Playstation（tim）】和【XPM（xpm）】等图像格式。

（2）Frame/Animation ext【帧 / 动画扩展名】参数组：该参数组提供了 name（Single Frame）【名称（单帧）】、name.ext（Single Frame）【名称.扩展名（单帧）】、name.#.ext【名称.#.扩展名】、name.ext.#【名称.扩展名.#】、name.#【名称.#】、name#.ext【名称#.扩展名】、name_#.ext【名称_#.扩展名】、name（Multi Frame）【名称（多帧）】和 name.ext（Multi Frame）【名称.扩展名（多帧）】9 种帧 / 动画扩展名方式，默认 name.ext（Multi Frame）【名称.扩展名（多帧）】帧 / 动画扩展名方式。

3. Frame Range【帧范围】卷展栏参数

Frame Range【帧范围】卷展栏参数，主要用来调节开始帧、结束帧、帧数、开始编号和帧数等参数，如图 12.4 所示。

图 12.3　File Output【文件输出】卷展栏

图 12.4　Frame Range【帧范围】卷展栏

4. Render able Cameras【可渲染摄影机】卷展栏参数

Render able Cameras【可渲染摄影机】卷展栏参数，主要用来选择可渲染摄影机，如图 12.5 所示。

Render able Camera【可渲染摄影机】参数组：该参数组在默认情况下，提供了 Persp【透视】、Front【前】、Side【侧】、Top【顶】和 Add Render able Camera【添加可渲染摄影机】5 个选项。

提示：在 Render able Camera【可渲染摄影机】参数组中，如果用户创建了摄影机，就会自动添加到该参数组选项中。

5. Image Size【图像大小】卷展栏参数

Image Size【图像大小】卷展栏参数，主要用来调节预设、宽度、高度、大小单位、分辨率、分辨率单位、设备纵横比例和像素纵横比等参数，如图 12.6 所示。

第 12 章　Render Settings【渲染设置】面板

图 12.5　Render able Cameras
【可渲染摄影机】卷展栏

图 12.6　Image Size
【图像大小】卷展栏

（1）Presets【预设】参数组：在该参数组中提供了 Custom【自定义】、320×240、640×480、1K Square【1K 平方】、2K Square【2K 平方】、3K Square【3K 平方】、4K Square【4K 平方】、CCIR PAL/Quantel PAL 和 CCIR 601/Quantel NTSC 等 33 种预设。

（2）Size units【大小单位】参数组：在该参数组中提供了 pixels【像素】、inches【英寸】、cm【厘米】、mm【毫米】、points【点】和 picas【派卡】6 种单位，默认 pixels【像素】单位。

（3）Resolution units【分辨率单位】参数组：在该参数组中提供了 pixels/inch【像素/英寸】和 pixels/cm【像素/厘米】2 种分辨率单位。

6. Scene Assembly【场景集合】卷展栏参数

Scene Assembly【场景集合】卷展栏参数，主要用来调节【渲染表现】参数，如图 12.7 所示。

Render Representation【渲染表现】参数组：在该参数组中提供了 Active Representation【活动表现】和 Custom【自定义】2 个选项，默认 Active Representation【活动表现】项。

7. Render Options【渲染选项】卷展栏参数

Render Options【渲染选项】卷展栏参数，主要用来调节渲染前/后 MEL、渲染层前/后 ME 和渲染帧前/后 MEL 等参数，如图 12.8 所示。

图 12.7　Scene Assembly【场景集合】卷展栏

图 12.8　Render Options【渲染选项】卷展栏

12.2　Render Settings【渲染设置】面板中的 Maya Software【Maya 软件】渲染器属性面板

在 Render Settings【渲染设置】面板中的 Maya Software【Maya 软件】渲染器属性面板中，有 Anti-aliasing【抗锯齿质量】、Field Options【场选项】、Ray tracing Quality【光线

跟踪质量、Motion Blur【运动模糊】、Render Options【渲染选项】、Memory and Performance Options【内存与性能选项】、IPR Options【IPR 选项】和 Paint Effects Rendering Options【Paint Effects 渲染选项】8 个卷展栏参数，如图 12.9 所示。

1. Anti-aliasing【抗锯齿质量】卷展栏参数

Anti-aliasing【抗锯齿质量】卷展栏参数，主要用来调节抗锯齿质量、采样数、多像素过滤和对比度阈值等参数，如图 12.10 所示。

图 12.9　Maya Software　　　　　图 12.10　Anti-aliasing
【Maya 软件】渲染器属性面板　　　【抗锯齿质量】卷展栏

（1）Quality【质量】参数组：在该参数组中提供了 Custom【自定义】、Preview Quality【预览质量】、Intermediate Quality【中间质量】、Production Quality【产品质量】、Contrast Sensitive Production【对比度敏感产品级】和 3D motion Blur Production【3D 运动模糊产品级】6 种渲染质量，默认 Custom【自定义】渲染质量。

（2）Edge anti-aliasing【边缘抗锯齿】参数组：在该参数组中提供了 Low Quality【低质量】、Medium Quality【中等质量】、High Quality【高质量】和 Highest Quality【最高质量】4 种边缘抗锯齿方式，默认 Low Quality【低质量】边缘抗锯齿方式。

2. Field Options【场选项】卷展栏参数

Field Options【场选项】卷展栏参数，主要用来调节渲染、场顺序、第零条扫描线、场扩展名等参数，如图 12.11 所示。

（1）Render【渲染】参数组：在该参数组中提供了 Frames【帧】、Both Fields（Interlaced）【两种场（交替）】、Both Fields Separate【两种场（单独）】、Odd Fields【奇场】和 Even Fields【偶场】5 种渲染方式，默认 Frames【帧】渲染方式。

（2）Zeroth Scanline【场顺序】参数组：在该参数组中提供了 Odd Field（NTSC）【奇场（NTSC）】和 Even field（PAL）【偶场（PAL）】两种场顺序，默认 Odd Field（NTSC）【奇场（NTSC）】场顺序。

3. Ray Tracing Quality【光线跟踪质量】卷展栏参数

Ray tracing Quality【光线跟踪质量】卷展栏参数，主要用来调节反射、折射、阴影和偏移等参数，如图 12.12 所示。

图 12.11　Field Options
【场选项】卷展栏

图 12.12　Ray Tracing Quality
【光线跟踪质量】卷展栏

4. Motion Blur【运动模糊】卷展栏参数

Motion Blur【运动模糊】卷展栏参数，主要用来调节运动模糊类型、模糊帧数、模糊长度、快门、平滑质量和 2D 模糊内存限制等参数，如图 12.13 所示。

5. Render Options【渲染选项】卷展栏参数

Render Options【渲染选项】卷展栏参数，主要用来调节后期处理、摄影机、灯光/阴影及颜色/合成等参数，如图 12.14 所示。

图 12.13　Motion Blur
【运动模糊】卷展栏

图 12.14　Render Options
【渲染选项】卷展栏

Shadow Linking【阴影连接】参数组：在该参数组中提供了 Shadows Obey Light Linking【阴影遵守阴影连接】、Shadows Obey Light Linking【阴影遵守灯光连接】和 Shadows Obey Light Linking【阴影忽略连接】3 种阴影连接方法，默认 Shadows Obey Light Linking【阴影遵守灯光连接】阴影连接方式。

6. Memory and Performance Options【内存与性能选项】卷展栏参数

Memory and Performance Options【内存与性能选项】卷展栏参数，主要用来调节细分、光线跟踪和多重处理等参数，如图 12.15 所示。

7. IPR Options【IPR 选项】卷展栏参数

IPR Options【IPR 选项】卷展栏参数，主要用来控制是否启用渲染着色、照明和辉光、渲染阴影贴图和渲染 2D 运动模糊等参数，如图 12.16 所示。

图 12.15　Memory and Performance Options
【内存与性能选项】卷展栏

图 12.16　IPR Options
【IPR 选项】卷展栏

8. Paint Effects Rendering Options【Paint Effects 渲染选项】卷展栏参数

Paint Effects Rendering Options【Paint Effects 渲染选项】卷展栏参数，主要用来控制是否启用笔划渲染、过采样、过采样后置过滤和仅渲染笔划等参数，如图 12.17 所示。

12.3　Render Settings【渲染设置】面板中的 Maya Hardware【Maya 硬件】渲染器属性面板

Render Settings【渲染设置】面板中的 Maya Hardware【Maya 硬件】渲染器属性面板，主要包括 Quality【质量】和 Render Options【渲染选项】2 个卷展栏参数，如图 12.18 所示。

图 12.17　Paint Effects Rendering Options
【Paint Effects 渲染选项】卷展栏

图 12.18　Maya Hardware
【Maya 硬件渲染器】属性面板

1. Quality【质量】卷展栏参数

Quality【质量】卷展栏参数，主要用来调节预设、采样数、帧缓冲区格式、透明度排序、颜色分辨率、凹凸分辨率和纹理压缩等参数，如图 12.19 所示。

（1）Presets【预设】参数组：在该参数组中提供了 Custom【自定义】、Preview Quality【预览质量】、Intermediate Quality【中间质量】、Production Quality【产品级质量】和 Production Quality with Transparency【带透明度的产品级质量】5 种质量预设方式，默认 Intermediate Quality【中间质量】质量预设方式。

（2）Number Of Samples【采样数】参数组：在该参数组中提供了 1 Sample【1 采样数】、

3 Sample【3 采样数】、4 Sample【4 采样数】、5 Sample【5 采样数】、7 Sample【7 采样数】、9 Sample【9 采样数】、16 Sample【16 采样数】、25 Sample【25 采样数】和 36 Sample【36 采样数】9 种采样数，默认 1 Sample【1 采样数】采样数。

（3）Frame buffer format【帧缓冲区格式】参数组：在该参数组中提供了 RGBA:8-bits fixed per channel【RGBA:每通道 8 位定点型】和 RGBA:16-bit fixed per channel【RGBA:每通道 16 位浮点型】2 种帧缓冲区格式，默认 RGBA:8-bits fixed per channel【RGBA:每通道 8 位定点型】帧缓冲区格式。

（4）Transparency Sorting【透明度排序】参数组：在该参数组中提供了 Per Object【每对象】和 Per Polygon【每多边形】两种透明度排序方式，默认 Per Object【每对象】透明图排序方式。

（5）Texture compression【纹理压缩】参数组：在该参数组中提供了 Disabled【已禁用】和 Enabled【已启用】两个选项，默认 Disabled【已禁用】选项。

2. Render Options【渲染选项】卷展栏参数

Render Options【渲染选项】卷展栏参数，主要用来调节消隐、图像大小的百分比、最大缓存大小（MB）、运动模糊帧数、曝光次数和阴影连接等参数，如图 12.20 所示。

图 12.19　Quality【质量】卷展栏　　　　图 12.20　Render Options【渲染选项】卷展栏

（1）Culling【消隐】参数组：在该参数组中提供了 Per Object【每对象】、All Double Sided【所有双面】和 All Single Sided【所有单面】3 种消隐方式，默认 Per Object【每对象】消隐方式。

（2）Shadow Linking【阴影连接】参数组：在该参数组中提供了 Shadows Obey Light Linking【阴影遵守阴影连接】、Shadows Obey Light Linking【阴影遵守灯光连接】和 Shadows Obey Light Linking【阴影忽略连接】3 种阴影连接方法，默认 Shadows Obey Light Linking【阴影遵守灯光连接】阴影连接方式。

12.4　Render Settings【渲染设置】面板中的 Maya Hardware 2.0【Maya 硬件 2.0】渲染器属性面板

Render Settings【渲染设置】面板中的 Maya Hardware 2.0【Maya 硬件 2.0】渲染器属性面板，主要包括 Performance【性能】、Screen-Space Ambient Occlusion【屏幕空间环境光

遮挡】、Hardware Fog【硬件雾】、Motion Blur【运动模糊】、Anti-aliasing【抗锯齿】、Floating Point Render Target【浮点渲染目标】和 Render Options【渲染选项】7 个卷展栏参数，如图 12.21 所示。

1. Performance【性能】卷展栏参数

Performance【性能】卷展栏参数，主要用来调节顶点动画缓存、灯光限制、透明度算法、透明度质量、最大纹理分辨率钳制和不支持的纹理类型的烘焙分辨率等参数，如图 12.22 所示。

图 12.21　Maya Hardware 2.0【Maya 硬件 2.0 渲染器】属性面板

图 12.22　Performance【性能】卷展栏

（1）Vertex Animation Cache【顶点动画缓存】参数组：在该参数组中提供了 Disable【禁用】、System【系统】和 Hardware【硬件】3 种顶点动画缓存，默认 Disable【禁用】顶点动画缓存。

（2）Transparency【透明度算法】参数组：在该参数组中提供了 Simple【简单】、Object Sorting【对象排序】、Weighted Average【加权平均】和 Depth Peeling【深度剥离】4 种透明度算法，默认 Object Sorting【对象排序】透明度算法。

2. Screen-Space Ambient Occlusion【屏幕空间环境光遮挡】卷展栏参数

Screen-Space Ambient Occlusion【屏幕空间环境光遮挡】卷展栏参数，主要用来调节屏幕空间环境光遮挡的数量、半径、过滤半径和采样数等参数，如图 12.23 所示。

Samples【采样数】参数组：在该参数组中提供了【8】、【16】和【32】3 种采样数，默认【16】采样数。

3. Hardware Fog【硬件雾】参数卷展栏

Hardware Fog【硬件雾】参数卷展栏，主要用来调节硬件雾的衰减、密度、开始、结束、颜色和 Alpha 等参数，如图 12.24 所示。

Falloff【衰减】参数组：在该参数组中提供了 Linear【线性】、Exponential【指数】和 Exponential squared【指数平方】3 种采样数，默认 Linear【线性】采样数。

第 12 章 Render Settings【渲染设置】面板

图 12.23　Screen-Space Ambient Occlusion
【屏幕空间环境光遮挡】卷展栏

图 12.24　Hardware Fog
【硬件雾】卷展栏

4. Motion Blur【运动模糊】卷展栏参数

Motion Blur【运动模糊】卷展栏参数，主要用来调节运动模糊的类型、快门和采样计数等参数，如图 12.25 所示。

（1）【类型】变换参数组：在该参数组中只提供了一个【变换】类型。

（2）Samples【采样数】参数组：在该参数组中提供了【4】、【8】、【16】和【32】4 种采样数，默认【8】采样数。

5. Anti-aliasing【抗锯齿】卷展栏参数

Anti-aliasing【抗锯齿】卷展栏参数，主要用来调节抗锯齿的采样技术，如图 12.26 所示。

图 12.25　Motion Blur【运动模糊】卷展栏

图 12.26　Anti-aliasing【抗锯齿】卷展栏

Samples【采样数】参数组：在该参数组中提供了【1】、【2】、【4】、【8】和【16】5 种采样数，默认【8】采样数。

6. Floating Point Render Target【浮点渲染目标】卷展栏参数

Floating Point Render Target【浮点渲染目标】卷展栏参数，主要用来调节浮点渲染目标，如图 12.27 所示。

图 12.27　Floating Point Render Target【浮点渲染目标】卷展栏

Format【格式】参数组：在该参数组中主要提供了【R32G32B32A32_FLOAT】、【R32G32B32_FLOAT】和【R16G16B16A16_FLOAT】3 种格式，默认【R32G32B32A32_FLOAT】格式。

7. Render Options【渲染选项】卷展栏参数

Render Options【渲染选项】卷展栏参数，主要用来调节渲染选项和对象类型过滤等参数,如图 12.28 所示。

(1) Lighting Mode【照明模式】参数组:在该参数组中提供了 Default【默认】、All【全部】、None【无】、Active【活动】和 Full Ambient【完整环境光】5 种照明模式,默认 All【全部】照明模式。

(2) Render Mode【渲染模式】参数组:在该参数组中提供了 Wire【线】、Shaded【着色】、Wire on Shaded【线框着色】、Default Material【默认材质】、Shaded And Textured【着色且带纹理】、Wire on Shaded And Textured【线框着色且带纹理】和 Bounding Box【边界框】7 种渲染模式,默认 Shaded and Textured【着色且带纹理】渲染模式。

(3) Render Override【渲染覆盖】参数组:在该参数组中只提供了 No Render Override【无渲染覆盖】模式。

12.5 Render Settings【渲染设置】面板中的 Maya Vector 【Maya 向量】渲染器属性面板

Render Settings【渲染设置】面板中的 Maya Vector【Maya 向量】渲染器属性面板,主要包括 Image Format Options(Maya)【图像格式选项(Maya)】、Appearance Options【外观选项】、Fill Options【填充选项】、Edge Options【边选项】和 Render Optimizations【渲染优化】5 个卷展栏参数,如图 12.29 所示。

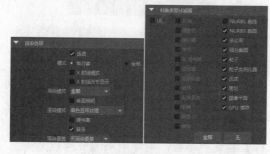

图 12.28　Render Options
【渲染选项】卷展栏

图 12.29　Maya Vector
【Maya 向量渲染器】属性面板

1. Image Format Options(Maya)【图像格式选项(Maya)】卷展栏参数

Image Format Options(Maya)【图像格式选项(Maya)】卷展栏在默认情况下,没有任何可调用的参数,如图 12.30 所示。

2. Appearance Options【外观选项】卷展栏参数

Appearance Options【外观选项】卷展栏参数,主要用来调节外观选项的曲线容差、细节级别预设和细节级别等参数,如图 12.31 所示。

第 12 章 Render Settings【渲染设置】面板

图 12.30 Image Format Options（maya）【图像格式选项（maya）】卷展栏

图 12.31 Appearance Options【外观选项】卷展栏

Detail Level preset【细节级别预设】参数组：在该参数组中提供了 Automatic【自动】、Low【低】、Medium【中等】、High【高】和 Custom【自定义】5 种细节级别预设，默认 Automatic【自动】细节级别预设。

3. Fill Options【填充选项】卷展栏参数

Fill Options【填充选项】卷展栏参数，主要用来调节填充选项的填充样式、高光级别和反射深度等参数，如图 12.32 所示。

4. Edge Options【边选项】卷展栏参数

Edge Options【边选项】卷展栏参数，主要用来调节边选项的边权重预设、边权重、边颜色和最小边角度等参数，如图 12.33 所示。

图 12.32 Fill Options【填充选项】卷展栏

图 12.33 Edge Options【边选项】卷展栏

（1）Edge weight preset【边权重预设】参数组：在该参数组中提供了 Hairline【细线】、0.5 pt【0.5 点】、1.0 pt【1.0 点】、1.5 pt【1.5 点】、2.0 pt【2.0 点】、3.0 pt【3.0 点】、4.0 pt【4.0 点】、5.0 pt【5.0 点】、6.0 pt【6.0 点】、7.0 pt【7.0 点】、8.0 pt【8.0 点】、9.0 pt【9.0 点】、10.0 pt【10.0 点】和 Custom【自定义】14 种边权重预设，默认 Hairline【细线】边权重预设。

（2）Edge style【边样式】参数组：在该参数组中提供了 Outlines【轮廓】和 Entire Mesh【整个网络】两种边样式，默认 Outlines【轮廓】边样式。

5. Render Optimizations【渲染优化】卷展栏参数

Render Optimizations【渲染优化】卷展栏只提供了渲染优化一个参数选项，如图 12.34 所示。

图 12.34　Render Optimizations【渲染优化】卷展栏

Render optimization【渲染优化】参数组：在该参数组中提供了 Safe【安全】、Good【良好】和 Aggressive【过度】3 种渲染优化参数。

第 13 章　Maya 命令中英文对照速查表

（续）

英文命令	中文命令	备注
A		
About Alembic	关于 Alembic	
Absolute Transform	绝对变换	
Accelerated Multi-sampling	加速的多重采样	
Acceleration	加速度	
Accumulate Opacity	累积不透明度	
Accuracy	精确度	
Activate/Deactivate Keys	激活/不激活关键帧	
Activate/Deactivate Modeling Toolkit	激活/关闭建模工具箱	
Active	激活	
Active Keying Feedback	激活关键帧反馈	
Active Only	仅激活	
Adaptive Camera	自适应摄影机	
Adaptive Mesh	自适应网络	
Add	添加	
Add Attribute	添加属性	
Add Boat Locator	添加木船定位器	
Add Divisions	添加细分	
Add Double Linear	添加双线	
Add Dynamic Attributes	添加动力学属性	
Add Dynamic Buoy	添加动力学浮标	
Add Dynamic Locator	添加动力学定位器	
Add Dynamic to Capsule	添加末端定位器到胶囊	
Add Frame Ext	添加帧扩展	
Add Holder	添加定位器	
Add In-between Target	添加中间目标	
Add Influence	添加影响	
Add Joint Labels	添加关节标签	
Add Key Tool	添加关键帧工具	
Add Light Name	添加灯光名称	
Add Matrix	添加矩阵	
Add Members	添加成员	

（续）

英文命令	中文命令	备注
Add New Objects to Current Layer	添加新的对象到当前层	
Add New Render Pass	添加新的渲染通道	
Add New Toon Outline	添加新的卡通轮廓线	
Add Ocean Surface Locator	添加海洋表面定位器	
Add Output Curves to Hair	将输出曲线添加到头发	
Add Paint Effects Output to Hair	将 Paint effects 笔刷指定给头发	
Add Point Tool	添加点工具	
Add Prefix/Suffix	添加前缀/后缀	
Add Preview Plane	添加预览平面	
Add Proxy	添加代理	
Add Scene Name	添加场景名称	
Add Scene Time Warp	添加场景时间弯曲	
Add Selected	添加所选	
Add Selected Objects	添加所选对象	
Add Selected to Graph	添加所选到图表	
Add Shot To Group	添加镜头到组	
Add Surfaces	添加曲面	
Add to a partition	添加到分区	
Add to Asset	添加到资源	
Add to Character Set	添加到角色集	
Add to Current scene	添加到当前场景	
Add to Current Selection	添加到当前选择	
Add to Selected Light	添加到选择的灯光	
Add to Selected	添加到当前选择	
Add Unselected	添加到未选	
Add/Edit Contents	添加/编辑内容	
Add/Remove In-between	添加/移除中间帧	
Adjust Pivot	调整枢轴	
Adjustment	调整	
Adobe[R Illustrator(R Object)]	Adobe[R Illustrator(R 对象)]	
Advanced	高级 标签	
Advanced Assets	高级资源	
Advanced Display Settings	高级显示设置	
Advanced Options	高级选项	
Advanced Raytracing Settings	高级光线追踪设置	
Advanced Settings	高级设置	
Advanced Twist Controls	高级扭曲控制卷展栏	
AdvPolygons	高级多边形	

(续)

英文命令	中文命令	备 注
Aero	阿诺德	
AE Display	显示属性编辑器	
Affect Selected Object	影响所有对象	
Affect Sliding	影响滑动	
Affect Sticky	影响黏连	
After Stroke cmd	绘制后的命令	
Aim	目标约束	
Aim Vector	目标向量	
Air	空气力场	
Alembic Cache	Alembic 缓存	
Algorithm	算法	
Align	对齐	
Align Curves	对齐曲线	
Align in	对齐轴	
Align Mode	对齐模式	
Align Objects	对齐对象	
Align Surfaces	对齐曲面	
Align to	对齐到	
Align Tool	对齐工具	
Align X to V	沿 X 对齐到 V	
Align Y to V	沿 Y 对齐到 V	
Align Z to V	沿 Z 对齐到 V	
All	全部	
All Affected Layers	所有受影响的层	
ALL Channels	所有通道	
ALL Characters	所有角色	
All Characters Retargeters	所有角色重定目标工具	
All Layers	所有的层	
All Points	所有点	
All Surface CVs	所有曲面 CVs	
Allow Multiple Bind Poses	允许多种绑定姿势	
Allow Neg Fat	允许方向脂肪	
Alpha	通道	
Alpha Channel	Alpha 通道	
Alpha Gain	阿尔法增益	
Alpha Mode	通道模式	
Alpha Offset	阿尔法偏移	
Alpha to Use	使用 Alpha	

（续）

英文命令	中文命令	备 注
Always	总显示	
Ambient	环境光	
Ambient Color	环境色	
Ambient Light	环境光	
Ambient Occlusion	环境遮挡	
Ambient Occlusion Settings	环境遮挡设置	
Ambient Shade	环境光明暗处理	
Amount	数量	
Amplify Tool	放大工具	
Amplitude	振幅	
Amplitude X/Y	振幅 X/Y	
Anchor Presets	锚点预设	
Angle	角度	
Angle Between	角度之间	
Angle Min/Max	最小/最大角度	
Angle Threshold	角度阈值	
Anim Clip Attributes	动画片段属性卷展栏	
Animate	动画 菜单	
Animate Camera Transitions	调节摄影机位移动画	
Animated	动画	
Animation	动画	
Animation Blending	动画融合	
Animation Clip	动画片段	
Animation Editors	动画编辑器	
Animation Layer	动画层编辑器	
Animation Layer Buttons	动画层按钮	
Animation Markers	动画标记	
Animation Menu Set	动画菜单设置	
Animation Start/end	动画起始/结束帧	
Anisotropic	各向异性	
Anisotropic Reflectivity	各向异性反射率	
Annotation	注释	
Anti-alias	抗锯齿	
Anti-aliasing	抗锯齿	
Anti-aliasing Contrast	抗锯齿对比度	
Anti-aliasing Quality	抗锯齿品质	
Appearance Options	外观选项	
Append Time Range	添加时间范围	

（续）

英文命令	中文命令	备 注
Append to Cache	附加到缓存	
Append to Polygon Tool	添加到多边形工具	
Applications	应用	
Apply Color	应用颜色	
Apply Default Weights	应用默认权重	
Apply Effect	应用效果	
Apply Muscle Multi Collide Deformer	应用肌肉多重碰撞变形器	
Apply Muscle Spline Deformer	应用肌肉样条变形器	
Apply Muscle Stretch Deformer	应用肌肉拉伸变形器	
Apply Muscle System Skin Deformer	应用肌肉系统皮肤变形器	
Apply Setting to Last Stroke	将设置应用于最后一个笔触	
Apply Setting to Selected Strokes	将设置应用于所选笔触	
Apply tessellation	应用镶嵌细分	
Approximation Editor	近似编辑器	
Arc Tools	圆弧工具	
Archive Scene	归档场景	
Area	区域	
Area Light	区域光	
Area tolerance	面积容差	
array functions	阵列函数	
Array Mapper	阵列映射	
Artisan	雕刻	
As Extra Large Swatches	作为超大样式	
As Lcons	以图表显示	
As Large Swatches	作为大样式	
As List	以文字列表显示	
As Medium Swatches	作为中等样式	
As Small Swatches	作为小样式	
Ascii File Compression Mode	Ascii 文件压缩模式	
Assembly Representations	集合表示	
Asset Editor	资源编辑器	
Asset Options	资源选项	
Assets	资源	
Assign Equal Weights	指定相等权重	
Assign Existing Bake Set	指定已有烘焙集	
Assign Existing Material	指定已有材质	
Assign Favorite Material	指定常用材质	
Assign Fill Shader	指定填充材质球	

（续）

英文命令	中文命令	备注
Assign Hair Constraint	指定头发约束	
Assign Hair System	指定头发系统	
Assign Invisible Faces	指定不可见的面	
Assign New Hotkey	指定新热键	
Assign New Material	指定新材质	
Assign New Vertex Bake Set	指定新顶点烘焙集	
Assign Offline File	指定脱机文件	
Assign Outline	指定轮廓线	
Assign Paint Effects Brush to Hair	指定画笔特效笔刷到头发	
Assign Paint Effects Brush to Toon Lines	将画笔特效笔刷应用于卡通轮廓线	
Assign Shader to Each Projection	为每个映射指定材质	
Assign Solver	指定解算器	
Assign Template	指定模板	
Assign unique name to child nodes	为子节点指定唯一的名称	
Assign/Edit Textures	指定/编辑纹理	
Assigned Surfaces	指定的曲面	
Associated Passes	关联通道	
Associated Passes Contribution Map	关联的通道成分贴图	
Assume Preferred Angle	采用优先角度	
Atmospheric Settings	大气层设置	
Atom file	Atom 文件	
Attach	连接	
Attach Brush to Curves	将笔刷连接到曲线	
Attach Curves	结合曲线	
Attach End	附加到末端	
Attach Existing Cache File	连接已有缓存文件	
Attach Fur Description	连接毛发描述	
Attach Hair System to Fur	将头发系统连接到毛发	
Attach method	结合方式	
Attach multiple output meshes	合并多个输出网络	
Attach Obj 1/Attache Obj 2	附加到对象1/附加到对象2	
Attach Start	附加到起始端	
Attach surfaces	连接曲面	
Attach to Motion Path	连接到运动路径	
Attach Without Moving	连接时不移动曲面	
Attenuation Settings	衰减设置	
Attract to Matching Mesh	吸引到匹配网络约束	
Attraction	吸引	

（续）

英文命令	中文命令	备注
Attraction Scale	吸引缩放	
Attractor Model	吸引模式	
Attractor Set	吸引设置	
Attribute Editor	属性编辑器	
Attribute Map Width/Attribute Map Height	属性贴图的宽/高	
Attribute Maps	属性贴图	
Attribute Name	属性名称	
Attribute Prefix	属性前缀	
Attribute Spread Sheet	属性总表	
Attribute to Paint	绘制的属性	
Attribute Type	属性类型	
Attributes	属性	
Attributes Selection	属性选择	
Attributes Settings	属性设置	
Attributes To Transfer	传递的属性	
Auto	自动	
Auto Closed Rail Anchor	自动关闭轨道的锚点	
Auto Create Curve	自动创建曲线	
Auto Create Rigid Body	自动创建刚体	
Auto Create Root Axis	自动创建根关节轴	
Auto Frame	自动帧	
Auto Ghost objects in Layer	自动重影层中的对象	
Auto Ghost Selected Objects	自动重影选择的对象	
Auto Key	自动关键帧	
Auto load	自动加载	
Auto Load Selected Objects	自动加载所选对象	
Auto Normal Dir	自动设置发线方向	
Auto Overrides	自动覆盖	
Auto Paint	自动绘制	
Auto Parent curve	自动父化曲线	
Auto Render Region	自动执行区域渲染	
Auto Resize	自动调整大小	
Auto Resolve	自动解析按钮	
Auto Reverse	自动翻转放样方向	
Auto Save	自动保存	
Auto Seams	自动接缝	
Auto Simplify Curve	自动简化曲线	
Auto Smooth	自动平滑	

(续)

英文命令	中文命令	备 注
Auto Weight Threshold	自动权重阈值	
Auto-Activate Modeling Toolkit	自动激活建模工具箱	
Auto Calculate	自动计算	
Auto-Fix Deleted/Missing	自动修整删除的/丢失的肌肉	
Auto-Fix Invalid Sticky	自动修整无效的黏连结合点	
Automatic	自动	
Automatic Baking	自动烘焙	
Automatic Mapping	自动映射	
Autopriority	自动优先权	
AutoSave	自动保存	
Auxiliary Nodes	辅助节点	
Available Node Types	可见节点类型	
Average Normals	平均化法线	
Average Vertices	平均化顶点	
Avg Value	平均值	
Axial Magnitude	轴向强度	
Axis	轴向	
Axis Divisions	轴向细分	
Axis preset	轴向预设	
B		
Back Shade Darkness	背面投影暗度	
Back Shade Factor	背面投影因数	
Backburner Options	Backburner 选项	
Backface Culling	消隐背面	
Background Color	背景色	
Background mode	背景模式	
Bake	烘焙	
Bake Alpha	烘焙通道	
Bake Animation	烘焙动画	
Bake Attribute	烘焙属性	
Bake Channel	烘焙通道	
Bake Color	烘焙颜色	
Bake Deformation to Skin Weights…	烘焙变形到蒙皮权重……	
Bake Layer	烘焙层	
Bake Layer Editor	烘焙层编辑器	
Bake Layers	烘焙层	
Bake Pivot	烘焙枢轴方向	
Bake Point Cloud	烘焙点状云	

第 13 章 Maya 命令中英文对照速查表

（续）

英文命令	中文命令	备注
Bake Resolution for Unsupported Texture Types	为不支持的纹理类型的烘焙分辨率	
Bake Selected Objects	烘焙所选对象	
Bake Shading Group Lighting	烘焙材质组灯光	
Bake Shadows	烘焙阴影	
Bake Simulation	烘焙模拟	
Bake Spring Animation	烘焙弹簧动画	
Bake to	烘焙到	
Bake Topology to Targets	烘焙拓扑结构到目标对象	
Bake Transparency	烘焙透明度	
Bake Ussing Virtual Plane	利用虚拟平面进行烘焙	
Bake View Dependent	烘焙取决于视图	
Baking	烘焙	
Baldness	秃度	
Bank	倾斜	
Bank limit	倾斜限制	
Bank Scale	倾斜比例	
Barm Doors	遮光板	
Base Ambient Color	底部环境色	
Base Color	底部颜色	
Base Curl	底部卷曲度	
Base Opacity	底部不透明度	
Base Width	底部宽度	
Basic	基本 标签	
Basic Emission Speed Attributes	基本发射速度属性	
Basic Emission Attributes	发射器基本属性	
Basics	基本	
Batch Bake	批量烘焙	
Batch Render	批渲染	
Batch Render Messages	批渲染消息	
Batch Render Options	批渲染选项	
Before Stroke cmd	绘制前的命令	
Benchmark	基准	
Bend	弯曲	
Bend Amount	弯曲量	
Best Plane	最佳平面	
Best Plane Texturing Tool	最佳平面纹理工具	
Bevel	倒角	
Bevel Cap Edge	倒角封边	

（续）

英文命令	中文命令	备 注
Bevel Corners	倒角曲面拐角	
Bevel Depth	倒角深度	
Bevel Inside Curves	曲线内部倒角	
Bevel Plus	倒角附加	
Bevel Tool	倒角工具	
Bevel Width	倒角宽度	
Bezier	贝塞尔	
Bezier Corner	贝塞尔拐角	
Bezier Curve to NURBS	贝塞尔曲线转换为 NURBS 曲线	
Bezier Curve Tool	贝塞尔曲线工具	
Bezier Curves	贝塞尔曲线	
Bias	偏移	
Bias Adjust	偏移调整	
Bilinear Filter	双线性过滤器	
Bind Fat on Muscle System	在肌肉系统上结合脂肪	
Bind Method	绑定方式	
Bind Skin	绑定蒙皮	
Bind to	绑定到	
Bins	排序箱	
Bins Sort Shading Node Only	排序箱仅管理材质节点	
Birail	双轨	
Birail 1 Tool	双轨 1 工具	
Birail 2 Tool	双轨 2 工具	
Birail 3+Tool	双轨 3+工具	
Bits per Channel	每通道的位数	
Blend	融合	
Blend Bias	融合偏移	
Blend Colors	颜色融合	
Blend Control	融合控制	
Blend Mode	融合模式	
Blend Shape	融合变形	
Blend Specular With Alpha	融合带 Alpha 的高光	
Blend Two Attr	融合两个属性	
Blender	融合器	
Blending	融合	
Blend Shape Deformers	融合变形	
Blend Shape node	融合变形节点	
Blinn	布林	

（续）

英文命令	中文命令	备注
Blur by Frame	模糊间距	
Blur for GI	全局照明模糊度	
Blur Iterations	模糊强度	
Blur Length	模糊长度	
Blur Sharpness	模糊锐化	
Bone Radius Settings	骨骼半径设置	
Bones	骨骼	
Bonus Rigging	额外装配	
Bookmark Current Objects	标记当前对象	
Bookmark Editor	书签编辑器	
Bookmark Selected Curves	标记所选曲线	
Bookmarks	书签	
Booleans	布尔运算	
Border Lines	边界线	
Border Target Shape	边界目标形状	
Border Width	边框宽度	
Boss	凸柱	
Boss Editor	凸柱编辑器	
Both Proxy and Subdiv Display	以代理和细分显示	
Bottom Barn Door	底端遮光板	
Bottom Half	下半部	
Bottom Left	左下	
Bottom Right	右下	
Boundary	边界	
Boundary Rules	边界规则	
Bounding Box	边界框	
Box Min Radius	长方体最小半径	
Box Radius	长方体半径	
Box Ratio	长方体比率	
Break	打断	
Break Light Links	断开灯光链接	
Break Links	打断链接	
Break Rigid Body Connections	打断刚体连接	
Break Shadow Links	打断阴影链接	
Break Tangent	打断切线	
Break/Unify Tangents	打断/统一切线	
Break Anchor Tangents	打断锚点切线	
Bridge	桥接	

（续）

英文命令	中文命令	备注
Bridge Tool	桥接工具	
Bright Spread	亮度扩展	
Brownian	布朗	
Brownian Attributes	布朗属性	
Brush	笔刷	
Brush Animation	笔刷动画	
Brush Profile	笔刷轮廓	
Brush Strength	笔刷强度	
Brush Tube Attributes	笔刷管属性	
Brush Type	笔刷类型	
Buffer Curve	缓冲曲线	
Build	构建	
Build/Update	创建/更新	
Built-in Noise	内置噪波	
Bulge	凸起纹理	
Bulge Tool	凸起工具	
Bulk A/B	体积 A/B	
Bulk Angular A/B	体积角 A/B	
Bulk Widen A/B	体积加宽 A/B	
Bump 2D	2D 凹凸	
Bump 3D	3D 凹凸	
Bump Mapping	凹凸贴图	
Bump Resolution	凹凸数值	
Buttons on Right	按钮在右边	
By Curvature	通过曲率	
By Cycle Increment	循环增量	
By Direction	通过方向	
By Name	按名称	
By Time	按时间	
By Type	按类型	
C		
C Muscle Shader	C 肌肉着色	
Cache	缓存	
Cache Density	缓存密度	
Cache Description	缓存描述卷展栏	
Cache Directory	缓存路径	
Cache File	缓存文件卷展栏	
Cache Format	缓存格式	

第 13 章　Maya 命令中英文对照速查表

（续）

英文命令	中文命令	备注
Cache In Direct Light	直接照明缓存	
Cache name	缓存名称	
Cache Path	缓存路径	
Cache Point Spacing	点空间缓存	
Cache Points	缓存点数	
Cache Time Range	缓存时间范围	
Caching	缓存	
Calculation Quality	计算质量	
Camera	摄影机	
Camera and Aim	摄影机和目标	
Camera Attributes	摄影机属性	
Camera Manipulators	摄影机操纵器	
Camera Panel	摄影机面板	
Camera Projection Attributes	摄影机映射属性	
Camera Properties	摄影机属性	
Camera Sequencer	摄影机序列	
Camera Aim and Up	摄影、目标和上方控制器	
Cameras	摄影机	
Cancel	撤销	
Cancel Batch Render	取消批渲染	
Canvas	画布菜单	
Canvas Undo	画布取消	
Cap Divisions	盖的细分	
Caps	盖	
Cascade Stitch Node	层叠缝合节点	
Cast Shadows	投射阴影	
Categories	类别	
Caustic and Global Illumination	焦散和全局照明	
Caustic Filter Kernel	焦散过滤器内核	
Caustic Filter Type	焦散过滤器类型	
Caustics	焦散	
Caustics Intensity	焦散强度	
Caustics Photo Map	焦散光子贴图	
Caustics Saturation	焦散饱和度	
Center Normal	中心法线	
Center Pivot	中心化轴心点	
Center U	中心 U	
Center V	中心 V	

（续）

英文命令	中文命令	备 注
Center View of Selection	当前选择的中心视图	
Chamfer Vertex	斜切顶点	
Change Curve Color	改变曲线颜色	
Change Rotation Interp	更改旋转插值	
Channel Control	通道控制	
Channel Offsets	通道偏移	
Channels	通道	
Character	角色	
Character Animation	角色动画	
Character Mapper	角色映射	
Character Set	角色集	
Checker	棋盘格	
Choice	选择	
Choose a Hotkey Set	选择热键集	
Choose a Menu Set	选择菜单集	
Choose a VP2.0 Preset	选择 VP2.0 预设	
Chooser	选择器	
Chord Height	弦高	
Chord Height Ratio	弦高比	
Chromatic Aberration	色差	
Circle	圆形	
Circle Radius	圆半径	
Circle Size Ratio	圆大小比率	
Circles	圆	
Circular arc Degree	圆弧度数	
Circular Fillet	圆形圆角	
Circularize	圆形圆角	
Clamp	夹具式	
Clamp Materials	夹具材质	
Clamp Min/Clamp Max	最小钳制/最大钳制	
Clamp Values	夹具值	
Clamped	钳制曲线	
Classic Hair	经典头发	
Clean Up Reference	清理引用	
Cleanup	清除	
Cleanup Effect	清除效果	
Clear	清除	
Clear All	清除所有	

第13章 Maya命令中英文对照速查表

（续）

英文命令	中文命令	备 注
Clear Before Graphing	清除以前图表	
Clear Clipboard	清空剪贴板	
Clear Graph	清除图表	
Clear History	清除历史	
Clear Initial State	清除初始状态	
Clear Input	清除输入	
Clip	片段	
Clip Edges	剪切边界	
Clip Effects Attributes	剪辑特效属性	
Clip Time	片段时间	
Clipboard Actions	剪贴板操作	
Clipboard Adjustment	剪贴板调整	
Clipping Planes	剪切平面	
Clone Brush Mode	仿制笔刷模式	
Clone Target Tool	克隆目标工具	
Close	关闭	
Close IPR File	关闭 IPR 文件	
Close IPR File and Stop Tuning	关闭 IPR 文件并且终止调整	
Close Point	最近的点	
Closest Position Locator	最近位置定位器	
Cloth	布料	
Cloud	云	
Cloud Attributes	云属性	
Cloud Import/Export…	元导入/导出……	
Clump and Hair Shape	发从和头发形状	
Clump Brush	成束笔刷	
Clump Shape	聚集的形状	
Clumping	丛	
Clumping Frequency	丛频率	
Cluster	簇	
Cluster Curve	簇化曲线	
cMuscieBuilderCamera Viewports	控制肌肉创建器摄影机视口	
Coil divisions	螺旋细分	
Coils	螺旋数	
Collapse	塌陷	
Collapse Asset	塌陷资源	
Collectively	一体化	
Collide Mode	碰撞模式	

英文命令	中文命令	备注
Collide with Effect	带效果碰撞	
Collision Attributes	碰撞属性	
Collision Blur Iterations	碰撞模糊迭代	
Collision Displace	碰撞置换	
Collision Ramps	碰撞渐变	
Collisions	碰撞	
Color	颜色	
Color A/B/C	颜色 A/B/C	
Color Balance	颜色平衡	
Color borders	颜色边界	
Color Channel Modifiers	颜色通道修改	
Color Channel Values	颜色通道值	
Color Curves	颜色曲线	
Color Feedback	颜色反馈	
Color Gain	颜色增益	
Color If False	颜色如果错误	
Color If True	颜色如果正确	
Color Management	颜色管理	
Color Material Channel	对材质通道上色	
Color mode	颜色模式	
Color Modifier Node	颜色修改节点	
Color Offset	颜色偏移	
Color Presets	颜色预设	
Color Profile	颜色配置	
Color Range	颜色范围	
Color Remap	颜色重映射	
Color Resolution	颜色分辨率	
Color Scale	颜色定位	
Color Set	颜色集	
Color Set Editor	颜色集编辑器	
Color Set Name	颜色集名称	
Color Set Sharing	颜色集共享	
Color Settings	颜色设置	
Color Swatch	调色板	
Color Trigger	颜色触发器	
Color Utilities	颜色工具节点	
Color Value	颜色值	
Color Wire Frame	颜色线框	

（续）

英文命令	中文命令	备注
Color Wire Frame 2	颜色线框 2	
Color with Alpha	带有通道的颜色	
Color/Compositing	颜色/合成	
Color Clip	颜色剪切	
Coloring	着色	
Colorize Skeleton	使骨骼变成彩色	
Colors	颜色	
Comb Brush	合并笔刷	
Combine	合并	
Combine Fills and Edges	合并填充和边	
Combine Mode	合并模式	
Command	命令	
Command Completion	命令完成	
Command Shell	命令编辑器	
Common End Points	公共结束点	
Common Material Attributes	材质通用属性	
Common Menu Set	常用菜单设置	
Common Options	常规选项	
Common Selection Options	常用选择选项	
Common Toon Attributes	通用卡通属性	
Component	组件	
Component Editor	组件编辑器	
Component Manipulators	组件操纵器	
Component to Component	组件到组件	
Component Type	组件类型	
Compositing Flag	合成标记	
Compress	压缩	
Compute from	以……来计算	
Compute Node	计算节点	
Concentric Ripple Attributes	通行波纹属性	
Condition	条件	
Cone	圆锥体	
Cone Angle	锥角	
Conform	统一	
Connect	关联	
Connect Border Edges	连接边界边	
Connect Breaks	连接断点	
Connect Components	连接组件	

（续）

英文命令	中文命令	备注
Connect IK/FK	连接 IK/FK	
Connect Joint	连接关节	
Connect Maps to	将贴图连接到	
Connect Maps to Shader	将贴图连接到材质球上	
Connect Muscles to KeepOut	将肌肉连接到 KeepOut	
Connect NURBs Curve to Muscle Displace	连接 NURBs 曲线到肌肉置换	
Connect Output Maps	连接输出贴图	
Connect Selected Muscle	连接所选肌肉方向	
Connect Selected Muscle Displace Nodes	连接所选肌肉置换节点	
Connect Selected Muscle Objects	连接所选肌肉对象	
Connect Selected Muscle Smart Collide Nodes	连接所选肌肉智能碰撞节点	
Connect to Time	连接到时间	
Connect Tool	连接工具	
Connect Density Range	连接密度范围	
Connection Editor	连接编辑器	
Connection Selected	连接所选	
Conservative Rasterization	保守的光栅化	
Constant	恒定	
Constrain	约束	
Construction	构建	
Construction Plane	构造平面	
Container Properties	容器特性	
Container（Asset）	容器（资源）	
Contents Details	内容细节	
Continuous	连续的	
Contour	轮廓	
Contour Contrast	轮廓对比	
Contour Output	轮廓输出	
Contour Shader	轮廓材质	
Contour Store	轮廓存储	
Contour Stretch	轮廓拉伸	
Contour Stretch Mapping Options	轮廓拉伸映射选项	
Contours	轮廓	
Contrast	对比度	
Contrast All Buffers	对比所有缓冲	
Contrast As Color	颜色对比度	
Contrast Threshold	对比度阈值	
Contribution	成分	

（续）

英文命令	中文命令	备注
Control	控制	
Control Curves	控制曲线	
Control For ALL Caches	控制所有缓存	
Control Points	控制点	
Control Rig	控制装配	
Convert	转换	
Convert nCloth OutPut Space	转换n布料输出空间	
Convert PSD to File Texture	转换PSD到文件纹理	
Convert PSD to Layered Texture	转换到PSD到层纹理	
Convert Selection	转换选项	
Convert Selection to Edges	转换选项到边	
Convert Selection to Faces	转换选择项到面	
Convert Selection to UV Edge Loop	转换选项到UV循环边	
Convert Selection to UVS	转换选项到UV	
Convert Selection to Vertices	转换选项到顶点	
Convert Smooth Skin to Muscle System	转换平滑皮肤到肌肉系统	
Convert Surface to Muscle/Bone	转换曲面到肌肉/骨头	
Convert Texture to Vertex Map	转换纹理到贴图	
Convert to File Texture (Maya Software)	转换为文件纹理（Maya软件）	
Convert to Key/Breakdown	转换为关键帧/受控帧	
Convert to Muscle	转换为肌肉	
Convert to Interactive Groom…	转化为交互式修饰……	
Convert Toon to Polygons	将卡通转换为多边形	
Convert Xgen Primitives to Polygons…	将Xgen基本体转化为多边形……	
Convert Vertex to Texture Map	转换顶点到纹理贴图	
Convexity	凸面	
Coordinate	坐标	
Coordinate Space	坐标空间	
Copy	复制	
Copy Attributes	复制属性	
Copy Channel Offstes	复制通道偏移	
Copy Colors, UV, and or Shaders from a Face to the Clipboard	复制面的颜色、UV及材质信息到剪贴板	
Copy Flexor	复制屈肌	
Copy Keys	复制关键帧	
Copy Layer	复制层	
Copy Skin Weights	复制蒙皮权重	
Copy UVs	复制UVs	

（续）

英文命令	中文命令	备注
Copy UVs to UV Set	复制 UVs 到 UV 集	
Copy Vertex Weights	复制顶点权重	
Copy/Paste	复制粘贴	
Copy/Paste Muscle Settings	复制/粘贴肌肉设置卷展栏	
Corner	拐角	
Cosine Power	余弦值	
Count	数量	
Coverage	覆盖	
Crack Shatter	分离破碎	
Crater	弹坑	
Crease Edges	折痕边	
Crease Lines	折痕线	
Crease Selection Sets	折痕选择集	
Crease Set Editor	折痕集编辑器	
Crease Tool	折痕工具	
Create	创建菜单	
Create 3D/2D Container	创建 3D/2D 容器	
Create 3D/2D Container with Emitter	创建带发射器的 3D/2D 容器	
Create Active Rigid Body	创建主动刚体	
Create Animated Sweep	创建动画扫描	
Create Animation Snapshot	创建动画快照	
Create Assembly Definition	创建集合定义	
Create Assembly Reference	创建集合引用	
Create Asset	创建资源	
Create Asset Effect	创建资源效果	
Create Asset with Transform	创建变换资源	
Create Asset with Transform's Effect	创建变换资源效果	
Create ATOM Template	创建 ATOM 模板	
Create Backburner Job	创建 Backburner 作业	
Create Bar	创建列表	
Create Barrier Constraint	创建屏障约束	
Create Base for Selected Muscle Objects	从所选肌肉对象创建基本对象	
Create Bevel	创建倒角	
Create Bin from Selected	从所选中创建排序箱	
Create Bookmake	创建书签	
Create Cache	创建角色集	
Create Cap	创建封口	
Create Character Set	创建角色集	

（续）

英文命令	中文命令	备 注
Create Clip	创建片段	
Create cMuscle Object Shape Node	创建 cMuscle Object 形状节点	
Create Constraint	创建约束	
Create Curve Flow	创建曲线流动	
Create Curve On Surface	在曲面上创建曲线	
Create Curves for	为以下项创建曲线	
Create Deformers	创建边形器菜单	
Create Dough Point Cloud	创建面团点状云	
Create Editable Motion Trail	创建可编辑的运动轨迹	
Create Emitter	创建发射器	
Create Empty Bin	创建空箱	
Create Empty Color Set	创建空白颜色集	
Create Empty Layer	创建空白层	
Create Empty UV Set	创建空的 UV 集	
Create Expression	创建表达式	
Create Fire	创建火焰	
Create Fireworks	创建烟火	
Create Flexor	创建屈肌	
Create LOD Group	创建 LOD 组	
Create Hair	创建头发	
Create Hinge Constraint	创建铰链约束	
Create Interactive Groom Splines	创建交互式修饰样条线	
Create Layer from Selected	为选择的对象创建层	
Create Layer Options	创建层选项	
Create Lightning	创建闪电	
Create MASH Network	创建 MASH 网络	
Create Marking Menu/Edit Marking Menu	创建/编辑标签菜单	
Create Maya Muscle Shader Network	创建 Maya 肌肉着色器网络	
Create Mirrored Muscle	创建肌肉镜像	
Create Modifier	创建修改器	
Create Muscle Displace	创建肌肉置换	
Create Muscle Smart Collide	创建肌肉智能碰撞	
Create Muscle Spline	创建肌肉样条	
Create Muscle Spline Rig	创建肌肉样条装配	
Create Nail Constraint	创建钉约束	
Create nCloth	创建 n 布料	
Create New	新建的形状	
Create New Cache	创建新缓存	

(续)

英文命令	中文命令	备 注
Create New Color Set	创建新的颜色集	
Create New Deformer	创建新边形器	
Create New Tabs	创建新的标签	
Create New UV Set	创建新的 UV 集	
Create Node	创建节点	
Create nParticles	创建 n 粒子	
Create Ocean	创建海洋	
Create Override Layer	创建覆盖层	
Create Override Layer form Selected	从选择的对象中创建覆盖层	
Create Particle Disk Cache	创建粒子磁盘缓存	
Create Passive Collider	创建被碰撞体	
Create Passive Rigid Body	创建被动刚体	
Create Pin Constraint	创建销约束	
Create Polygon Tool	创建多边形工具	
Create Pose	创建姿势	
Create Pose Interpolator	创建姿势插值器	
Create PSD Network	创建 PSD 网络	
Create Reference	创建引用	
Create Render Node	创建渲染节点	
Create Rigid Body Solver	创建刚体解算器	
Create Shatter	创建破碎	
Create Smoke	创建烟	
Create Soft Body	创建柔体	
Create Spring Constraint	创建弹簧约束	
Create Springs	创建弹簧	
Create Subcharacter Set	创建子角色集	
Create Surface Flow	创建曲面流动	
Create Texture Reference Object	创建纹理参考对象	
Create Unattached	创建独立毛发	
Create UVs Based on Camera	基于摄影机创建 UVs	
Create Vertex Face Normals	创建顶点面面法线	
Create View	创建视图	
Create Wake	创建尾流	
Create/Delete Entry	创建/删除入口	
Cross Section	截面	
Cross Section List	截面列表	
Crossing Effect	交叉效果	
Cube	立方体	

（续）

英文命令	中文命令	备 注
Culling	消隐	
Cur Len	当前长度	
Curl	卷曲	
Curl Amount	卷曲量	
Curl Frequency	卷曲频率	
Current Character	当前角色	
Current Character Retargeters	当前角色重定位目标工具	
Current Hotkeys	当前热键	
Current Project	当前项目	
Current Rigid Body Solver	当前刚体解算器	
Current Solver	当前解算器	
Current State	当前状态	
Current Tab	当前标签	
Current Time	当前时间	
Current Track（Snap to Previous）	捕捉到前一个（当前轨道）	
Curvature	曲率	
Curvature Based Width Scaling	基于宽度缩放比例的曲率	
Curvature Scale	曲率缩放	
Curvature Scale First/Second	第 1/2 条曲线曲率缩放	
Curvature Tolerance	曲率容差	
Curve Degree	曲率度	
Curve Editing Tool	曲线编辑工具	
Curve Fillet	曲线倒角	
Curve Fit Checkpoints	曲线匹配检查点	
Curve Functions	曲线函数	
Curve Info	曲线信息	
Curve Ordering	曲线顺序	
Curve Presets	曲线预设	
Curve Radius	曲线半径	
Curve Range	曲线范围	
Curve Smoothness	曲线平滑	
Curve Tolerance	曲线容差	
Curve Utilities	曲线特效	
Curve Warp	曲线扭曲	
Curves	曲线	
Curves Per Fur	每个毛发曲线	
Curves Type	曲线类型	
Custom	自定义	

（续）

英文命令	中文命令	备 注
Custom Command Options	自定义命令选项	
Custom Data Locations	自定义数据位置	
Custom Entities	自定义实体	
Custom equalizer	自定义均衡器	
Custom Globals	自定以全局参数	
Custom Multi Rig	自定义多重装置	
Custom Muscle Shapes	自定义肌肉形状	
Custom Rig	自定义装备	
Custom Scene Text	自定义场景文本	
Custom Stereo Rig	自定义立体装置	
Custom string	自定义字符串	
Cut	剪切	
Cut along	沿……剪切	
Cut Brush	修剪笔刷	
Cut Curve	剪切曲线	
Cut Keys	剪切关键帧	
Cut Mode	修剪模式	
Cut UV Edges	剪切 UV 边	
Cut/Copy/Paste/Delete	剪切/复制/粘贴/删除	
Cutting Radius	剪切半径	
CV Curve Tool	CV 曲线工具	
CV Hardness	CV 硬度	
Cycle	循环	
Cycle Frames	循环帧	
Cycle Min/Cycle Max	最小循环/最大循环	
Cycle Start/Cycle Mid/Cycle End	循环起始端/中部/末端	
Cycle UVs	循环 UVs	
Cycle with Offset	偏移循环	
Cylinder	圆柱体	
Cylindrical Mapping	圆柱体映射	
D		
Dampen on Squash	挤压阻尼	
Dampen on Stretch	拉伸阻尼	
Data Conversion	数据转换	
Data Type	数据类型	
Debug	调试	
Decay Rate	衰减率	
Decay Regions	衰减区域	

（续）

英文命令	中文命令	备 注
Decompose Matrix	分解矩阵	
Def	变形	
Default	默认	
Default Cameras	默认摄影机	
Default Color	默认颜色	
Default Object Manipulator	默认对象操纵器	
Default Scene	默认场景	
Default Time Slider Settings	默认时间滑块设置	
Default Transparency	默认透明度	
Default ViewCube Settings	默认视图导航器设置	
Default Weight	默认权重	
Default Working Units	默认工作单位	
Define	定义	
Deformer	变形器	
Deformer Node	变形器节点	
Deformers	变形	
Degree	度数	
Delete	删除	
Delete All by Type	按类型全部删除	
Delete All Type	删除所有类型	
Delete Attribute	删除属性	
Delete Backup Files	删除备份文件	
Delete Baked Channels	删除烘焙通道	
Delete by Type	按类型删除	
Delete Cache	删除缓存	
Delete Cache Frame	删除缓存帧	
Delete Current Color Set	删除当前颜色集	
Delete Current UV Set	删除当前 UV 集	
Delete Curve Attractor Set	删除曲线吸引设置	
Delete Duplicate Shading Networks	删除复制的材质网络节点	
Delete Edge/Vertex	删除边/顶点	
Delete Empty Layers	删除空层	
Delete Entire Hair System	删除整个头发系统	
Delete Envelopes On Bake	烘焙时删除封套	
Delete existing caches	继承修改	
Delete Farmes	删除帧	
Delete Group and Contents	删除组合内容	
Delete Hair	删除头发	

（续）

英文命令	中文命令	备 注
Delete History	删除历史	
Delete History Ahead of Cache	删除缓存之前的历史	
Delete inputs	删除输入	
Delete Keys	删除关键帧	
Delete Layer	删除层	
Delete Muscle Jiggle Cache	删除肌肉抖动缓存	
Delete Node Cache	删除节点缓存	
Delete Nodes Ahead of Cache	删除缓存之前的节点	
Delete Pass Contribution Map	删除通道成分贴图	
Delete Per-Point Skin Jiggle Cache	删除每点蒙皮抖动缓存	
Delete Scene Time Warp	删除场景时间弯曲	
Delete Selected Muscle Objects	删除所选肌肉对象	
Delete Surface Flow	删除曲线流动	
Delete Tab	删除选项卡	
Delete Texture Reference Object	删除纹理参考对象	
Delete time	删除时间	
Delete Unused Nodes	删除不使用的节点	
Delete UVs	删除 UVs	
Density	密度	
Density Brush	密度笔刷	
Dependency Graph Evaluation	附属计算表	
Depth	深度	
Depth divisions	沿深度方向细分	
Depth Map Shadow Attributes	深度贴图阴影属性	
Depth Max	最大深度	
Depth of Field	景深	
Desaturate	去饱和	
Description	描述	
Deselect All	全部取消选择	
Detach	打断	
Detach Components	分离组件	
Detach Curves	分离曲线	
Detach Hair System form Fur	从毛发上断开头发系统连接	
Detach Selected Joints	断开所选关节	
Detach Skeleton	断开骨骼	
Detach Skin	断开蒙皮	
Detach Surfaces	分离曲面	
Detail Level	细节级别	

(续)

英文命令	中文命令	备注
Detail Level Preset	细节级别预设	
Developer Kit	开发工具包	
Devices	设备	
Diagnose Bsp	诊断 BSP	
Diagnose Final Gather	诊断最终聚集	
Diagnose Grid	诊断栅格	
Diagnose Photon	诊断光子	
Diagnose Sample Rate	诊断采样率	
Diagnose Samples	诊断采样	
Difference	差集	
Difference Tool	差集工具	
Diffuse	漫反射	
Diffuse Boost	漫反射加速	
Dim Image	灰暗的图像	
Dir Axis	方向坐标轴	
Direct Illumination Shadow Effects	直接照明阴影效果	
Direction	方向	
Direction Light	平行光	
Direction Occlusion Map	方向遮挡贴图	
Directory	路径	
Disable All Caches On Selected	禁用选定对象的所有缓存	
Disable All On Selected	禁用所有集合体缓存	
Disable Collision	禁用碰撞约束	
Disable Implicit Control	禁用隐式控制	
Disable Selected IK Handles	禁用所选 IK 手柄	
Disable Weight Normalization	禁用权重规格化	
Disable/Enable All Caches On Selected	禁用/启用所选对象的所有缓存	
Disc	圆盘	
Disconnect all Muscle Directions	断开所有肌肉方向	
Disconnect all Muscle Displaces	断开所有肌肉置换	
Disconnect all Muscle Objects	断开所有肌肉对象	
Disconnect all Muscle Smart Collides	断开所有肌肉智能碰撞	
Disconnect Joint	断开关节	
Disconnect Muscles form KeepOut	从 KeepOut 断开肌肉	
Disconnect NURBs Curve to Muscle Displace	从肌肉置换中断开 NURBs 曲线	
Disconnect selected Muscle Direction	断开所选肌肉方向	
Disconnect Selected Muscle Displace nodes	断开所选肌肉置换节点	
Disconnect Selected Muscle Objects	断开所选肌肉对象	

（续）

英文命令	中文命令	备 注
Disconnect Selected Muscle Smart Collide nodes	连接所择的肌肉智能碰撞节点	
Discrete Scale	离散缩放	
Disk Based Dmaps	保存深度贴图	
Disk Cache Options	磁盘缓存选项	
Displacement	置换	
Displacement Map	置换贴图	
Displacement Shader	置换着色器	
Displacement to Polygons	置换转换为多边形	
Displacement to Polygons with History	置换转换为带有历史记录的多边形	
Display	显示	
Display Alpha Channels	显示 Alpha 通道	
Display Borders	显示边框	
Display Center	显示中心点	
Display Colors	显示颜色	
Display Current Mesh	显示当前网络	
Display Edges	显示边	
Display Fog	显示雾效	
Display HVD	显示 HVD	
Display Image	显示图像	
Display Input Mesh	显示输入网络	
Display Intermediate Objects	显示中间对象	
Display Invisible Faces	显示不可见的面	
Display Layer	显示层	
Display Layer Editor	显示层编辑器	
Display Non Planar	显示非平面的面	
Display Normal	显示法线	
Display Normalized	规格化显示	
Display Options	显示选项	
Display Quality	显示质量	
Display RGB Channels	显示 RGB 通道	
Display Settings	显示设置	
Display Size	显示大小	
Display Tangent	显示切线	
Display Triangles	显示三角形	
Display Unfiltered	显示未过滤的	
Display UVs	显示 UVs	
Display Vertices	显示顶点	
Display Work Area Only	仅显示工作区域	

第 13 章　Maya 命令中英文对照速查表

（续）

英文命令	中文命令	备　注
Distance	距离	
Distance Attributes	距离属性	
Distance Based on	距离基于	
Distance Between	求解距离	
Distance Cutoff	距离截断值	
Distance/Direction Attributes	距离/方向属性	
Dither	振动	
Dither Final Color	抖动最终颜色	
Dithered	抖动	
Division Levels	细分级别	
Divisions	细分	
Documentation	文档	
Domains	区域	
Dope Sheet	信息清单	
Dope Sheep Summary	信息清单概要	
Double Click Command	双击命令标签	
Double Sided	双面	
Double Switch	双通道转换	
Drag	拖拽立场	
Draw	绘制	
Draw as Mesh	作为网络绘制	
Draw Brush	绘制笔刷	
Draw Brush Affected Vertices	绘制时显示受影响的顶点	
Draw Brush Feedback	绘制笔刷反馈	
Draw Brush Tangent Outline	绘制笔刷切线轮廓	
Draw Brush While Painting	绘制时显示笔触范围	
Draw Method	绘制方法	
Draw Order	绘制顺序	
Driven	被驱动对象	
Driven channels	驱动通道	
Driver	驱动对象	
Drop off	衰减	
Drop off	衰减	
Drop off Distance	衰减距离	
Drop off Rate	衰减率	
Duplicate	复制	
Duplicate Input Connections	复制输入连接	
Duplicate Input Graph	复制收入图标	

（续）

英文命令	中文命令	备 注
Duplicate NURBS Patches	复制 NURBS 面片	
Duplicate Reference	复制引用	
Duplicate Special	特殊复制	
Duplicate Surface Curves	复制曲面曲线	
Duplicate With Transform	变换复制	
Duration	持续时间	
Dynamic Constraint Attributes	动态约束属性	
Dynamic Constraints	动力学约束	
Dynamic Photon Map	动态光子贴图	
Dynamic Properties	动力学特性	
Dynamic Relationships	动力学关系	
Dynamic Simulation	动力学模拟	
Dynamics	动力学	
Dynamics and nDynamics	动力学和 n 动力学	
Dynamics MenuSet	动态菜单设置	
E		
Each Face separately	每个面单独执行	
Eccentricity	离心率	
Echo All Commands	回显所有命令	
Echo Collision Commands	回声崩溃命令	
Edge	边	
Edge Anti-aliasing	边界抗锯齿	
Edge Color	边颜色	
Edge Detail	边细节	
Edge Dilation	边界扩张	
Edge Options	边选项	
Edge Snapping	边捕捉	
Edge Style	边类型	
Edge Swap	边缘交换	
Edge U/Edge V	U 向边界/V 向边界	
Edge Weight	边权重	
Edge Weight Preset	边权重预设	
Edge Weights	边权重	
Edge Either Side	任何边	
Edge with Zero Length	零长度边	
Edit	编辑	
Edit Area	编辑区域	
Edit Assigned Bake Set	编辑指定的烘焙集	

第13章 Maya命令中英文对照速查表

（续）

英文命令	中文命令	备 注
Edit Attribute	编辑属性	
Edit Cross Section	编辑截面	
Edit Curve Attractor Set	编辑曲线吸引设置	
Edit Curves	编辑曲线	
Edit Deformers	编辑变形器	
Edit Edge Flow	编辑边流	
Edit Fluid Resolution	编辑流体分辨率	
Edit Fur Description	编辑毛发描述	
Edit Layer	编辑层	
Edit Layouts	编辑布局	
Edit Membership Tool	编辑成员工具	
Edit Mesh	编辑网络	
Edit Normalization Groups	编辑规划组	
Edit NURBS	编辑 NURBS	
Edit Oversampling or Cache Settings	编辑采用或缓存设置	
Edit PSD Network	编辑 PSD 网络	
Edit Rigid Skin	编辑刚性蒙皮	
Edit Rocket Burst Colors	编辑发射爆炸颜色	
Edit Rocket Positions	编辑发射位置	
Edit Rocket Trail Colors	编辑发拖尾颜色	
Edit Smooth Skin	编辑柔性蒙皮	
Edit Template Brush	编辑模板笔刷	
Edit Texture	编辑纹理	
Edit UVs	编辑 UVs	
Edit in Main Window	在主窗口中的编辑	
Editors	编辑	
Effect	效果	
Emission Attributes (see also emitter tabs)	发射属性（也可参见发射器标签）	
Emission Random Stream Seeds	发射随机流种子	
Emission Region	发射区域	
Emission Speed Attributes	发射速度属性	
Emission Scale	发射缩放	
Emit Diffuse	发射漫反射	
Emit from Object	从对象发射粒子	
Emit Specular	发射镜面反射	
Emitter	发射器	
Emitter Name	发射器名称	
Empty	清空	

（续）

英文命令	中文命令	备注
Empty Group	空群组	
Empty Pass Contribution Map	清空通道成分贴图	
Empty Render Layer	清空渲染层	
Enable	启用	
Enable All Caches on Selected	启用选定对象的所有缓存	
Enable Caustics	开启焦散	
Enable Default Scene	开启默认场景	
Enable Displace	开启置换	
Enable Force	启用力度	
Enable Fur	启用毛发	
Enable Geometry Mask	启用几何体蒙版	
Enable Global Illumination	开启全局照明	
Enable IK Handle Snap	启用 IK 手柄捕捉	
Enable IK Solver	启用 IK 解算器	
Enable IK/FK Control	启用 IK/FK 控制	
Enable Input Gamma Correction	开启输入伽马修正	
Enable Jiggle	开启抖动	
Enable Map Visualizer	启用贴图可视化器	
Enable Quality Limits	启用质量限制	
Enable Relax	启用松弛	
Enable Scene Time Warp	启用场景时间弯曲	
Enable Selected IK Handles	启用所选 IK 手柄	
Enable Shaping	启用变形	
Enable Sliding	启用滑动	
Enable Smooth	启用平滑	
Enable Sound	激活音频	
Enable Sticky	启用黏连	
Enable Stroke Rendering	启用笔触渲染	
Enable Vertex Color Filter	启用顶点颜色过滤器	
Enable Weight Normalization	启用权重规格化	
Enable Weight Post Normalization	启用权重后期规格化	
Enable/Disable	开启/关闭	
Enabled Clips	启用片段	
Encoding	编码	
End Bounds	结束边界	
End Cycle Extension	结束循环扩展	
End frame	结束帧	
End point tolerance	结束点容差	

（续）

英文命令	中文命令	备 注
End sweep angle	终止扫掠角度	
End Time	结束时间	
Engine Settings	引擎设置	
Enum Names	枚举名称	
Env Chrome	环境金属	
Env Cube	环境立方	
Env Fog	环境雾	
Env Sky	环境天空	
Env Sphere	环境天球	
Env Textures	环境纹理	
Env Rays	环境光线	
Env Scale	重建	
Envelope	封套	
Envelope Surfaces	包裹曲面	
Environment Ball Attributes	环境球属性	
Environment Chrome Attributes	环境金属属性	
Environment Shader	环境着色器	
Environment Sky Attributes	环境天空属性	
Environment Textures	环境纹理	
Environments	环境	
EP Curve Tool	EP 曲线工具	
Equalization	平衡	
Erase Surface	擦出曲面	
Erase Target Tool	擦除目标工具	
Error	错误	
Error Cutoff	误差中止	
Escape Radius	逃逸半径	
Euler Filter	Euler 过滤器	
Evaluate every	求值频率	
Evaluate every frame(s)	每……帧进行一次采样	
Evaluate Nodes	解算节点	
Even Anchor Tangents	平均化锚点切线	
Even Fields	偶数场	
Even Actions	事件作用	
Even Type	事件类型	
Exclude Collide Pairs	排除碰撞对约束	
Exclusive	专用	
Exclusive Bind	排除绑定	

（续）

英文命令	中文命令	备注
Exclusive Partition	专用分割区	
Execute	执行	
Existing Nodes	已存在的线节点	
Exit	退出	
Exit on Completion	完成后退出	
Expand Asset	开展资源	
Exponential Controls	指数控制	
Export	导出	
Export All	导出全部	
Export All to Alembic	将所有内容导出到 Alembic	
Export Animation	导出动画	
Export Animation Clip	导出动画片段	
Export as Preset…	导出为预设……	
Export Branch	导出子成级	
Export Cache	导出缓存	
Export Character Mapper	导出角色映射	
Export Deformer Weights	导出变形器权重	
Export Editorial	导出可编辑的文件	
Export Exact Hierarchy	到出精确层次	
Export Fluids	到出流体	
Export Full Dag Path	导出完整 Dag 路径	
Export Hair	导出头发	
Export Layer	导出层	
Export Particle Instances	导出粒子实例	
Export Particles	导出粒子	
Export Post Effects	导出后期效果	
Export Pre-Compositing	导出预合成	
Export Proxy Asset	导出代理资源	
Export Selected Network	导出选择的材质网络	
Export Selection	导出所选对象	
Export Skin Weight Maps	导出蒙皮权重贴图	
Export Textures First	首先导出纹理	
Export to Offline File	导出为脱机文件	
Export Value	导出值	
Export Vertex Colors	导出顶点颜色	
Expression	表达式	
Expression Clip	表达式片段	
Expression Editor	表达式编辑器	

（续）

英文命令	中文命令	备注
Extend	延伸	
Extend Curve at	延伸曲线在	
Extend Curve on Surface	延伸曲面曲线	
Extend Fluid	扩展流体	
Extend Method	延伸方式	
Extend Seam Color	扩展缝隙颜色	
Extend Side	延伸侧面	
Extend Surfaces	延伸曲面	
Extension Type	延伸类型	
External Communication	外部通信	
Extra Attributes	附加属性	
Extract Non-layered Animation for All Objects	为所有对象提取无分成动画	
Extract Non-layered Animation for Selected Objects	为所选对象提取无分层动画	
Extract Selected Objects	提取选择的对象	
Extrude	挤出	
Extrude Distance	挤出距离	
Extrude Height	挤出高度	
Extrude Length	挤出长度	
Extrude Tool	挤出工具	
F		
Face	面	
Faces Tool	切面工具	
Faces with Zero Geometry Area	包含零几何体区域的面	
Faces with Zero Map Area	具有零贴图区域的面	
Faces Ration	面向比率	
Falloff	衰减	
Falloff Around Selection	环绕选择衰减	
Falloff Based on	基于……衰减	
Falloff Color	衰减颜色	
Falloff Curve	衰减曲线	
Falloff Inner/Falloff Outer	内部衰减/外部衰减	
Falloff Mode	衰减模式	
Falloff Radius	衰减半径	
Falloff Start/Stop	衰减开始/停止	
Fast	快速	
Fast Interaction	快速交互	
Fat	脂肪	
Favorites	收藏夹	

英文命令	中文命令	备注
Fdynamic Properties Map	动力学特性贴图	
Feature Preservation	性能保存	
Features	功能	
Fidelity Keys Tolerance	关键帧精确度容差	
Field Chart	区域图	
Field Dominance	场优先	
Field Extension	场扩展名	
Field Options	场选项	
File	文件	
File Clean-up	文件清空	
File Content	文件内容	
File Dialog	文件对话框	
File Distribution	文件分布	
File Format	文件格式	
File Name	文件名	
File Output	文件输出	
File Particulars	文件细节	
File Path Editor	文件路径编辑器	
File References	文件引用	
File Referencing	文件引用	
File Textures	文件纹理	
File Type Specific Options	文件类型特定选项	
Files on Disk	硬盘中的文件	
Files/Projects	文件/工程	
Fill Gaps in Selection	在选择之间填充间隙	
Fill Hole	补洞	
Fill Object	填充对象	
Fill Options	填充选项	
Fill Style	填充类型	
Fill Texture Seams	填充纹理的裂缝	
Fill Tool	填充工具	
Filler Color	填充颜色	
Fillet Blend Tool	圆角融合工具	
Film Back	底片	
Film Back Properties	胶片背属性	
Film Gate	胶片门	
Filter	过滤器	
Filter Attributes	过滤属性	

第 13 章　Maya 命令中英文对照速查表

（续）

英文命令	中文命令	备注
Filter Field	过滤去区域	
Filter Normal Deviation	过滤器偏离	
Filter Offset	过滤器偏移	
Filter Shape	过滤器形状	
Filter Size	过滤器大小	
Filter Type	过滤器类型	
Filter Width/Filter Height	过滤器宽度/过滤器高度	
Final Gather	最终聚集	
Final Gather Mode	最终聚集模式	
Final Gather Quality	最终聚集质量	
Final Gather Reflect	最终聚集反射	
Final Gathering	最终聚集	
Final Gathering Quality	最终聚集质量	
Final Gathering Tracing	最终聚集跟踪	
Finalize	最终	
Find Intersections	查找交叉点	
Fine Offset U	精细偏移 U	
Fine Offset V	精细偏移 V	
Fireworks Display Type	烟火显示类型	
Fireworks Sparks Attributes	烟火火花属性	
First Term	第 1 项	
Fit B-spline	适配 B 样条曲线	
Fit Fill	适合填充	
Fit projection to	匹配到	
Fit Time Range	适配时间范围	
Fit to BBox	适配到边界	
Fit Type	适配类型	
Fitting tolerance	适配容差	
Fix by Tesselation	嵌套修复	
Fix Invalid Muscle Object Nodes	修正无效肌肉对象节点	
Fix Surface Attach to allow for Poly Smooth	修正曲面依附到允许的平滑多边形	
Fixed	固定式曲线	
Flare	扩展	
Flash Version	Flash 版本	
Flat	平坦式曲线	
Flat Shade All	平直显示所有对象	
Flat Shade Selected Items	平直显示所选对象	
Flatten A/B	平整 A/B	

（续）

英文命令	中文命令	备 注
Flatten Tool	展开工具	
Flip	翻转	
Flip Triangle Edge	翻转三角边	
Flip Tube Direction	翻转管的方向	
Flip U/Flip V	沿 U/V 向翻转	
Flipped Normal	翻转法线	
Float	浮点	
Float Selected Objects	漂浮所选对象	
Float Point Render Target	浮点渲染目标	
Floating Window	浮动窗口	
Flood Erase	覆盖擦除	
Flood Paint	覆盖绘制	
Floor Attributes	地面属性	
Flow Animation	生长动画	
Flow Path Object	对象跟随路径	
Flow Speedmulti	生长速度控制	
Fluid Attributes	流体属性	
Fluid Emission Turbulence	流体发射扰乱	
Fluid nCache	流体 n 缓存	
Fluid Shape	流体形状	
Fluid Texture 3D	3D 流体纹理	
Fluid Texture 2D	2D 流体纹理	
Fluid to Polygons	流体转换为多边形	
Fluids	流体	
Foam	泡沫	
Foamy Tool	泡沫工具	
Foam Mask	泡沫遮罩	
Focus	关注	
Fog Shadow Intensity	雾效阴影强度	
Fog Shadow Samples	雾效阴影采样	
Fog Spread	雾扩展	
Follicles	毛囊	
Follow	跟随	
Font	字体	
Force editor texture rebake	强制编辑器纹理再烘焙	
Force Field	力场约束	
Force Field Generation	产生力场	
Force Normalize	强制标准化	

第13章 Maya 命令中英文对照速查表

（续）

英文命令	中文命令	备 注
Forces	力	
Format	格式	
Forward	前进	
Fractal	不规则碎片	
Frame	显示帧	
Frame All	全部显示	
Frame All in All Views	在所有窗口中显示所有对象	
Frame All Scene Clips	当前时间中心	
Frame Buffer	帧缓存	
Frame Buffer Format	帧缓存格式	
Frame Cache	帧缓存	
Frame Image	以适合比例显示图像	
Frame in/out	入帧/出帧	
Frame Padding	帧填充	
Frame Playback Range	显示播放范围帧	
Frame Range	帧范围	
Frame Rate	帧速率	
Frame Region	以合适比例显示区域	
Frame Selection	显示选择	
Frame Selection in All Views	在所有窗口中显示所有选择对象	
Frame Selection with Children	框显当前选择（包括子对象）	
Frame Selection with Children in Views	在所有视图中框显当前选择（包含子对象）	
Frame Buffer	帧缓冲区	
Frames to Display	要显示的帧	
Free Image Plane	可用的图像平面	
Freeform Fillet	自由圆角	
Freeform Type	自由类型	
Freeze	冻结	
Freeze Brush	冻结笔刷	
Freeze Tool	冻结工具	
Freeze Transformations	冻结变换	
Freeze/Unfreeze	冻结/解冻	
Frequency	频率	
Frequency Ratio	频率比	
Fresnel Index	菲列尔率	
From	从	
From Channel Box	从通道盒	

（续）

英文命令	中文命令	备注
From Current	从当前位置	
From Rest	从静止位置	
From Start	从起始位置	
From->to	从->到	
Front Axis	前轴	
Full Square	完整方形	
Fur	毛发	
Fur Attribute	毛发的属性	
Fur Description	毛发描述	
Fur Description（more）	毛发描述（更多）	
Fur Render Options	毛发渲染选项	
Fur Render Settings	毛发渲染设置	
Fur Shading Type	毛发的投影类型	
Fur Shadowing Attributes	毛发阴影属性	
G		
Gamma	伽马	
Gamma Correct	Gamma 校正	
Gap Color	间隙颜色	
Gaps	空隙	
Gear	齿轮	
General	常用功能	
General Control Attributes	常规控制属性	
General Editors	常用编辑器	
General Options	常规选项	
General Utilities	常规工具节点	
General LOD Group	生成 LOD 网络	
Generate Polygon Cylinders from Capsules	从胶囊生成多边形圆柱	
Geometry	几何体	
Geometry Type	几何体类型	
Geometry Cache	几何体缓存	
Geometry to Bounding Box	几何体到边界框	
Get Brush	获取笔刷	
Get Effects Asset	获取效果碰撞	
Get Fluid Example	获取流体实例	
Get Hair Example	获取头发实例	
Get nCloth Example	获取 n 布料实例	
Get nParticle Example	获取 n 粒子实例	
Get Ocean/Pond Example	获取海洋/池塘实例	

第13章 Maya命令中英文对照速查表

(续)

英文命令	中文命令	备注
Get Settings from Selected Stroke	从所选笔触上获取设置	
Get Tool Example	获取卡通实例	
Ghost Selected	重影选择	
Ghost/Color Layer	重影/层颜色	
Ghosts	重影	
GI Environment	全局照明环境	
GI Transparency	全局照明透明度	
Gird	网络	
Global Illumination	全局照明	
Global Illumination/Caustics	全局照明/焦散	
Global Photon Map	全局光子贴图	
Global Scale	全局缩放	
Global Stitch	全局缝合	
Glow	辉光	
Glow Intensity	辉光强度	
Go to Bind Pose	恢复绑定姿势	
Go to Line	转到行	
Go to Next	下一个驱动关键帧	
Go to Previous	前一个驱动关键帧	
Goal	目标	
Goal Weight and Objects	目标权重与对象	
GPU Cache	GPU 缓存	
Grab Brush	抓取笔刷	
Grab Color	吸取颜色	
Grab Icon	获取图标	
Grab Swatch to Hypershade/Visor	从材质编辑器/遮板中获取样板	
Grab Tool	抓取工具	
Gradient	渐变	
Gradients	梯度渐变	
Granite Attributes	花岗岩属性	
Graph	图表	
Graph Anim Curves	曲线编辑器中的动画曲线	
Graph Editor	曲线编辑器	
Graph Materials on Selected Objects	所选对象的图标材质	
Gravity	重力场	
Gravity Strength	重力强度	
Gravity X/Y/Z	重力 X/Y/Z	
Gray Option	灰度选项	

（续）

英文命令	中文命令	备注
Grease Pencil	油脂铅笔	
Grease Pencil Frames	油脂铅笔标记帧	
Grid	网络	
Grid Placement	网络布局	
Grid Size	规格大小	
Grid Smear	网络涂抹	
Grid U/V	U/V 网络	
Grid Cache	网络缓存	
Group	群组	
Group Pivot	群组轴心点	
Group under	群组类型	
Group with Original	和原始曲线成组	
Grouping	成组	
Grouping Threshold	组阈值	
Grow	生长	
Grow Along Loop	沿循环方向扩大	
Grow CV Selection	扩大当前选择的 CV	
Grow Selection Region	扩展选择区域	
Guide	导向	
H		
Hair Color	头发颜色	
Hair Systems	头发系统	
Hair Tube Shader	头发管状着色器	
Handle Selection	手柄选择	
Hard Edges	硬边	
Harden Edge	硬化边缘	
Hardware Environment Lookup	硬件环境查找	
Hardware Geometry Cache	硬件几何缓存	
Hardware Information	硬件信息	
Hardware Render	硬件渲染	
Hardware Renderer	硬件渲染器	
Hardware Shading	硬件材质	
Hardware Texturing	硬件纹理	
Heads up Display	平视显示仪	
Heatmap Falloff	热量贴图衰减	
Height	高度	
Height divisions	高度细分	
Height Field	高度区域	

第 13 章　Maya 命令中英文对照速查表

（续）

英文命令	中文命令	备注
Height Scale	高度缩放	
Height/Width	高度/宽度	
Helix	螺旋体	
Help	帮助	
Help Images	帮助图像	
Help Location	帮助位置	
Hidden Edges	隐藏边	
Hide	隐藏	
Hide All	全部隐藏	
Hide All Image Planes	隐藏所有图像平面	
Hide All Labels	隐藏所有标签	
Hide Deformers	隐藏变形器	
Hide Geometry	隐藏几何体	
Hide Intermediate Objects	隐藏中间对象	
Hide Kinematics	隐藏运动学对象	
Hide Selection	隐藏所选	
Hide Source	隐藏原始对象	
Hide Strokes	隐藏笔触	
Hide Unselected CVs	隐藏未选择的 CVs	
Hide Unselected Objects	隐藏未选择的对象	
Hierachy	层级	
Hierachy below/none	层级之下/无	
High Dynamic Range Image Preview Options	高动态范围图像预览选项	
High quality lighting	高质量灯光	
Highlight	高亮	
Highlight Backfaces	高亮显示背面	
Highlight Level	高亮级别	
Highlight Size	高亮显示区的大小	
Highlight What's New	高亮新功能	
Highlight Shaded	高亮实体显示	
History	历史	
Hold	保持	
Hold Current Keys	保持当前关键帧	
Holders	固定器	
Hold-Outs	透底	
Horizontal	水平	
Hotbox Controls	热盒控制器	
Hotkey Editor	热盒编辑器	

（续）

英文命令	中文命令	备 注
HSV Color Key	HSV 颜色键	
HSV Color Noise	HSV 转换为 RGB	
HUD Poly Count	题头显示多边形数量	
Humanlk Example	Humanlk 示例	
Hybrid	混合	
Hypergraph	超图	
Hypershade tab filter	材质编辑器标签过滤器	
I		
IBL Light Rig Editor	IBL 灯光装备编辑器	
Icon Only	仅图标	
Icon/Text Below	在下面显示图标/文本	
Icon/Text Beside	在旁边显示图标/文本	
Ignore Glows	忽略辉光	
Ignore Hardware Shader	忽略硬件着色器	
Ignore Light Links	忽略灯光链接	
Ignore Mirrored Faces	忽略镜像面	
Ignore Names	忽略名称	
Ignore Shadows	忽略阴影	
IK Handle Attributes	IK 手柄属性	
IK Handle Settings	IK 手柄设置	
IK Handle Tool	IK 手柄工具	
IK Solver Attributes	IK 解算器属性	
IK Spline Handle Tool	IK 样条手柄工具	
IK/FK Keys	IK/FK 关键帧	
Illuminates by Default	默认照明	
Illumination	照明	
Image	图像	
Image Based Lighting	基于图像的照明	
Image Based Lighting Attributes	基于图像照明属性	
Image Editing Applications	图像编辑应用程序	
Image File	图像文件	
Image Format	图像格式	
Image Format Options	图像格式选项	
Image Options	图像选项	
Image Plane	图像平面	
Image Plane Attributes	图像平面属性	
Image Properties	图像属性	
Image Range	图像范围	

（续）

英文命令	中文命令	备 注
Image Settings	图像设置	
Image Size	图像大小	
Image Viewing Applications	图像查看应用程序	
Implode	向心爆炸	
Implode Center	向心爆炸中心	
Import	导入	
Import Alembic	导入 Alembic	
Import Animation	导入动画	
Import Animation Clip	导入动画片段	
Import Animation Clip to Characters	导入动画片段到角色	
Import Audio	导入音频	
Import Cache	导入缓存	
Import Deformer Weights	导入变形器权重	
Import Editorial	导入可编辑的文件	
Import Method	导入方法	
Import Objects from Reference	从引入中导入对象	
Import Preset…	导入预设……	
Import Sequencer Audio	导入序列音频	
Import Shot Audio	导入镜头音频	
Import Skin Weight Maps	导入蒙皮权重贴图	
Import Value	导入值	
Importons	重要性粒子	
Imprint Tool	压印工具	
In Alpha	在透明度中	
In Color	在颜色中	
In Color/Out Color	内线框颜色/外线框颜色	
In Connections	输入连接	
In Direction X/Y/Z	在 X/Y/Z 方向	
In Double	双面	
In HSV	输入 HSV	
In Quad	四面	
In RGB	输入 RGB	
In Shape	形状	
In Single	单面	
In Triple	3 面	
In/Out Tangent	入/出切线	
Inactive	未激活	
In-between Weight	中间权重	

（续）

英文命令	中文命令	备注
Incandescence	自发光	
Inclination	倾斜度	
Include	包含	
Include Connected	包括已连接的节点	
Include Edges	包含边	
Include External Files of Unloaded References	包含卸载引用的外部文件	
Include Hierachy	包含层级	
Include Options	包含选项	
Include Placement with Textures	包含带有纹理的置换	
Include Shaders	包括着色器	
Include Shading Group with Materials	包括带有材质的着色组	
Increase Fidelity	提高精确度	
Increment	增量	
Increment & Save	递增并保存	
Indent Selection/Unindent selection	缩进当前选择/取消缩进当前选择	
Independent Euler	独立 Euler	
Indirect Lighting	间接照明	
Indirect Passes	间接过程	
Individual Cache Status	单独缓存状态	
Infinity	无限	
Inflection	变形	
Influence	影响	
Influence Association	联合影响	
Influence List	影响列表	
Info.	信息	
Inherit modifications	继承修改	
Inherit Smooth Mesh Preview	继承平滑网络预览	
Initial Placement	初始值换	
Initial Settings	初始设置	
Initial State	初始状态	
Initial States	初始化妆态	
Inner Bevel Style	内倒角样式	
Inner Falloff Start/Mid/End	内部衰减起始端/中部/末端	
Input	输入	
Input and Output Connections	输入和输出连接	
Input and Output Ranges	输入和输出范围	
Input Connections	输入连接	
Input Curves	输入曲线	

第 13 章　Maya 命令中英文对照速查表

（续）

英文命令	中文命令	备注
Input Gamma Correction	输入伽马修正	
Input Position Locator	输入位置定位器	
Input 1/Input 2	输入 1/输入 2	
Insert Breakdowns	插入受控关键帧	
Insert Clip	插入片段	
Insert Edge Loop Tool	插入循环边工具	
Insert Isoparms	插入等参线	
Insert Joint Tool	插入关节工具	
Insert key Tool	插入关键帧工具	
Insert Knot	插入结	
Insert Location	插入位置	
Insert New Layers After Current	在当前选择的层之后插入新层	
Insert parameter	插入参考	
Insert Pose	插入姿势	
Insert Projection Before Deformers	在变形器之前插入投影	
Instance Leaf Nodes	实例化叶节点	
Instance to Object	实例转换为对象	
Instancer (Geometry Replacement)	替换（几何体替代）	
Instancer (Replncement)	替换（替代）	
Int.	整数	
Intensity	强度	
Intensity Curve	强度曲线	
Intensity Multiplier	强度倍增	
Intensity Scale	强度缩放	
Interaction Mode	交互模式	
Interactive Creation	交互式创建	
Interactive Groom Editor	交互式修饰编辑器	
Interactive Groom Tool	交互式修饰工具	
Interactive Playback	交互播放	
Interactive Sequence Caching Options	交互式序列缓存选项	
Interactive Shading	交互着色	
Interactive Skin Bind	交互式蒙皮绑定	
Interactive Skin Bind Tool	交互蒙皮绑定工具	
Interactive Split Tool	交互分离工具	
Interface	界面	
Interpoints	插值点数量	
Interpolate	插值	
Interpolate Frames	内插帧	

(续)

英文命令	中文命令	备 注
Interpolate Samples	插值采样	
Interpolation Points	插值点	
Interset	交叉	
Intersect Curves	交叉曲线	
Intersect Surfaces	交叉曲面	
Intersection	交集	
Intersection Lines	交叉线	
Intersection Tool	交叉工具	
Inverse Front	翻转前方向	
Inverse Kinematics	方向运动学	
Invert	反转	
Invert Reference Vector	反向参考向量	
Invert Selection	反选	
Invert Shown	反向显示	
In-view Messages	视图中消息	
IPR	交互式渲染菜单	
IPR Options	交互式渲染选项	
IPR Render	IPR 渲染	
IPR Render Current Frame	IPR 渲染当前帧	
IPR Tuning Options	IPR 调节选项	
Irradiance Particles	辐照度粒子	
Isolate Curve	隔离曲线	
Isolate Select	隔离选择	
J		
Jiggle	抖动	
Jiggle Collisions	抖动碰撞	
Jiggle Deformer	抖动变形器	
Jiggle Disk Cache	抖动磁盘缓存	
Jiggle Impact	抖动影响	
Jiggle Impact Start/Jiggle Impact Stop	抖动影响开始/抖动影响停止	
Jiggle Min/Jiggle Max	最小抖动/最大抖动	
Jiggle Presets	抖动预设 卷展栏	
Jiggle Start/Jiggle Mid/Jiggle End	抖动起始端/中部/末端	
Jiggle X/Y/Z	抖动 X/Y/Z	
Jitter	抖动	
Job Settings	作业设置	
Join	连接	
Join to Original	连接原始曲线	

(续)

英文命令	中文命令	备 注
Joint	关节	
Joint Labeling	关节标签	
Joint Orient Settings	关节方向设置	
Joint Rotation Limit Damping	关节旋转限制阻尼	
Joint Settings	关节设置	
Joint Tool	关节工具	
K		
Keep	保留	
Keep Aspect Ratio	保持长宽比	
Keep Construction History	保持构建历史	
Keep Face Group Border	保持面组边界	
Keep Face Together	保持面合并	
Keep Hard Edges	保留硬边	
Keep Image in Render View	在渲染窗口中保留图像	
Keep Image Width/Height Ratio	保持图像宽/高比例	
Keep in Render View Image	保持当前渲染窗口中的图像	
Keep Motion Vectors	保持运动矢量	
Keep New Faces Planar	保持新建平面共面	
Keep Original	保留原始曲线	
Keep Originals	保留原始曲线（复数）	
Keep Swatches at Current Resolution	在当前分辨率下保留样本	
Keep Texture Border	保留文理边界	
Keep Unbaked Keys	保留未烘焙的关键帧	
Key	设置关键帧	
Key in Last Active Layer	在上一个激活的动画层中设置关键帧	
Key in Selected Layer	在选择的层中设置关键帧	
Key Size	关键帧大小	
Key Ticks	关键帧标记	
Key Ticks Size	关键帧标记的大小	
Keyboard/Mouse	键盘/鼠标	
Keys	关键帧	
Kill Field	禁用场	
Kinematics	运动学	
Knife Tool	刀工具	
Knot Spacing	结间距	
L		
Label	标签	
Label Based on Joint Names	基于关节名称设置标签	

（续）

英文命令	中文命令	备 注
Lamina Faces (faces sharing all edges)	薄面	
Lasso Select Tool	套索选择工具	
Lattice	晶格	
Lattice Around	晶格环绕	
Lattice Deform Keys Tool	晶格变形关键帧工具	
Layer	图层	
Layer Mode	层模式	
Layered Shader	分层着色器	
Layered Textures	层纹理	
Layout	布局	
Layout Rectangle	排布矩形	
Layout Settings	布局设置	
Layout UVs Options	排布 UVs 选项	
Leaf Effect	叶效果	
Leather	皮革	
Leather Attributes	皮革属性	
Left Barn Door	左侧遮光板	
Left half	左半部	
Legacy Options	旧版设置	
Len Default/Squash/Stretch	默认长度/挤压长度/拉伸长度	
Length	长度	
Length Brush	长度笔刷	
Length Out	延伸长度	
Length Tolerance	长度容差	
Lens Properties	镜头属性	
Lenses	镜头	
Level	级别	
Level Max	最大级别	
Level min	最小级别	
Level of Detail	细节层级	
Libraries	库（复数）	
Library	库	
Lifespan Attributes（see also per-particle tab）	生命周期属性（也可参见每粒子标签）	
Light Absorbance	灯光吸收率	
Light Angle	灯光角度	
Light Direction Only	仅灯光方向	
Light Effects	灯光特效	
Light Emission	灯光发射	

（续）

英文命令	中文命令	备注
Light Fog	灯光雾	
Light Fog Attributes	灯光雾属性	
Light Info	灯光信息	
Light Linking Editor	灯光连接编辑器	
Light Manipulators	灯光操纵器	
Light Maps	灯光贴图	
Light Model	灯光模式	
Light Rig Name	灯光装备名称	
Light Shape	灯光形状	
Light-Centric	以灯光为中心	
Lighting	灯光（菜单）	
Lighting Mode	照明模式	
Lighting/Shading	灯光材质 菜单	
Lights	灯光	
Lights and Shadows	灯光和阴影	
Limit Information	限制信息	
Limit Number of Edges Selected	选定边数的数量限制	
Limit Shell Size	限制UV面片大小	
Limit the Number of Points	限制点数量	
Line Blending	线混合	
Line Color	线框颜色	
Line Focus	线焦点	
Line Modifier Attributes	线条修改器的属性	
Line Numbers in Errors	出错行号	
Line Offset Ratio	线偏移比率	
Line Offset U	线偏移U	
Line Offset V	线偏移V	
Line Resampling	线条采样	
Linear Resampling	线性采样	
Linear Controls	线性控制	
Link to Camera	连接摄影机	
Link to Layer Set	连接层集	
Liquid Simulation	液体模拟	
List	列表	
List Reference Edits	引用编辑列表	
List Unknown Edits	未知编辑列表	
Live Update	实时更新	
Load	加载	

（续）

英文命令	中文命令	备注
Load On Startup	启动时加载	
Load Options	加载选项	
Load Projection	加载投影	
Load Related Reference	加载相关引用	
Load Render Pass	加载渲染通道	
Load Script	加载脚本	
Load Selected	加载所选	
Load Selected Objects	加载所选对象	
Load Source	加载源角色	
Load Target	加载目标角色	
Load/Save Weights	加载保存权重	
Loaded	已加载	
Lobes	光锥	
Local	局部	
Local Effect	局部效果	
Local Influence	局部影响	
Location	定位	
Locator	定位器	
Lock	锁定	
Lock Camera	选择摄影机	
Lock Dir Wt	锁定方向权重	
Lock Layer	锁定层	
Lock Length	锁定长度	
Lock Muted Layer	锁定屏蔽层	
Lock Normals	锁定法线	
Lock Reference	锁定引用	
Lock Sliding Wt	锁定滑动权重	
Lock Smart Wt	锁定智能权重	
Lock Sticky Wt	锁定黏连权重	
Lock Unpublished Attributes	锁定未发布的属性	
Lock/Free Tangent Weight	锁定/释放切线权重	
Lock/Unlock Render Layer and Render Pass Rendering	渲染时锁定/解除锁定渲染层和渲染过程	
LOD（Level of Detail）	LOD（详细功能）	
Loft	放样	
Long name	长名称	
Loop Brush Animation	循环笔刷动画	
Loop Cutting	线圈剪切	

第13章 Maya 命令中英文对照速查表

（续）

英文命令	中文命令	备 注
Looping	循环播放	
Luminance	亮度	
Luminance/Alpha Channel	亮度/Alpha 通道	
M		
Maintain Max Influences	保持最大影响	
Maintain Offset	保持偏移	
Maintain Pic Aspect Ratio	保持像素纵横比	
Make	生成	
Make Attribute	使属性成为	
Make Boats	创建木船	
Make Brush Spring	生成笔刷弹簧	
Make Capsule	创建胶囊	
Make Capsule with End Locator	创建带有末端定位器的胶囊	
Make Collide	创建碰撞	
Make Hole Tool	创建洞工具	
Make Light Links	生成灯光链接	
Make Links	创建链接	
Make Live	激活	
Make Motion Field	创建运动场	
Make Motor Boats	创建摩托艇	
Make Muscle Direction	创建肌肉方向	
Make New Layers Current	激活当前新创建的动画层	
Make Paintable	使对象可绘制	
Make Pressure Curve	创建压力曲线	
Make Selected Curves Dynamic	使所选曲线成为动力学曲线	
Make Selected Exclusive	使选择作为专用	
Make Shadow Links	对阴影做链接	
Make Soft	生成柔体	
Mandelbrot Inside Method	Mandelbrot 内部方法	
Mandelbrot Shade Method	Mandelbrot 着色方法	
Mandelbrot Type	Mandelbrot 类型	
Manip Mode	操作模式	
Manipulator Sizes	操纵器大小	
Manipulators	操纵器	
Manual Scale	手动缩放	
Manual Squish	手动挤压	
Map	映射	
Map borders	贴图边界	

（续）

英文命令	中文命令	备注
Map File	贴图文件	
Map Height	贴图高度	
Map Item	映射项目	
Map name	贴图名称	
Map Options	贴图选项	
Map size presets	贴图尺寸预设	
Map size X/Y	X/Y 轴贴图尺寸	
Map UV Border	UV 边界贴图	
Map width	贴图宽度	
Map Width/Height	贴图宽/高	
Mapped	已映射	
Mapped Area	映射区域	
Mapping method	映射方法	
Mapping Settings	映射设置卷展栏	
Marble	大理石	
Marble Attributes	大理石属性	
Marking Menu Editor	标签菜单编辑器	
Marking Menu Item Editor	标签菜单项目编辑器	
Mask Target Tool	遮罩目标工具	
Mask Unselected	屏蔽未选择的对象	
MASH (Multi Stage Noise Shaping)	多级噪声整形技术	
MASH Editor	MASH 编辑器	
Match All Transforms	匹配所有变换	
Match Clips	匹配片段	
Match Method	匹配方法	
Match Pivots	匹配枢轴	
Match Render Tessellation	匹配渲染镶嵌细分	
Match Rotation	匹配旋转	
Match Scaling	匹配缩放	
Match Transformations	匹配变换	
Match Translation	匹配平移	
Match Using	使用以下对象匹配	
Material Alpha Gain	材质 Alpha 增益	
Material Attributes	材质属性	
Material Blend	材质混合	
Materials	材质	
Math Functions	数学函数	
Matte Opacity	蒙版透明度	

(续)

英文命令	中文命令	备 注
Matte Opacity Mode	蒙版透明度类型	
Max	最大	
Max Color	最高色值	
Max Curvature	最大曲率	
Max Depth	最大深度	
Max Displacement	最大值换值	
Max Distance	最大距离	
Max Edge Length	最大边长度	
Max Glossy Rays	最大高光光线	
Max Influences	最大影响	
Max Photon Depth	最大光子深度	
Max Playback Speed	最大播放速度	
Max Ray Depth	最大光线深度	
Max Ray Length	最大光线长度	
Max Sample Level	最高采样级别	
Max Sample Rate	最小采样率	
Max Search Depth（%）	最大搜索深度（%）	
Max Separation	最大间隔	
Max Shadow Rays	最大阴影光线	
Max Squash	最大挤压	
Max Stretch	最大拉伸	
Max Subd	最大细分	
Max Subdivision Density	最大细分密度	
Max Time	最大时间	
Max Trace Depth	最大追踪深度	
Max Triangles	最大三角形	
Max UV	最大 UV	
Max Value	最大值	
Maximum 2D/3D Angle	最大二维/三维角度	
Maximum Base Mesh Faces	基于网络的最大面数	
Maximum Cache Size (MB)	最大缓存大小（MB）	
Maximum Edges per Vertex	每个顶点转换为边的最大值	
Maximum Number Polygons	多边形的最大数量	
Maximum Texture Resolution Clamping	最大纹理分辨率钳制	
Maximum U Value	捕捉所选 UVs 到 U 值最大的边上	
Maximum V Value	捕捉所选 UVs 到 V 值最大的边上	
Maya Classic	Maya 经典	
Maya Common Output	Maya 通用输出	

(续)

英文命令	中文命令	备 注
Maya Hardware 2.0	Maya 硬件 2.0	
Maya Muscle	Maya 肌肉	
Maya Start Frame	Maya 起始帧	
Maya Vector	矢量渲染	
Measure Tools	测量工具	
MEL Profile Window	MEL 分析窗口	
Memory and Performance	存储和性能	
Memory and Performance Options	内存和性能选项	
Memory Caching	内存缓存	
Mental Ray Common Output	Mental Ray 通道输出	
Mental Ray Preferences	Mental Ray 参数	
mental Ray User Data Shaders	Mental Ray 用户数据着色器	
MentalRay Lights	MentalRay 灯光	
Merge	缝合	
Merge Caching	合并缓存	
Mental Caching Clips	合并缓存片段	
Merge Caches	合并缓存	
Merge Character Sets	合并缓存集	
Merge Common Inputs	合并公用输入	
Merge Connections	合并连接	
Merge Distance	合并距离	
Merge Edge Tool	缝合边工具	
Merge Layers	合并层	
Merge to Center	缝合到中心	
Merge to one map	合并到一张贴图	
Merge Tolerance	合并容差	
Merge UVs	合并 UVs	
Merge Vertex Threshold	缝合顶点阈值	
Merge Vertex Tool	缝合顶点工具	
Merge Vertices	缝合顶点	
Merge with the Original	与原对象缝合	
Mesh	网络	
Mesh Borders	网络边界	
Mesh Output	网络输出	
Mesh Quality Attributes	网络质量属性	
Method	方法	
Mib_lookup_cylindrical	柱状环境材质	
Min	最小	

（续）

英文命令	中文命令	备 注
Min Alpha/Max Alpha	最小 Alpha 通道/最大 Alpha 通道	
Min Color	最低色值	
Min Edge Length	最小边长度	
Min Sample Level	最低采样级别	
Min Sample RATE	最小采样率	
Min Samples/Max Samples	最大采样数/最大采样数	
Min Screen	最小屏幕	
Min Shadow Rays	最小阴影光线	
Min Value	最小值	
Min/Max	最小/最大	
Min/Max Color	最小/大颜色	
Min/Max Radius	最小/大半径	
Min/Max Samples	最小/大采样率	
Min/Max Value	最小/最大值	
Minimize Application	最小化应用程序	
Minimum U Value	捕捉所选 UVs 到 U 值最小的边上	
Minimum V Value	捕捉所有 UVs 到 V 值最小的边上	
Minimum Weight	最小权重	
Minor Radius	辅助半径	
Minor Sweep Angle	次要扫描角度	
Mirror	镜像	
Mirror Across	镜像平面	
Mirror Axis Position	镜像轴位置	
Mirror Cut	镜像剪切	
Mirror Deformer Weights	镜像变形器权重	
Mirror Direction	镜像方向	
Mirror Function	镜像功能	
Mirror Geometry	镜像几何体	
Mirror Joint	镜像关节	
Mirror Skin Weights	镜像蒙皮权重	
Mirror Weights	镜像权重	
Miscellaneous	杂项	
Miss_fast_skin_maya	sss 快速皮肤材质	
Mocap Example	Mocap 示例	
Mode	模式	
Mode U，Mode V	模式 U，模式 V	
Model View Hardware Visualization	模型视图硬件可视化	
Modeling	建模	

（续）

英文命令	中文命令	备注
modeling MenuSet	建模菜单设置	
Modeling Toolkit	建模工具包	
Modeling Toolkit Preferences	建模工具箱参考	
Modification Falloff	改变衰减	
Modification Resistance	修改阻力	
Modification Type	修改类型	
Modify	修改	
Modify Boundary	修改边界	
Modify Constrained Axis	修改约束轴向	
Modify Current Color Set	修改当前颜色集	
Modify Curves	修改曲线	
Modify Existing Rigs	修改现有装置	
Modify Position	修改位置	
Modify Texture	修改纹理	
Modules	模块	
Monte Carlo	蒙特·卡洛	
Motion Blur	运动模糊	
Motion Blur by Frame	每帧运动模糊	
Motion Blur Optimization	运动模糊优化	
Motion Blur type	运动模糊类型	
Motion Blurred	运动模糊	
Motion Field	运动场	
Motion Paths	运动路径	
Motion Steps	运动步数	
Mountain	山脉	
Move	移动	
Move and Rotate Objects	移动并旋转对象	
Move and Sew UV Edges	移动并缝合 UV 边	
Move Down	向下移动	
Move IK to FK	移动 IK 到 FK	
Move Keys Tool	移动关键帧工具	
Move Normal Tool	法线移动工具	
Move Objects Only	只移动对象	
Move Seam	移动接缝	
Move Selection down in List	在列表中向下移动选择的层	
Move Selection up in List	在列表中向上移动选择的层	
Move Settings	移动设置	
Move Shell Settings	移动壳设置	

第13章　Maya命令中英文对照速查表

（续）

英文命令	中文命令	备　注
Move Skinned Joints Tool	移动蒙皮关节工具	
Move Tab Down	下移标签	
Move Tab Left	左移标签	
Move Tab Right	右移标签	
Move Tab UP	上移标签	
Move Tool	移动工具	
Move UV Shell Tool	移动UV壳工具	
Move UVs to	移动UV到	
Move Weights to Influences	移动权重到影响	
Move/Rotate/Scale Tool	移动/旋转/缩放工具	
Movement Options	运动选项	
Movie	影片	
Movie File	影片文件	
Multi Double Linear	多重双线	
Multi Processing	多线程处理	
Multi Stereo Rig	多立体装备	
Multi Streaks	头发（条纹）补充	
Multi-Camera Output	多个摄影机输出	
Multi-Component	多重组件	
Multi-Cut Tool	多项剪切工具	
Multi-Pixel Filtering	多像素过滤	
Multiple Knots	多节点	
Multiplicity	多重值	
Multiply Divide	乘除法	
Muscle	肌肉	
Muscle Builder	肌肉构建器	
Muscle Creator	肌肉创建器	
Muscle Name	肌肉名称	
Muscle Object Settings	肌肉对象设置	
Muscle Objects	肌肉对象	
Muscle Parameters	肌肉参数	
Muscle Stretch Advanced Attributes	肌肉拉伸高级属性	
Muscle Stretch Jiggle Attributes	肌肉拉伸抖动属性	
Muscle Stretch Scaling Attributes	肌肉拉伸缩放属性	
Muscles/Bones	肌肉/骨头	
Mute	屏蔽	
Mute Key/Unmute Key	禁用/取消禁用关键帧	
Mute Layer	屏蔽层	

（续）

英文命令	中文命令	备 注
Mute/Unmute Channel	屏蔽/取消屏蔽通道	
N		
N Seg	N 段数	
N Sides	N 侧面段数	
Name	名称	
Name Fragment	名称关键帧	
Name Match	名称匹配视图	
Name Suffix	名称后缀	
Namespace	命名空间	
Namespace Editor	命名空间编辑器	
Namespace Options	命名空间选项	
nCache	n 缓存	
nCloths	n 布料	
nConstraint Membership Tool	n 约束成员工具	
nDynamics	n 动力学	
Neighbors	相邻	
Never	从未	
New	新建	
New Geometry Settings	新几何体设置	
New Image	新建图片	
New Layer	新建层	
New Name	新建名称	
New Panel	新建面板	
New Scence	新建场景	
New Start/End Times	新的起始/结束时间	
New Tab	新建选项卡	
New Tab Name	新标签的名称	
New UV SetName	新 UV 集的名称	
Newton	牛顿立场	
Nice Name	友好命名	
No Bounding Box	无边界框	
No Geometry	无几何体	
No Interpolation	无内插	
Node	节点	
Node Behavior	节点行为	
Node Editor	节点编辑器	
Node Publishing	节点发布	
Node State	节点状态	

（续）

英文命令	中文命令	备 注
Node Types That Are Hidden in Editors	曲线编辑中被隐藏的节点类型	
Node Unpublishing	节点取消发布	
Noise	噪波	
Noise Attributes	噪波属性	
Noise Brush	噪波笔刷	
Noise Type	噪波类型	
Noise UV	噪波UV	
Non-Deformer History	非变形历史	
None	无	
Nonlinear	非线性	
Non-manifold Geometry	不可展开几何体	
Non-particle Expressions	非粒子表达式	
Non-power of-two Texture	无二次幂的纹理	
Non-Uniform	非均匀	
Normal	法线	
Normal-Based	基于法线	
Normal Camera	法线摄影机	
Normal Direction	法线方向	
Normal Format	法线格式	
Normal Lengths	法线长度	
Normal Map	法线贴图	
Normal Options	法线选项	
Normal Size	法线长度	
Normal Tolerance	法线容差	
Normal Type	法线类型	
Normalization	规格化	
Normalize	规格化	
Normalize Normal	标准化法线	
Normalize Output	标准输出	
Normalize Weights	规格化权重	
Normals	法线	
nParticle to Polygons	n粒子转换为多边形	
nParticle Tool	n粒子工具	
nParticles	n粒子	
nRigids	n刚体	
nSegs	n段数	
nSolver	n解算器	
nSpans	n跨度	

（续）

英文命令	中文命令	备 注
Num Controls	控制数目	
Num Driven&Type	驱动数量&类型	
Num Insertion Controls&Type	插入的控制器数量&类型	
Num Waves	波纹数量	
Num Controls/Controls&Type	控制器数/截面数	
Num Segments Around	细分段数	
Num of Copies	复制份数	
Number of Exposures	曝光数量	
Number of Faces	面片数	
Number of Frames	帧数	
Number of LOD Levels	LOD 级别的数量	
Number of processors to use	使用的处理器的数目	
Number of Samples	采样数量	
Number of Sections	曲面片段数	
Number of Sides in Base	底的边数	
Number of Spans	段数	
Number U, Number V	U 向数目，V 向数目	
Numeric Attribute Properties	数字属性的参数值	
NURBS Curve to Bezier	NURBS 曲线转换为贝塞尔曲线	
NURBS Display	NURBS 显示	
NURBS Interaction	NURBS 交互	
NURBS Output	NURBS 输出	
NURBS Primitives	NURBS 基本体	
NURBS Samples	NURBS 采样值	
NURBS Texture Placement Tool	NURBS 纹理放置工具	
NURBS to Polygons	NURBS 转多边形	
NURBS to Subdiv	NURBS 转换为细分曲面	
O		
Obey Max Influences	按照最大影响	
Object	对象	
Object Display	对象显示	
Object Path Completion	对象路径完成	
Object Shading	对象材质菜单	
Object Type	对象类型	
Object Type Filter	对象类型过滤器	
Object/Component	对象/组件	
Object-Centric	以对象为中心	
Occlusion falloff	遮挡衰减	

第 13 章　Maya 命令中英文对照速查表

（续）

英文命令	中文命令	备　注
Occlusion rays	遮挡光线	
Occlusion Volume	占用体积	
Ocean	海洋	
Ocean Attributes	海洋属性	
Ocean Shader	海洋着色器	
Odd fields	奇数场	
Off	禁用	
Off/Wireframe/Rendered	关闭/线框/渲染	
Offset	偏移	
Offset Curve	偏移曲线	
Offset Curve On Surface	偏移曲面曲线	
Offset Distance	偏移距离	
Offset Edge Loop Tool	偏移循环边工具	
Offset Fur Direction by	偏移毛发方向	
Offset Scale	偏移缩放	
Offset Surfaces	偏移曲面	
Old Geometry Options	原有几何体选项	
Old Min/Old Max	源最小值/源最大值	
On Active Keys	在激活的关键帧上显示	
Once	一次	
Only Render Strokes	仅渲染笔触	
Only Scale Specified Keys	只缩放指定的关键帧	
Opacity	不透明度	
Opaque Assets	不透明的资源	
Open Alembic	打开 Alembic	
Open Hypergraph	打开超图	
Open Image	打开图像	
Open in Browser	在浏览器中打开	
Open IPR File	打开 IPR 文件	
Open Scence	打开场景	
Open the Graph Editor	打开曲线编辑器	
Open the Trax Editor	打开非线性编辑器	
Open/Close Curves	打开/关闭曲线	
OpenGL-Based Selection Options	OpenGL 的选择选项	
Operation	操作	
Optical FX Attributes	Optical FX 属性	
Optimization Threshold	优化阈值	
Optimize Animations for Motion Blur	优化运动模糊动画	

（续）

英文命令	中文命令	备注
Optimize for	优化	
Optimize Hierarchy	优化层级	
Optimize Scene Size	优化场景大小	
Options	选项	
Order	顺序	
Orient	方向	
Orient Children of Selected Joints	确定选定关节子对象的方向	
Orient Joint	关节定向	
Orientation	方向	
Orientation Settings	方向设置	
Original Object	原始对象	
Orthogonal Reflection	正交反射	
Orthographic Views	正交视图	
Oscillate	摇摆	
Other Animation Tools	其他动画工具	
Other Settings	其他设置	
Other Speed	其他速度	
Other Textures	其他纹理	
Other Color	输出颜色	
Out Connections	输出连接	
Out Glow Color	输出辉光颜色	
Out HSV	输出 HSV	
Out Matte Opacity	输出蒙版不透明度	
Out Transparency	输出透明度	
Outer Bevel Style	外侧倒角样式	
Outer Falloff Start/Mid/End	外部衰减起始端/中部/末端	
Outer/Inner Bevel Style	外/内倒角类型	
Outliner	大纲	
Outlines at Intersections	在交叉点的轮廓线	
Output Connections	输出连接	
Output Correction	输出修正	
Output Gamma Correction	输出伽马修正	
Output Geometry	输出几何体	
Output Geometry Type	输出几何体类型	
Output Maps	输出贴图	
Output Mesh	输出网络	
Output Settings	输出设置	
Output Shader	输出着色器	

(续)

英文命令	中文命令	备 注
Output Type	输出类型	
Output UV Set	输出 UV 集	
Output Verbosity	输出冗余	
Outputs	输出	
Overlay Label	叠加标签	
Override Mesh UV Set Assignments	覆盖网络 UV 集分配	
Override Nice Name	覆盖友好命名	
Overrides	覆盖	
Oversample	过采样	
Oversample Post Filter	过采样后期过滤	
Overwrite Color Set	复写颜色集	
P		
Package	软件包	
Package into Assets	打包到资源	
Packing Region	紧缩区域	
Paint	绘制	
Paint Attributes	绘制属性	
Paint Attributes Tool	绘制属性工具	
Paint Blend Shape Weights Tool	绘制融合变形权重工具	
Paint Cache Weights Tool	绘制缓存权重工具	
Paint Cluster Weights Tool	绘制簇权重工具	
Paint Effects	画笔特效	
Paint Effects Globals	画笔特效全局设置	
Paint Effects Panel	画笔特效面板	
Paint Effects Rendering Options	画笔特效渲染选项	
Paint Effects to Curves	画笔特效转换为曲线	
Paint Effects to NURBS	画笔特效转换为 NURBS	
Paint Effects to Polygons	画笔特效转换为多边形	
Paint Effects Tool	画笔特效工具	
Paint Fluids Tool	绘制流体工具	
Paint Fur Attributes Tool	绘制毛发属性工具	
Paint Grid	绘制网络	
Paint Hair Follicles	绘制头发毛囊	
Paint Hair Textures	绘制头发纹理	
Paint Jiggle Weights Tool	绘制抖动权重工具	
Paint Line Attributes	绘制卡通线属性	
Paint Muscle Weights	绘制肌肉权重	
Paint Nonlinear Weights Tool	绘制非线性权重工具	

（续）

英文命令	中文命令	备 注
Paint on Paintable Objects/View Plane	在可绘制的对象上/视图平面进行绘制	
Paint operation	绘制方式	
Paint Operations	绘制操作	
Paint Properties by Texture Map	通过纹理贴图绘制属性	
Paint Properties by Vertex Map	通过顶点贴图绘制属性	
Paint Random	随机绘制	
Paint Reduce Weights Tool	绘制精简权重工具	
Paint Scene/Paint Canvas	绘制场景/绘制画布	
Paint Scripts Tool	绘制脚本工具	
Paint Selection Tool	绘制选择工具	
Paint Set Membership Tool	绘制集成员工具	
Paint Skin Weights Tool	绘制蒙皮权重工具	
Paint Soft Body Weights Tool	绘制柔体权重工具	
Paint Texture Properties	绘制纹理特性	
Paint Transfer Attributes Weights Tool	绘制传递属性权重工具	
Paint Vertex Color Tool	绘制顶点颜色工具	
Paint Vertex Properties	绘制顶点特性	
Paint Wire Weights Tool	绘制线权重工具	
Paint, Smear, Blur, Erase	绘制、涂抹、模糊、擦除	
Panel Configurations	面板配置	
Panel Editor	面板编辑器	
Panel Toolbar	面板工具栏	
Panels	面板	
Parameter Range	参数范围	
Parameterization	参数化	
Parameters	参数	
Parametric Distance	参数化距离	
Parametric Length	参数长度	
Parent	创建父化关系	
Parent Base Wire	父化基础线	
Parent Controller	父控制器	
Parent Method	父化方式	
Part Brush	分隔笔刷	
Particle	粒子	
Particle Cloud	粒子云	
Particle Collision Even Editor	粒子碰撞事件编辑器	
Particle Disk Cache	粒子磁盘缓存	
Particle Size	粒子大小	

第 13 章 Maya 命令中英文对照速查表

（续）

英文命令	中文命令	备注
Particle Tool	粒子工具	
Particle Utilities	粒子工具节点	
Partition	分区	
Pass Contribution Map name	通道成分贴图名称	
Pass Contribution Maps	通道成分贴图	
Pass Custom Alpha Channel	传递自定义 Alpha 通道	
Pass Custom Depth Channel	传递自定义深度通道	
Pass Custom Label Channel	传递自定义标签通道	
Pass Used by Contribution Map	通过成分贴图使用的通道	
Passes	通道	
Paste	粘贴	
Paste Attributes	粘贴属性	
Paste Keys	粘贴关键帧	
Paste Method	粘贴方式	
Paste U Value to Selected UVs	将 U 值粘贴到所选 UVs	
Paste UVs	粘贴 UVs	
Paste UVs to Selected Face	将 UVs 粘贴到所选择的面上	
Paste V Value to Selected UVs	将 V 值粘贴到所选 UVs	
Paste Vertex Weights	粘贴顶点权重	
Path Tracer	路径追踪器	
Pause IPR Tuning	暂停 IPR 调整	
Pause IPR tuning	暂停 IPR 渲染	
Pct Squash/Stretch	挤压/拉伸百分比	
Pencil Curve Tool	铅笔曲线工具	
Penumbra	半影	
Penumbra angle	半影角	
Per Instance Sharing	每实例共享	
Per Particle (Array Attributes)	每粒子（阵列属性）	
Percent	百分比	
Percent of Image Size	图像大小的百分比	
Percentage Space	间隔百分比	
Performance	性能	
Performance Attributes	性能属性	
Performance Settings	性能设置	
Per-point Emission Rates	每个点的发射速率	
Photon Auto Volume	光子自动体积	
Photon Density	光子密度	
Photon Emission	光子发射	

（续）

英文命令	中文命令	备注
Photon Map	光子贴图	
Photon Map File	光子贴图文件	
Photon Reflections	光子反射	
Photon Refractions	光子折射	
Photon Tracing	光子追踪	
Photon Volume	光子体积	
Photon Volumetric Materials	光子体积材质	
Photon Materials	光子材质	
Physical Fog	物理雾	
Physical Sun and Sky	物理太阳和天空	
Pick/Marquee	拾取/框选	
Pin/Unpin Channel	固定通道/取消固定通道	
Pinch Tool	收缩工具	
Pinning	锁定	
Pipe	管状体	
Pipeline Cache	流水线缓存	
Pivot	枢轴	
Pivot Point	轴心点坐标	
Pixel Center	像素中心	
Pixel Snap	像素吸附	
Place 2D Texture	放置2D纹理节点	
Placement	放置	
Placement Extras	放置附加	
Placement Settings	布局设置	
Planar	平面	
Planar Mapping	平面映射	
Planarity	平面化	
Plateau	平顶式曲线	
Platonic Solid	柏拉图多面体	
Platonic Solids	柏拉图多面体	
Platonic Type	柏拉图类型	
Play Every Frame，Free	逐帧播放，不限制	
Play Every Frame，Max Real-time	逐帧播放，最大实时	
Playback	播放	
Playback by	播放方式	
Playback Looping	循环播放测试	
Playback Speed	播放测试速度	
Playback start/end	播放起始/结束帧	

（续）

英文命令	中文命令	备 注
Playblast	播放测试	
Playblast All Shots	播放测试所有镜头	
Playblast Selected Shots	播放测试选择的镜头	
Playblast Sequence	播放测试序列	
Playlist	播放列表	
Plug-in Manager	插件管理器	
Point	点	
Point Camera	摄影机空间点坐标	
Point Cloud Bake Editor	点状云烘焙编辑器	
Point Density	点密度	
Point Interpolation	点插值	
Point Light	点光源	
Point Obj	对象空间点坐标	
Point on Curve	曲线上的点	
Point on Poly	多边形上的点约束	
Point Size	点大小	
Point to Point	点对点	
Point to Surface	点到面约束	
Point Weight	点权重	
Point World	世界空间点坐标	
Point per Curve	每条曲线上的点	
Points to 2 Points	两点对两点	
Points/Balls/Cloud/Thick Cloud/Water	点/球体/云/厚云/水	
Poke Face	凸起面	
Polar	极性	
Pole Axis	极轴	
Pole Vector	极向量约束	
Poly limit	多边形限制	
Polygon Edges to Curve	多边形边转换为曲线	
Polygon Primitives	多边形基本几何体	
Polygon Selection	多边形选择	
Polygon Smoothness	多边形平滑度	
Polygon Tangent Space	多边形切线空间	
Polygons	多边形	
Polygons Display	多边形显示	
Polygons to Subdiv	多边形转换为细分曲面	
Polynomial Texture Maps	多项纹理贴图	
Pond	池塘	

（续）

英文命令	中文命令	备 注
Popup Help	弹出帮助	
Popup Menu Items	弹出菜单项	
Pose	姿势	
Pose Editor	姿势编辑器	
Pose Space Deformation	姿势空间变形	
Position Along Curve	沿曲线放置	
Post Frames	渲染后帧	
Post Processing	后期处理	
PO Weight	位置和方向权重	
Power	幂	
Pre frames	渲染前帧	
Pre/Post Infinity	无限向前/向后	
Precalc Irradiance	预计算发光	
Precalc Step	预计算步骤	
Precalculate Irradiance	发光预计算	
Pre-Compositing	预合成	
Pre-Compositing Scene Anchor	预合成场景锚点	
Precompute Photon Lookup	预计算光子查找	
Preferences	参数	
Preferences and Settings	参数和设置	
Prefix	前缀	
Prefix Hierarchy Names	为层级名称添加前缀	
Pre-Illumination Controls	预照明控制	
Prelight	预亮	
Premultiply	预乘	
Premultiply Alpha	预乘阿尔法	
Premultiply Threshold	预乘阈值	
Prep for Sculpt	准备雕刻	
Prepass	预通道	
Pre-Select Highlight	预选高亮	
Preserve	保持	
Preserve Aspect Ratio	保持外观比例	
Preserve Child Transform	保持子对象变换	
Preserve History	保持历史记录	
Preserve Length	保持长度	
Preserve Length ratio	保持长度比	
Preserve Normals	保留法线	
Preserve Original shape	保持原状	

第 13 章　Maya 命令中英文对照速查表

（续）

英文命令	中文命令	备　注
Preserve Position	保持位置	
Preserve Quads	保持四边形	
Preserve Seam	保持接缝	
Preserve Skin Groups	保持蒙皮群组	
Preserve UV Ratios	保留 UV 比	
Preserve UVs	保持 UVs	
Preset Blending	预设融合	
Presets	预设	
Pressure	压力	
Pressure Mapping	压力映射	
Pressure Mappings	压力映射（复数）	
Preview	预览	
Preview Animation	预览动画	
Preview Calculation Pass	预览计算通道	
Preview Convert Tiles	预览转化分片	
Preview Final Gather Tiles	预览最终聚集分片	
Preview Light Direction	预览光线方向	
Preview Loop/Ring	预览循环/环形	
Preview Motion Blur	预览运动模糊	
Preview Render Tiles	预览渲染分片	
Preview Tone Map Tiles	预览色调映射分片	
Primary Diffuse Scale	主漫反射比例	
Primary Final Gather File	主最终聚集文件	
Primary GI	初级全局照明	
Primary Intensity	初级强度	
Primary Project Locations	主项目位置	
Primary Saturation	初级饱和度	
Primary Scale	初级值	
Primary Visibility	笔触可见	
Primary Winding Order	原始弯曲顺序	
Priority	优先权	
Prism	棱柱体	
Production Start Time	产品级开始时间	
Profile	轮廓	
Profile Blend Value	轮廓融合值	
Profile Lines	轮廓线	
Profiling	分析	
Progress	进程	

（续）

英文命令	中文命令	备 注
Progressive Mode	渐进式模式	
Project Type	投射方式	
Project	投射	
Project Along	沿下一项投影	
Project Curve on Mesh	在网络上投影曲线	
Project Curve on Surface	在曲面上投影曲线	
Project From	投射方向	
Project Settings	工程设置	
Project Tangent	投射切线	
Project Window	项目窗口	
Projection	投影	
Projection Attributes	投影属性	
Projection Geometry	投影几何体	
Projection Object	投影对象	
Propagate Edge Hardness	增加边的硬度	
Propagation	扩展	
Proportional Modification Tool	比例修改工具	
Proxy	代理	
Proxy Options	代理选项	
Prune Below	精减小于……的权重	
Prune Membership	删减成员	
Prune Small Weights	精减细微权重	
Prune Weights	精简权重	
Prune Weights if less than…in value	如果值小于……就精简权重	
PSD	PSD 文件	
Publish	发布	
Publish as Selection Transform	发布为选择变换	
Publish Attributes	发布属性	
Publish Child Anchor	发布子锚点	
Publish Connections	发布连接	
Publish Node	发布节点	
Publish Options	发布选项	
Publish Parent Anchor	发布父锚点	
Push Mode	推动模式	
Pyramid	棱锥体	
Q		
Quad Draw Tool	四边形绘制工具	
Quad Output	四边形输出	

第13章 Maya命令中英文对照速查表

（续）

英文命令	中文命令	备 注
Quad Switch	四通道转换	
Quadrangulate	四边化	
Quality	质量	
Quality Settings	质量设置	
Quality Cubic	四元数立方	
Quaternion Slerp	四元数球面线性	
Quaternion Tangent Dependent	四元数切线从属	
Quick Rig	快速装备	
Quick Select Set	快速选择集	
R		
Radial	放射力场	
Radial Branch Amt	辐射状分支数量	
Radial Branch Depth	辐射状分支深度	
Radiosity Normal Maps	辐射法线贴图	
Radius	半径	
Radius Scale	半径缩放	
Radius (L)	半径（L）	
Radius (U)	半径（U）	
Raise Application Windows	激活应用窗口	
Raise Main Window	激活主窗口	
Ramp	渐变	
Ramp Attributes	渐变属性	
Ramp Shader	渐变着色器	
Random	随机	
random number functions	随机函数	
Random offset	随机偏移	
Randomize Follides	随机化毛囊	
Randomness	随机性	
Range	范围	
Rasterizer	光栅化器	
Rasterizer use opacity	栅格化使用不透明度	
Rate	采样率	
Ratio	比率	
Ray Direction	光线方向	
Ray Selection	光线选择	
Ray Tracing	光线追踪	
Raycast	投射	
Raycast/Marquee	投射/框选	

(续)

英文命令	中文命令	备注
Rays	光线数	
Raytrace Options	光线追踪选项	
Raytrace Shadow Attributes	光线追踪阴影属性	
Raytracing	光线追踪	
Raytracing Quality	光线追踪品质	
Read this Depth File	读取此深度文件	
Real Size	实际大小	
Real-time	实时	
Rearrange Graph	对齐图表	
Reassign Attribute Maps	重新指定属性贴图	
Reassign Bone Lattice Joint	重新指定骨骼晶格关节	
Reattach Selected Joints	重新连接所选关节	
Reattach Skeleton	重新连接骨骼	
Re-Bind Sticky for selected Muscle Objects	为所选肌肉对象重新结合粘连	
Rebuild	重建	
Rebuild Curve	重建曲线	
Rebuild Method	重建方法	
Rebuild Photon Map	重建光子贴图	
Rebuild Surfaces	重建曲面	
Rebuild Type	重建类型	
Recent Commands	最近使用命令	
Recent Commands List	最近的命令列表	
Recent Files	最近的文件	
Recent History Size	最近历史记录数量	
Recent Increments	最近的递增文件	
Recent Projects	最近的项目	
Recently Replaced Files	最近替换的文件	
Recursion Depth	递归深度	
Reduction Method	减少方法	
Red Channel/Green Channel/Blue Channel	红色通道/绿色通道/蓝色通道	
Red Channel/Green Channel/Blue Channel/All Channels	红色通道/绿色通道/蓝色通道/所有通道	
Redirect	重定向	
Redo	重做	
Redo Previous IPR Render	重复上一次的 IPR 渲染	
Redo Previous IPR render	重新执行前一次 IPR 渲染	
Redo Previous Render	重复上一次的渲染	
Redo View Change	重做视图更改	

（续）

英文命令	中文命令	备 注
Reducation Method	减少方式	
Reduce	精减	
Reference	引用	
Reference Editor	引用编辑器	
Reference Options	引用选项	
Reference Vector	参考向量	
Reference Paths	引用路径	
Referencing	引用	
Referencing Options	引用选项	
Reflect	反射	
Reflected Color	反射颜色	
Reflection about Origin	基于原点的镜像	
Reflection Axis	反射轴向	
Reflection Depth	反射深度	
Reflection Limit	发射限制	
Reflection Settings	映射设置	
Reflection Space	映射空间	
Reflection Specularity	镜面反射	
Reflection/Refraction Blur Limit	反射/折射模糊限制	
Reflections	反射	
Reflectivity	反射率	
Refraction Limit	折射限制	
Refractions	折射	
Refractive Index	折射率	
Refresh Graph Defaults	同步节点编辑/视口选择	
Refresh IPR Image	刷新 IPR 图像	
Refresh the current UV values	刷新当前 UV 值	
Region Keys Tool	区域关键帧工具	
Register a new rig	注册一个新的装置	
Re-Initialize Setup Data on Muscle System	重新初始化肌肉系统的设置数据	
Relationships	关系	
Relative	相对坐标	
Relative Sticky	相对黏连	
Relative transform	相对变换	
Relax	松弛	
Relax Collisions	松弛碰撞	
Relax Compress	松弛压缩	
Relax Expand	松弛扩展	

（续）

英文命令	中文命令	备 注
Relax Iterations	松弛迭代	
Relax Mode	松弛模式	
Relax Strength	松弛强度	
Relax Tool	松弛工具	
Reload File Textures	重新加载文件纹理	
Reload Left and Reload Right	重新载入左面板和重新载入右面板	
Reload Proxy as	将代理重新加载为	
Reload Reference	重新加载引用	
Remake Color Palette	修改颜色调色板	
Remap Color	重映射颜色	
Remap HSV	重映射 HSV	
Remap Value	重映射值	
Remove	移除	
Remove All Bookmarks	移除所有书签	
Remove all Creases	移除所有折痕	
Remove All Images from Render View	从渲染预览窗口中移除所有的图像	
Remove Asset	移除资源	
Remove Brush Sharing	移除笔刷共享	
Remove Current Toon Outlines	移除当前卡通轮廓线	
Remove Curve Color	移除曲线颜色	
Remove duplicate	移除重复项	
Remove Dynamic Constraint	移除动力约束	
Remove empty	移除空项目	
Remove Empty Tracks	移除空轨道	
Remove from Asset	从资源移除	
Remove from Character Set	从角色集中移除	
Remove from Selected Light	从选择的灯光中移除	
Remove Geometry	移除几何体	
Remove Image	移除图像	
Remove Image from Render View	从渲染窗口中移除图像	
Remove Influence	移除影响	
Remove Invalid	移除无效值	
Remove Item	移除项目	
Remove Joint	移除关节	
Remove Lattice Tweaks	移除晶格扭曲	
Remove Members	移除成员	
Remove Multiple Knots	删除多个节点	
Remove nCloth	移除 n 布料	

第13章 Maya命令中英文对照速查表

（续）

英文命令	中文命令	备 注
Remove Pass Contribution Map from Layer	从层中移除通道成分贴图	
Remove Proxy	移除代理	
Remove Reference	移除引用	
Remove Selected	移除所选	
Remove Selected Creases	移除所选折痕	
Remove Selected from Graph	从图表中移除所选择的节点	
Remove Selected Objects	移除选择的对象	
Remove Shot From Group	从组内移除镜头	
Remove Shot Gaps	移除镜头间隙	
Remove Shot Gaps and Overlaps	移除镜头间隙和重叠	
Remove Shot Overlaps	移除镜头重叠	
Remove Subdiv Proxy Mirror	移除细分代理镜像	
Remove Tab	移除标签	
Remove Target	移除目标	
Remove temporary Files	移除临时文件	
Remove Unused	移除未使用的值	
Remove Unused Influences	移除无用的影响	
Remove Frame Renaming	已移除帧的重命名	
Rename	重命名	
Rename Current Color Set	重命名当前颜色集	
Rename Current UV Set	重命名当前UV集	
Rename Interpolated Frames	重命名内插帧	
Rename Joints From Labels	以标签重命名关节	
Rename Tab	重命名标签	
Render	渲染	
Render 2D Motion Blur	渲染2D运动模糊	
Render All Layers	渲染所有层	
Render Attributes	渲染属性	
Render Current Frame	渲染当前帧	
Render Diagnostics	渲染诊断	
Render Influence Objects	渲染影响对象	
Render Info	渲染信息	
Render Layer	渲染层	
Render Mode	渲染模式	
Render Offscreen	在后台检测	
Render Optimizations	渲染优化	
Render Options	渲染选项	
Render Override	渲染覆盖	

（续）

英文命令	中文命令	备 注
Render Pass	渲染过程	
Render Pass Mode	渲器通道模式	
Render Pass	渲染通道	
Render Prepass Only	仅渲染预通道	
Render Region	渲染区域	
Render Selected Objects Only	只渲染选择的对象	
Render Sequence	渲染序列	
Render Settings	渲染设置	
Render shading,lighting and glow	渲染阴影、阴影和辉光	
Render shading shadow maps	渲染阴影贴图	
Render Stats	渲染状态	
Render Target	渲染目标	
Render Texture Range	渲染纹理范围	
Render Type	渲染类型	
Render Using	选择渲染器	
Render View	渲染视图	
Renderable Cameras	可渲染的摄影机	
Renderer	渲染器	
Rendering and Render Setup	渲染和渲染设置	
Rendering Features	渲染特性	
rendering MenuSet	渲染菜单设置	
Rendering Threads	渲染线程	
Renormalize	重新规格化显示	
Reorder Vertices	对顶点重新排序	
Reorient the local scale axes	重新确定局部缩放轴方向	
Repath Files	重新指定文件路径	
Repeat	重复	
Repeat Tool	重复工具	
Repeat UV	重复 UV	
Replace	替换	
Replace Alembic	替换 Alembic	
Replace Cache	替换缓存	
Replace Cache Frame	替换缓存帧	
Replace Members	替换成员	
Replace Objects	替换对象	
Replace References	替换引用	
Replace Region	替换区域	
Replacement Names for Duplicated Joints	替换复制关节的名称	

第13章 Maya命令中英文对照速查表

(续)

英文命令	中文命令	备注
Re-Playblast Selected Shot	再次播放测试选择的镜头	
Represent Hierachy as	表示层次为	
Reroot Skeleton	重设根关节	
Resample Curve	重采样曲线	
Resample Type	重采样类型	
Reset	重置	
Reset Base Pose for Muscle Spline Deformer	重置肌肉样条变形器的基本姿势	
Reset Base Pose for Selected Muscle Objects	为所选肌肉对象重置基本姿势	
Reset Brushes	重置笔刷	
Reset Frame	重置帧	
Reset Lattice	重置晶格	
Reset Region Marquee	重置区域选框	
Reset Template Brush	重新设置模板笔刷	
Reset Transformations	重设变换	
Reset Weights to Default	恢复默认权重	
Resolution	分辨率	
Resolution Gate	分辨率门	
Resolution Status	解析状态	
Resolved Name	重新解算的名称	
Rest	静态	
Rest Min/Rest Max	最小静止/最大静止	
Rest Shape	静止形状	
Rest Start/Rest Mid/Rest End	静态起始端/中部/末端	
Result Position	结果位置	
Retime Tool	重定时工具	
Reverse	翻转	
Reverse Curvature	反转曲率	
Reverse Curve Direction	翻转曲线方向	
Reverse Direction	翻转方向	
Reverse Direction Left/Right	翻转方向左侧/右侧	
Reverse Fur Normals	翻转毛发法线	
Reverse Layer Stack	翻转层列表	
Reverse normal Left/Right	向左翻转法线/向右翻转法线	
Reverse Normals	翻转法线	
Reverse Normals on	在……翻转法线	
Reverse Order	翻转顺序	
Reverse Primary/Secondary Surface Normals	翻转主要/次要曲面法线	
Reverse Surface Direction	翻转曲面方向	

（续）

英文命令	中文命令	备注
Reverse Surfaces	反转曲面	
Revert Selected Swatches	还原选择的样本	
Revert to Default Tabs	恢复默认标签	
Revolve	旋转	
Re-weighting Distance Tolerance	重新设置权重距离容差	
RGB Blending/Alpha Blending	RGB 融合/阿尔法融合	
RGB Min/Max	RGB 最小/最大	
RGB Scale/Alpha Scale	RGB 大小/阿尔法大小	
RGB to HSV	RGB 转换为 HSV	
Rig Selection for Keepout	为 Keepout 装配选项	
Rigging	装配	
Rigging MenuSet	操作菜单设置	
Right Barm Dorr	右侧遮光板	
Right Half	右半部	
Rigid A/B	刚性 A/B	
Rigid Bind	刚性绑定	
Rigid Bodies	刚体	
Rigid Body Attributes	刚体属性	
Rigid Body Solver Attributes	刚体解算器属性	
Rigid Skins	刚性蒙皮	
Rigid Solver Attributes	刚体解算器属性	
Rigid Solver Display Options	刚体解算器显示选项	
Rigid Solver Method	刚体解算器方式	
Rigid Solver States	刚体解算器状态	
Ripple Edit	涟漪编辑	
Ripples	波纹	
Rock	岩石	
Rock Color	岩石颜色	
Rock Roughness	岩石粗糙度	
Rocket Attributes	发射属性	
Rocket Trail Attributes	发射拖尾属性	
Roll	旋转	
Root on Curve	根关节在曲线上	
Root Twist Mode	根关节扭曲模式	
Rotate	旋转	
Rotate Frame	旋转框架	
Rotate Settings	旋转设置	
Rotate to Stroke	旋转笔触	

（续）

英文命令	中文命令	备 注
Rotate Tool	旋转工具	
Rotate UV	旋转 UV	
Rotation	旋转	
Rotation Accumulation	旋转积累	
Rotation Angle	旋转角度	
Rotation Interpolation	旋转内插值	
Roughness	粗糙度	
Round Cap	圆形的盖	
Round Tool	圆化工具	
Run Render Diagnostics	运行渲染诊断	
Run up from	从……运行	
Run up to Current Time	当前帧准备运算	
S		
Safe Action	安全动作	
Safe Delete History	安全删除历史	
Safe Title	安全标题	
Same as Outer Style	与外倒角类型相同	
Same as Outer Style	相同于外部风格	
Sample by	取样值	
Sample Density	采样密度	
Sample Look	采样锁定	
Sample Options	采样选项	
Sample Space	采样空间	
Sample Info	采样信息	
Samples	采样数	
Samples Along Edge	沿边采样数	
Samples Per Object	每对象采样数	
Sampling	采样	
Sampling Controls	采样控制	
Sampling Density	采样密度	
Sampling Mode	采样模式	
Sampling Options	采样选项	
Sampling Quality	采样质量	
Save	保存	
Save Actions	保存动作	
Save Automatically	保存自动设置	
Save Brush Preset	保存笔刷预设	
Save Depth as Grayscale	保存深度为灰度级	

(续)

英文命令	中文命令	备注
Save Every	保存频率	
Save Every Evaluation	每……样本进行保存	
Save Image	保存图像	
Save IPR File	保存 IPR 文件	
Save Only on Request	仅在请求时保存	
Save Output to File	将输出保存到文件	
Save Preferences	保存参数	
Save Preset	保存预设	
Save Reference Edits	保存引用编辑	
Save Scene	保存场景	
Save Scene as	场景另存为	
Save Script	保存脚本	
Save Script to Shelf	将脚本保存至工具架	
Save Snapshot	保存快照	
Save Startup Cache for Particles	为粒子保存启动缓存	
Save State as	将状态保存为	
Save Texture on Stroke	绘制时保存纹理	
Save Textures	保存纹理	
Save to Color Set	保存到颜色集	
Save to File	保存到文件	
Save to Render View	保存到渲染视图	
Save/Save As	保存/另存为	
Saved Mel Scripts	保存的 Mel 脚本	
Scalar Utilities	标量工具节点	
Scale	缩放	
Scale Accumulation	缩放积累	
Scale and Texture Sizes	尺寸和纹理大小	
Scale Curvature	缩放曲率	
Scale Factor	缩放因数	
Scale Hair Tool	缩放头发工具	
Scale Image from Center	从中心缩放图像	
Scale Image from Side	从一侧缩放图像	
Scale Keys	缩放关键帧	
Scale Keys Tool	缩放关键帧工具	
Scale Mode	缩放模式	
Scale Pivot	缩放轴心点	
Scale RGBA	缩放 RGBA	
Scale Tool	缩放工具	

（续）

英文命令	中文命令	备 注
Scanline	扫描线	
Scene	场景	
Scene Assembly	场景集合	
Scene Passes	场景通道	
Scene Pre-Compositing Notes	场景预合成注释	
Scene Summary	场景概要	
Scene Time Warp	场景时间弯曲	
Scope	范围	
Scraggle	起伏度	
Scraggle Correlation	起伏相关度	
Scraggle Frequency	起伏频率	
Scrape Tool	刮擦工具	
Screen Pixels	屏幕像素	
Screen Projection	屏幕投影	
Screen-space Ambient Occlusion	屏幕空间环境光遮挡	
Screen-space Width Control	屏幕空间宽度控制	
Script Editor	脚本编辑器	
Sculpt	雕刻	
Sculpt Deformer	雕刻变形器	
Sculpt Geometry Tool	雕刻几何体工具	
Sculpt Parameters	雕刻参数	
Sculpting	雕刻	
Sculpting Tools	雕刻工具	
Seam Falloff	接缝衰减	
Seam Tolerance	接缝容差	
Seam/Pole Tolerance	接缝/极容差	
Search and Replace	搜索和替换	
Search and Replace Names	搜索并替换名称	
Search Envelop（%）	搜索封套（%）	
Search Method	搜索方法	
Search/Replace	搜索/替换	
Second Term	第2项	
Secondary Curve Fitting	次级曲线匹配	
Secondary Diffuse Bounces	次漫反射反弹数	
Secondary Diffuse Scale	次漫反射比例	
Secondary Final Gather File	次最终聚集文件	
Secondary Intensity	次级强度	
Secondary Project Locations	次项目位置	

（续）

英文命令	中文命令	备 注
Secondary Saturation	次级饱和度	
Secondary Scale	次级值	
Section Radius	截面半径	
Section Spans	跨度段数	
Sections	段数	
Seed	种子	
Segment Delection Options	分段监测选项	
Segment Meshing Options	分段网络选项	
Segments	分段数	
Select	选择	
Select All	选择全部	
Select All by Type	按类型全部选择	
Select Asset Contents	选择资源内容	
Select Border Edge Tool	选择边界边工具	
Select Branch	选择子层级	
Select Brush	选择笔刷	
Select Brush/Stroke Names Containing	选择含有某名称的笔刷/笔触	
Select Brushes	选择笔刷	
Select by Name	按名称选择	
Select Caches to Delete	选择要删除的缓存	
Select Character Set Members	选择角色集成员	
Select Character Set Node	选择角色集节点	
Select Children Under Collapsed Parent	在塌陷父层下选择子层	
Select Connected Faces	选择连接的面	
Select Connected Muscle Directions from Selected muscle Objects	从所选肌肉系统中选择连接的肌肉系统	
Select Connected Muscle Directions from Selected muscle systems	从所选肌肉系统中选择连接的肌肉方向	
Select Connected Muscle Objects from Selected muscle systems	从所选肌肉系统中选择连接的肌肉对象	
Select Connected Muscle Systems from Muscle Directions	从所选肌肉方向中选择连接的肌肉系统	
Select Contained Faces	选择包含的面	
Select Content	选择内容	
Select Contiguous Edges	选择连续的边	
Select Curve CVs	选择曲线上的 CVs	
Select Curve Nodes	选择曲线节点	
Select CV Selection Boundary	选择 CV 选择边界	

（续）

英文命令	中文命令	备 注
Select Edge Loop Tool	循环边选择工具	
Select Edge Ring Tool	环形边选择工具	
Select File Contents	选择文件内容	
Select First CV on Curve	选择曲线上的初始 CV	
Select First/Last U/V	选择第一个/最后一个 U/V	
Select Hierarchy	选择层级	
Select Keyframe Tool	选择关键帧工具	
Select Last CV on Curve	选择曲线上的终止 CV	
Select Layer Node	选择动画层节点	
Select Lights Illuminating Object	选择照明对象的灯光	
Select Lights Shadowing Object	选择阴影对象的灯光	
Select Materials from Objects	从对象选择材质	
Select Members	选择成员	
Select Mode	选择模式	
Select Objects	选择对象	
Select Objects Illuminated by Light	选择灯光照明的对象	
Select Objects in Layer	在层中选择对象	
Select Objects in Pass Contribution Map	在通道成分贴图中选择对象	
Select Objects Shadowed by Light	选择灯光阴影的对象	
Select Objects with Materials	选择所有材质的对象	
Select Previously Sent Objects	选择以前发送的对象	
Select Scene Time Warp	选择场景时间弯曲	
Select Selection Boundary	选择所选边界	
Select Shell	选择壳	
Select Shell Border	选择壳边界	
Select Shortest Edge Path Tool	选择最短边路径工具	
Select Shot	选择镜头	
Select Similar	选择类似	
Select Strokes	选择笔触	
Select Surface Border	选择曲面边界	
Select Surfaces Attached to	选择毛发附着的面	
Select Tab	选择选项卡	
Select Texture Reference Object	选择纹理参考对象	
Select Tool	选择工具	
Select Unsnapped	选择未吸附的关键帧	
Select Unsorted Content	选择无序内容	
Select Using Constrains	选择使用约束	
Selected	选定	

（续）

英文命令	中文命令	备 注
Selected Attributes	选择属性	
Selected Color	选择的颜色	
Selected Keys Override Specified Time Range	所选关键帧覆盖指定时间范围	
Selected Only	仅所选择的笔触	
Selected State	选择状态	
Selection	选择	
Selection Style	选择类型	
Self Attraction and Repulsion	自吸附和排斥	
Self Blur Iterations	自身模糊迭代	
Self Collision	自碰撞	
Self Collision Groups	自碰撞组	
Self Falloff	自衰减	
Self Relax Iterations	自身松弛迭代	
Self Relax Strength	自身松弛强度	
Self Shade	自投影	
Self Shade Darkness	自投影的暗度	
Self Smooth Hold	保持自身平滑	
Self Smooth Iterations	自平滑迭代	
Self Smooth Strength	自平滑强度	
Self Tolerance	自身容差	
Self Volumize	自身体积膨胀	
Self/Multi Collision	自身/多重碰撞	
Send as New Scene	作为新场景发送	
Send to 3ds Max	发送到 3ds Max	
Send to Adobe(R) After Effects(R)	发送到 AE	
Send to Print Studio	发送到 Print Studio	
Send to Unity	发送到 Unity	
Send to Unreal	发送到 Unreal	
Separate	分离	
Separate Extracted Faces	分离提取的表面	
Sequence Viewing Applications	序列查看应用程序	
Set	设置	
Set Active Key	设置主动关键帧	
Set Active Shot	设置激活的镜头	
Set Blend Shape Target Weight Keys	设置融合变形目标权重关键帧	
Set Breakdown	设置受控关键帧	
Set Camera Background Color	设置摄影机背景颜色	
Set Clip Ghost Root	设置片段重影根	

(续)

英文命令	中文命令	备注
Set Clone Source	设置仿制来源	
Set Current Asset	设置当前资源	
Set Current Character Set	设置当前角色集	
Set Current Time to Start Frame	将当前时间设置为开始帧	
Set Current UV Set	设置当前 UV 集	
Set Curve Color	设置曲线颜色	
Set Driven Key	设置驱动关键帧	
Set Erase Image	设置擦出图像	
Set Full Body IK Keys	设置全身 IK 关键帧	
Set IK/FK Key	设置 IK/FK 关键帧	
Set Initial State	设置初始状态	
Set Key	设置关键帧	
Set Keyframe for Vertex Color	为顶点颜色设置关键帧	
Set Keys at	在……设置关键帧	
Set Keys on	设置关键帧	
Set Layer to Override	设置图层为覆盖	
Set Location of File Cache	设置文件缓存位置	
Set Max Influence	设定最大影响	
Set Modifier Fill Object	设置修改器填充对象	
Set Motion Path Key	设置运动路径关键帧	
Set Normal Angle	设置法线角度	
Set NURBS Tessellation	设置 NURBS 镶嵌细分	
Set Passive Key	设置被动关键帧	
Set Preferred Angle	设置优先角度	
Set Project	设置项目	
Set Range	设置范围	
Set Range to	设置范围到	
Set Rest Position	设置静态位置	
Set Rigid Body Collision	设置刚体碰撞	
Set Rigid Body Interpenetration	设置刚体穿透	
Set Selected Muscles/Bones as Not Relative	以非相关的方式设置所选肌肉/骨头为不相关的	
Set Selected Muscles/Bones as Relative	以非相关的方式设置所选肌肉/骨头为相关的	
Set Size	设置大小	
Set Start Position	设定初始位置	
Set Start Position To	设置开始位置	
Set Stroke Control Curves	设置笔触控制曲线	

（续）

英文命令	中文命令	备注
Set Time Code	设置时间代码	
Set to Face	设置到面	
Set Transform Keys	设置变形关键帧	
Set Vertex Normal	设置顶点法线	
Set Weight to 1 and Key Layer	将权重设置为1并对层设置关键帧	
Set Weight/Flood	设置权重/覆盖	
Sets	集	
Settings	设置	
Setup	设置	
Setup for Relative Sticky Deformation	设置相关黏连变形	
Setup Master Muscle Control	建立主肌肉控制	
Sew UV Edges	缝合UV边	
Shade Color	实体颜色	
Shade UVs	着色UVs	
Shaded	显示模式	
Shader	着色器	
Shader Outputs	着色器输出	
Shading	着色	
Shading Group Attributes	阴影组属性	
Shading Groups	阴影组	
Shading Map	材质贴图	
Shading Map Attributes	材质贴图属性	
Shading Network	材质网络	
Shading Quality	阴影质量	
Shading Samptles	着色采样数	
Shadow Attenuation	阴影衰减	
Shadow Color	阴影颜色	
Shadow Control	阴影控制	
Shadow Linking	阴影连接	
Shadow Map	阴影贴图	
Shadow Map File Name	阴影贴图文件名称	
Shadow Maps	阴影贴图	
Shadow Mask	阴影遮罩	
Shadow Method	阴影方式	
Shadow Rays	阴影光线	
Shadow Shaders	阴影材质	
Shadows	阴影	
Shadows Color	阴影颜色	

第13章 Maya 命令中英文对照速查表

(续)

英文命令	中文命令	备注
Shape	形状	
Shaping Blend	变形融合	
Share One Brush	共享一个笔刷	
Share UVs	共享 UVs	
Shared Reference Options	共享引用选项	
Sharpness	锐度	
Shelf Editor	工具架编辑器	
Shell	壳	
Shell Layout	壳布局	
Shell Pre-Rotation	壳前期旋转	
Shell Pre-Scaling	壳前缩放	
Shell Spacing	壳空间	
Shell Stacking	壳堆积	
Shelves	工具架标签	
Shift	偏移	
Shot	镜头	
Show	显示	
Show All	显示全部	
Show All Image Planes	显示所有图像平面	
Show All Labels	显示所有标签	
Show Auxiliary Nodes	显示辅助节点	
Show Back Faces	显示背面	
show Base Wire	显示基础线变形器	
Show Batch Render	显示批渲染	
Show Both	两者都显示	
Show Bottom Tabs Only	只显示底部标签	
Show Buffer Curves	显示缓冲曲线	
Show Deformers	显示变形器	
Show Directories Only	只显示路径	
Show Files Only	只显示文件	
Show Frame Numbers	显示帧数	
Show Geometry	显示几何体	
Show Kinematics	显示运动学对象	
Show Last Hidden	显示上次隐藏	
Show Layer Weight	显示层权重	
Show Line Numbers	显示行号	
Show Manipulator	显示操纵器	
Show Manipulator Tool	显示操纵器工具	

（续）

英文命令	中文命令	备注
Show Modeling Toolkit	显示建模工具箱	
Show Namespace	显示命名空间	
Show Next Graph	显示下一个材质编辑器的图表	
Show Ornaments	显示提示	
Show Previous Graph	显示上一个材质编辑器的图表	
Show Quick Help	显示快速帮助	
Show Region Marquee	显示区域选框	
Show Relationship Connections	显示关联连接	
Show Results	显示结果	
Show Selected Type	显示所选类型	
Show Selection	显示选择	
Show Stack Trace	显示堆栈跟踪	
Show the ViewCube	显示视图导航器	
Show Ticks	显示标记	
Show Tooltip Help	显示工具提示帮助	
Show Top And Bottom Tabs	显示顶部和底部标签	
Show Top Tabs Only	只显示顶部标签	
Show Warnings	显示警告	
Show Wireframe	显示线框	
Show/Hide	显示/隐藏	
Show/Hide Backfaces	显示/隐藏背面	
Show/Hide Face Triangles	显示/隐藏面上的三角面	
Shrink Along Loop	沿循环方向收缩	
Shrink CV Selection	缩减当前选择的 CV	
Shrink Selection Region	收缩选择区域	
Shrink Surface	收缩曲面	
Shrink Wrap	收缩包裹	
Shrink Wrap Selection	收缩包裹选项	
Shutter Open/Close	快门打开/关闭	
Side Length	边长	
Similarity Tolerance	相似度容差	
Simple Fog	简单雾	
Simple Muscles	简单肌肉	
Simplify Curve	简化曲线	
Simplify Method	简化方式	
Simplify Stroke Path Curves	简化笔触路径曲线	
Sine	正弦	
Single Edge	单边	

（续）

英文命令	中文命令	备 注
Single Pixel Brush	单一像素笔刷	
Single Switch	单通道转换	
Single-sided Lighting	单面照明	
Size	尺寸	
Size Radius	半径大小	
Size Rand	随机尺寸	
Size Units	大小单位	
Skeleton	骨骼	
Skeleton Info	骨骼信息	
Skin	蒙皮	
Skin Binding Settings	蒙蔽绑定设置	
Skin Setup	蒙皮设置	
Skinning Method	蒙皮方式	
Sky Attributes	天空属性	
Sky Light	天空颜色	
Slide A/B	滑动 A/B	
Slide Angular A/B	滑动角 A/B	
Slide Angular Rear A/B	后向滑动角 A/B	
Slide Edge Tool	滑动工具	
Slide on Surface	面上滑动	
Slide Rear A/B	先后滑动 A/B	
Sliding Strength	滑动强度	
Small Object Culling threshold	小对象消隐阈值	
Smart Bake	智能烘焙	
Smart Collision	智能碰撞	
Smart Transform	智能变换	
Smear Frequency	涂抹频率	
Smear Tool	涂抹工具	
Smoke Effect Attributes	烟雾特效属性	
Smoke Emission Attributes	烟雾发射属性	
Smoke Motion Attributes	烟雾运动属性	
Smooth	平滑	
Smooth Alpha/Color	平滑 Alpha/颜色	
Smooth Anchor Tangents	平滑锚点切线	
Smooth Bind	柔性绑定	
Smooth Brush	平滑笔刷	
Smooth Collision	智能碰撞	
Smooth Collisions	平滑碰撞	

（续）

英文命令	中文命令	备 注
Smooth Compress	平滑压缩	
Smooth Curve	平滑曲线	
Smooth Edge Ratio	平滑边缘比	
Smooth Expand	平滑扩展	
Smooth Factor	平滑因数	
Smooth Hold	保持平滑	
Smooth Hold Post	后期平滑保持	
Smooth Hold Pre	平滑保持预设	
Smooth Iterations	平滑迭代	
Smooth Iterations Post	后期平滑迭代	
Smooth Iterations Pre	平滑迭代预设	
Smooth Mesh Preview to Polygons	平滑网络预览转换为多边形	
Smooth Shade All	平滑显示所有对象	
Smooth Shade Selected Items	平滑显示所选对象	
Smooth Skin Weights	平滑蒙皮权重	
Smooth Skins	平滑蒙皮	
Smooth Strength	平滑强度	
Smooth Strength Post	后期平滑强度	
Smooth Strength Pre	平滑强度预设	
Smooth Tangent	平滑切线	
Smooth Target Tool	平滑目标工具	
Smooth Tool	平滑工具	
Smooth UV Tool	平滑 UV 工具	
Smooth UVs	平滑 UVs	
Smooth Value	平滑值	
Smooth Wireframe	平滑线框	
Smoothing	平滑	
Smoothing angle	平滑角度	
Smoothness	平滑程度	
Smooth-Pre attributes	平滑属性预设	
Snap	捕捉	
Snap Align Objects	捕捉对齐对象	
Snap Curve to root	吸附曲线到根关节	
Snap Enable	启用吸附	
Snap Keys	捕捉关键帧	
Snap Shot	快照	
Snap Times to Multiple of	吸附到时间的 X 倍数	
Snap to Next	捕捉到下一个	

（续）

英文命令	中文命令	备 注
Snap to Next(Current Track)	捕捉到下一个（当前轨道）	
Snap to Polygon Face	捕捉到多边形面上	
Snap to Previous	捕捉到前一个	
Snap Together Tool	捕捉聚集工具	
Snap Type	捕捉类型	
Snap Values to Multiple of	吸附到数值的 X 倍数	
Snapping	捕捉	
Snapshot	快照	
Snow	雪	
Snow Altitude	雪高度	
Snow Color	雪颜色	
Snow Dropoff	雪衰减	
Snow Roughness	雪粗糙度	
Snow Slope	雪斜率	
Soccer Ball	足球体	
Soft Body	柔体	
Soft Body Attributes	柔体属性	
Soft Modification	软修改	
Soft Modification Tool	柔性修改工具	
Soft Select	软件选择	
Soft Selection	软选项	
Soften Edge	软化边缘	
Soften/Harden Edges	软化/硬化边	
Software Render	软件渲染	
Software Renderer	软件渲染器	
Solid Fractal	固体碎片	
Solid Shatter	固体破碎	
Solo	孤立	
Solo Layer	单独显示层	
Solver Enable	启用解算器	
Solver Settings	解算设置	
Solvers	解算器	
Sound	声音	
Sound Length	声音长度	
Sounds	声音	
Source	源	
Source Clip Attributes	源片段属性	
Source in/out	源入/源出	

（续）

英文命令	中文命令	备注
Source Meshes	源网络	
Source Script	源化脚本	
Source Surfaces	源曲面	
Source UV Set	源 UV 集	
Spacing Presets	间隔预设	
Spans U/Spans V	U 向段数/V 向段数	
Sparse Curve Bake	稀疏曲线的烘焙	
Special Effects	特效	
Specify Node	指定节点	
Specular Color	高光颜色	
Specular Roll Off	镜面反射衰减	
Specular Scale	高光值	
Specular Shading	高光着色	
Specular Sharpness	高光锐度	
Sphere	球体	
Spherical Harmonics	球形谐波	
Spherical Mapping	球形映射	
Spin Edge Forward/Backward	向前/后旋转边	
Spline	样条曲线	
Spline Length Settings	样条长度设置卷展栏	
Split	分离	
Split Mesh with Projected Curve	使用投影的曲线分割网络	
Split Shots	分离镜头	
Split UVs	分离 UVs	
Spot Light	聚光灯	
Spottyness	斑点	
Spray Tool	喷射工具	
Spreadsheet	数据表	
Spring Attributes	弹簧属性	
Spring Damp	弹簧阻尼	
Spring Methods	弹簧方式	
Spring Stiffness	弹簧硬度	
Spring Travel	弹簧跨度	
Springs	弹簧	
Sprite Attributes	精灵属性	
Sprite Wizard	魔术精灵	
Square	正方形	
Squash	挤压	

(续)

英文命令	中文命令	备注
Squash Start/Squash Mid/Squash End	挤压起始端/中部/末端	
Squash X End/Squash Z End	在 X/Z 轴向挤压肌肉的末端	
Squash X Mid/Squash Z Start	在 X/Z 轴向挤压肌肉的中部	
Squash X Start/Squash Z Start	在 X/Z 轴向挤压肌肉的起始端	
Stacked Curves	二叠曲线	
Stagger	交错	
Stalks U	茎 U	
Stalks V	茎 V	
Stamp Depth	图章深度	
Stamp Spacing	图章间距	
Stance	初始	
Stance Pose	初始姿势	
Start	开始	
Start Curve Attract	起始曲线吸引	
Start Cycle Extension	起始循环扩展	
Start Frame	起始帧	
Start Sweep Angle	起始扫掠角度	
Start Time	开始时间	
Start Time/End Time	起始时间/结束时间	
Start/End	开始/结束	
Start/End Delete Frame	起始/结束删除帧	
Start/End interpolation Frame	起始/结束内插帧	
Start/End Length	开始/结束长度	
Start/End Replace Frame	起始/结束替换帧	
Start/End Sweep Angle	起始扫掠角度	
Start/End Time	起始/结束时间	
Static Channels	静态通道	
Stencil	图案	
Stencil Attributes	图案属性	
Step Size	步长	
Stepped	阶梯式曲线	
Stepped Next	下级阶梯式曲线	
Steps after Current Frame	当前帧之后的步数	
Steps before Current Frame	当前帧之前的步数	
Steps Size	步长	
Stereo	立体	
Stereo Camera	立体摄影机	
Stereo Display	立体显示	

（续）

英文命令	中文命令	备 注
Sticky	黏性	
Sticky A/B/C	黏性 A/B/C	
Sticky Strength	黏性强度	
Stiffness Scale	刚度比例	
Stitch	缝合	
Stitch Corners	缝合角点	
Stitch Edges	缝合边	
Stitch Edges Tool	缝合边工具	
Stitch Partial Edges	缝合部分边	
Stitch Smoothness	缝合平滑度	
Stitch Surface Points	缝合曲面点	
Storage Options	存储选项	
Store Points as	存储精度	
Straighten	拉直	
Straighten UV Border	拉直 UV 边界	
Straightness	拉直度	
Strength	强度	
Strength Dropoff	强度衰减	
Stretch Start/Stretch Mid/Stretch End	拉直起始端/中部/末端	
Stretch Volume Presets	拉直提及预设	
Stretch X End/Stretch Z End	在 X/Z 轴向拉伸肌肉的末端	
Stretch X Mid/Stretch Z Mid	在 X/Z 轴向拉伸肌肉的中部	
Stretch X Start/Stretch Z Start	在 X/Z 轴向拉伸肌肉的起始端	
String	字符串	
Stroke	笔触	
Stroke Attributes	笔触属性	
Stroke Length	笔触长度	
Stroke Refresh	笔触刷新	
Strokes per Span	每段上的笔触	
Stucco	灰泥	
Stucoo Attributes	灰泥属性	
Style	类型	
Stylus	压感笔	
Stylus Pressure	手写笔压力	
Subcharacters	子角色	
Subdiv Proxy	细分代理	
Subdiv to NURBS	细分曲面转换为 NURBS	
Subdiv to Polygons	细分曲面转换为多边形	

（续）

英文命令	中文命令	备 注
Subdivision Surface Mode	细分面模式	
Subdivs	细分	
Subdivs Display	细分显示	
Subsample Size	子采样大小	
Substance	物质	
Substance Output	物质输出	
Substance Parameters	物质参数	
Substitute Geometry	替换几何体	
Sub-Surface Scattering	次表面散射	
Sun Attributes	太阳属性	
Super Ellipse	超椭圆	
Super Shapes	超形状	
Suppress Command Results	抑制命令结果	
Suppress Error Messages	抑制错误消息	
Suppress Info Messages	抑制信息消息	
Suppress Stack Window	抑制堆栈窗口	
Suppress Warning Messages	抑制警告消息	
Surf Luminance	曲面亮度	
Surface	曲面	
Surface Association	曲面连接	
Surface Attach	曲面依附	
Surface Degree	曲面度数	
Surface Direction	曲面方向	
Surface Editing	曲面编辑	
Surface Editing Tool	曲面编辑工具	
Surface Fillet	曲面圆角	
Surface Fitting	曲面匹配	
Surface Info.	曲面信息	
Surface Offset	曲面偏移	
Surface Offset Min/Surface Offset Max	表面最小/大偏移	
Surface Output Options	曲面输出选项	
Surface Shader	表面着色器	
Surface Shading Properties	曲面材质特性	
Surface Shatter	曲面破碎	
Surface Thickness	曲面厚度	
Surface	曲面菜单	
SVG	SVG	
SVG animation	SVG 动画	

（续）

英文命令	中文命令	备注
Swap	交换	
Swap Buffer Curve	交换缓冲曲线	
Swap Y<->Z	交换 Y、Z 轴	
Switch Tag for Active Proxy to	交换激活代理的标记	
Switch Tag for Proxy	切换代理的标记	
Switch Utilities	转换工具节点	
Symmetrize	对称	
Symmetry	对称	
Symmetry Plane	对称平面	
Symmetry Tolerance	对称容差	
Synchronized Euler	同步 Euler	
T		
Tab Toolbar	标签工具栏	
Tab Type	标签类型	
Tabs	标签	
Tag as Controller	标记为控制器	
Tangent	切线约束	
Tangent Align Direction	切线对齐方向	
Tangent Length	切线长度	
Tangent Manip Size	切线操纵器大小	
Tangent Mode	切线模式	
Tangent Options	切线选项	
Tangent Rotation	切线旋转	
Tangent Scale	切线缩放	
Tangent Scale First/Second	第 1/2 条曲线切线缩放	
Tangent Space	切线空间	
Tangent Space UV Set	切线空间 UV 集	
Tangent U Camera	U 向切线摄影机	
Tangent V Camera	V 向切线摄影机	
Tangents	切线菜单	
Target	目标	
Target Index	目标索引	
Target Meshes	目标网络	
Target shape options	目标形状选项	
Target Surfaces	目标曲面	
Target UV Set	目标 UV 集	
Target Weld Tool	目标焊接工具	
Tearable Surface	可撕裂表面	

第13章 Maya命令中英文对照速查表

（续）

英文命令	中文命令	备 注
Template	模板	
Template Brush Settings	模板笔刷设置	
Template Curves of Locked layers	锁定的动画层的模板曲线	
Template/Untemplate Channel	模板/取消模板通道	
Tessellation	细分	
Tessellation Method	镶嵌细分方式	
Tessellation Mode	镶嵌细分模式	
Test Resolution	测试分辨率	
Test Texture	测试纹理	
Text	文本	
Text Only	仅文本	
Texture	纹理	
Texture Attributes	纹理属性	
Texture Bake Settings	纹理烘焙设置	
Texture Cache	纹理缓存	
Texture Compression	纹理压缩	
Texture Filter	纹理过滤器	
Texture Mapping	纹理贴图	
Texture Map Size	纹理贴图大小	
Texture Placements	纹理放置	
Texture Resampling Editor	纹理重采样编辑器	
Texture Resolution	纹理分辨率	
Texture Space	纹理空间	
Texture to Geometry	纹理转换为几何体	
Texture Channel	纹理通道	
Texturing	纹理	
Thickness	厚度	
Thin Line Multi Streaks	细线多重流线	
Three Point Circular Arc	三点圆弧	
Threshold	阈值	
Threshold Length	长度阈值	
Tick Span	帧跨度	
Tile Scheme	平铺方法	
Tile Size	平铺大小	
Time	时间	
Time Attributes	时间属性	
Time Offset	时间偏移	
Time Range	时间范围	
Time Ratio	时间比	

（续）

英文命令	中文命令	备注
Time Samples	时间采样	
Time Scale/Pivot	时间缩放/轴心点	
Time Slider	时间滑块	
Time Step	时间步长	
Time Tolerance	时间容差	
Time Warp	时间弯曲	
Tip Ambient Color	顶端环境色	
Tip Color	顶色	
Tip Curl	顶端蜷曲度	
Tip Opacity	顶端不透明度	
Tip Width	顶端宽度	
to	到	
to Contained Edges	到包含的边	
to Contained Faces	到包含的面	
to Edge Loop	到循环边	
to Edge Ring	到环形边	
to Edges	到边	
to Face Path	到面路径	
to Shell	到壳线	
to Shell Border	到壳边界	
to UV Border	到 UV 边界	
to UV Edge Loop	到 UV 循环边	
to UV Shell	到 UV 壳线	
to UVs	到 UV	
to Vertex Faces	到顶点面	
to Vertices	到顶点	
Toggle	关联切换	
Toggle All	全部切换	
Toggle Copy/Paste for Faces/UVs	切换复制/粘贴选项	
Toggle Display Colors Attribute	切换显示颜色属性	
Toggle Local Axes Visibility	切换局部轴可见性	
Toggle Proxy Display	切换代理显示	
Toggle Selected Labetls	切换显示所选标签	
Toggle Show/Hide	切换显示/隐藏	
Toggle the Create Bar on/off	创建列表的显示/隐藏	
Toggle the Display of Texture Borders for the Active mesh	切换对激活网络纹理边界的显示	
Toggles the mode of the UV entry fields between	在隔离 UV 点和相关变形工具值之间	

(续)

英文命令	中文命令	备 注
absolute UV position and relative transformation tool values	来切换 UV 输入区域的模式	
Tolerance	容差	
Tolerance Value	容差值	
Tonemap Scale	色调映射比例	
Tool	工具	
Tool Behavior	工具作用	
Tool Settings	工具设置	
Toolbar	工具栏	
Toon	卡通	
Top and Bottom Tabs	顶部和底部标签切换	
Top Barn Door	顶端遮光板	
Top half	上半部	
Top Left	左上	
Topology	拓扑	
Top Right	右上	
Torus	圆环	
Total Photons	光子总数	
Trail Thickness	轨迹厚度	
Transfer All Strokes to New Object	将所有笔触传递到新的对象	
Transfer Attribute Values	传递属性值	
Transfer Attributes	传递属性	
Transfer Cache to Input Mesh	传递缓存到输入网络	
Transfer from	从……传递	
Transfer in	传递方式	
Transfer Maps	传递贴图	
Transfer Settings	传递设置	
Transfer Shading Sets	传递着色集	
Transfer to	传递到	
Transfer Verterx Order	传递顶点属性	
Transfer Weights	传递权重	
Transform	变换约束	
Transform Attributes	变换属性	
Transform Component	变换组件	
Transform Control	变换控制	
Transform Display	变换显示	
Transformation Tools	变换工具	
Translate	平移	

（续）

英文命令	中文命令	备 注
Translate Frame	位移框架	
Translate/Rotate	平移/旋转	
Translation	转换	
Translator Data Locations	转换数据位置	
Translucence	半透明	
Translucence Depth	半透明深度	
Translucence Focus	半透明聚焦	
Transparency	透明度	
Transparency Balance	透明度平衡	
Transparency Gain	透明度增益	
Transparency Offset	透明度偏移	
Transparency Sorting	透明度排列	
Transparent Shadow Maps	透明阴影贴图	
Transplant Hair	移植头发	
Traverse	穿越	
Trax Editor	非线性编辑器	
Triangulate	三角化	
Trim	修剪	
Trim after	修剪之后	
Trim before	修剪之前	
Trim Tool	修剪工具	
Triple Switch	三通道转换	
Truncate Cache	缩减缓存	
Tube Direction	管道方向	
Tubes	管	
Turbulence	扰乱场	
Turbulence Speedmulti	扰动速度控制	
Turn Light Dome	翻转灯光穹顶	
Turn on Ghosts Manually	手动打开重影选项	
Turntable	可旋转的	
Turtle	海龟渲染器	
Tweak/Marquee	微调/框选	
Twist	扭曲	
Twist Type	扭曲类型	
Two Point Circular Arc	两点圆弧	
Two Sided Lighting	双面照明	
Type	类型	
Type of Ghosting	重影类型	

(续)

英文命令	中文命令	备 注
U		
U Angle	U 角度	
U Color	U 向颜色	
U Coordinate,V Coordinate	U 向坐标，V 向坐标	
U Division factor,V Division Factor	U 向细分因数，V 向细分因数	
Ultra Shape	Ultra 形状	
U Min/Max,V Min/Max	U 最小值/最大值，V 最小值/最大值	
U Min/U Max/V Min/V Max	U 向最小值/U 向最大值/V 向最小值/V 向最大值	
U patches	U 向曲面数	
U Wave	U 向波纹	
U Width	U 向宽度	
U Width/V Width	U/V 向宽度	
U/V Min/Max	U/V 最大/最小	
U/V Patches	U/V 向面片数	
U/V Width	U/V 宽度	
UI Elements	用户界面元素	
Undo	撤销	
Undo View Change	取消视图更改	
Undo/Redo	取消/重做	
Uneven Anchor Tangents	不均衡锚点切线	
Unfold	展开	
Unghost All	取消所有重影	
Unghost Selected	取消选择对象的重影	
Ungroup	解组	
Ungroup LOD Group	解组 LOD 组	
Ungroup under	解组类型	
Unified Sampling	统一采样	
Uniform	统一力场	
Union	并集	
Union Tool	并集工具	
Unit Conversion	单位转换	
Unitize	单元化	
Universal Manipulator	万能操纵器	
Universal Manipulator Settings	万能操纵器设置	
Unload Reference	卸载引用	
Unload Related References	卸载相关引用	
Unlock Length	解锁长度	

（续）

英文命令	中文命令	备注
Unlock Normals	解锁法线	
Unlock Reference	解除对引用的锁定	
Unlock Unpublished Attributes	取消锁定未发布的属性	
Unmapped	未映射	
Unparent	解除父化关系	
Unparent method	解除父化方式	
Unpublish Attributes	取消发布属性	
Unpublish Child Anchor	取消发布子锚点	
Unpublish Node	取消发布节点	
Unpublish Parent Anchor	取消发布父锚点	
Unpublish Selection Transform	取消发布选择变换	
Unresolved Name	未解算的名称	
Unselect	不选择	
Unselect All	取消全部选择	
Untemplate	取消模板	
Untrim	取消剪切	
Untrim Surfaces	取消剪切曲面	
Up Axis	上方轴	
Up Vector	向上向量	
Update	更新	
Update Current Scene	更新当前场景	
Update Fur Maps	更新毛发贴图	
Update Image Planes/Background	更新图像平面/背景	
Update on Stroke	绘制时更新	
Update PSD Network	更新 PSD 网络	
Update Shadow Maps	更新阴影贴图	
Update Snapshot	更新快照	
Update View	更新视图	
Use 2d Blur Memory Limit	使用 2d 模糊存储限制	
Use All Available Processors	使用所有可用的处理器	
Use All Lights	使用所有灯光	
Use All Surface Isoparms	使用所有曲面等参线	
Use Auto Focus	使用自动焦距	
Use Background	使用背景	
Use Bind	使用绑定	
Use Blending	使用出血	
Use Cache	使用缓存	
Use Cache Name as Prefix	将缓存名称作为前缀	

第 13 章 Maya 命令中英文对照速查表

（续）

英文命令	中文命令	备注
Use Chord Height	使用弦高	
Use Chord Height Ratio	使用弦高比	
Use Color Ramp	使用颜色渐变	
Use Default Lighting	使用默认灯光	
Use Depth Map Shadows	使用深度贴图阴影	
Use FG Map File	使用最终聚集贴图文件	
Use Ghost Driver	使用重影驱动	
Use Hardware Shader	使用硬件着色器	
Use Hardware Texture Cycling	使用硬件纹理循环	
Use Image Ratio	使用图像比率	
Use Instancing	使用实例	
Use Light Position	使用灯光位置	
Use Macro	使用宏指令	
Use Max Distance	使用最大距离	
Use Mid Dist	使用中间距离	
Use Min Photon Depth	使用最小光子深度	
Use Min screen	使用最小屏幕	
Use One Thread per CPU	每个 CPU 使用一个线程	
Use Only Single Dmap	仅使用当个深度贴图	
Use Path Tracer Map File	使用路径跟踪器贴图文件	
Use Photon Map File	使用光子贴图文件	
Use Selected as Source of Field	使用所选对象作为场源	
Use Selected Emitter	使用所有选择的发射器	
Use Selected Lights	使用选择灯光	
Use Sequence Time	使用序列时间	
Use Shutter Open/Close	使用快门打开/关闭	
Use Smooth Edge	使用平滑边	
Use SSE	使用 SSE	
Use Stylus Pressure	使用压感笔	
Use Tabs for Indent	使用制表符进行缩进	
Use Tolerance	使用容差	
Use Trax Sounds	使用非线声音	
User Normals	用户法线	
User Scale	用户缩放	
User Scale x/y/z	用户缩放 x/y/z	
Utilities	工具	
UV borders	UV 边界	
UV Chooser	UV 选择器	

（续）

英文命令	中文命令	备注
UV Coord	UV 坐标	
UV Coordinates	UV 坐标	
UV Editor	UV 编辑器	
UV Gap Tolerance	UV 间隙容差	
UV Lattice Tool	UV 晶格工具	
UV Range	UV 范围	
UV Set Editor	UV 集编辑器	
UV Set Name	UV 集名称	
UV Set Sharing	共享 UV 集	
UV Sets	UV 集	
UV Settings	UV 设置	
UV Size	UVs 大小	
UV Smudge Tool	UV 柔化工具	
UV Snapshot	UV 快照	
UV Space	UV 空间	
UV Texture Editor	UV 纹理编辑器	
UV Texture Editor Baking	UV 纹理编辑烘焙	
UV Texture Editor Snapping	UV 纹理编辑器捕捉	
V		
V Angle	V 向角度	
V Color	V 向颜色	
V Patches	V 向曲面数	
V Wave	V 向波纹	
V Width	V 向宽度	
Value	属性值	
Value Offset	数值偏移	
Value Scale/pivot	数值缩放/轴心点	
Value Tolerance	数值容差	
Values	值	
Vector	向量	
Vector Displacement	向量置换	
Vector Encoding	向量编码	
Vector Functions	向量函数	
Vector Product	矢量乘积	
Vector Renderer	矢量渲染器	
Vector Renderer Control	矢量渲染控制	
Vector Space	向量空间	
Verbose	赘言	

（续）

英文命令	中文命令	备 注
Version	版本	
Vertex	顶点	
Vertex Backface Culling	消隐背面顶点	
Vertex Bake Settings	顶点烘焙设置	
Vertex Bias	顶点偏移	
Vertex Color Filtering	顶点颜色滤镜	
Vertex Color Mode	顶点着色模式	
Vertex Face	顶点面	
Vertex Index Map	顶点索引贴图	
Vertex Merge Threshold	顶点合并阈值	
Vertex Normal	顶点法线	
Vertex Normal Edit Tool	顶点法线编辑工具	
Vertex Size	顶点大小	
Vertical	垂直	
Vertices	顶点	
View	视图	
View（Radii in Pixel Size)	视图（半径以像素大小为单位）	
View All References	查看所有引用	
View Along Axis	沿轴查看	
View Connected Faces	显示连接面	
View Contained Faces	显示包含的面	
View Faces of Selected Images	显示选择的图片面	
View Image	查看图像	
View Selected References	查看所选引用	
View Sequence	查看序列	
ViewCube	视图导航器	
Viewport 2.0	视图 2.0 版	
Viewport Color	视口颜色	
Viewport 2.0 Font Size	视图 2.0 版字体大小	
Visible in Background	背景中可见	
Visible in Reflections	反射中可见	
Visible in Refractions	折射中可见	
Visible Surface Isoparms	可视的曲面参数线	
Visible UI Elements	可见的 UI 元素	
Visor	遮板	
Visor Directory	遮板路经	
Visualize	可视化	
Visualize in Model View	在模型视图中可视化	

（续）

英文命令	中文命令	备注
Visualize Sticky Bind Distance for Selected Muscle Objects	为所选肌肉对象可视化黏连结合距离	
Volume	体积	
Volume Axis	体积轴场	
Volume Curve	体积曲线场	
Volume Emitter Attributes	体积发射器属性	
Volume Fog	体积雾	
Volume Fog Attributes	体积雾属性	
Volume Light	体积光	
Volume Noise	体积噪波	
Volume Noise Attributes	体积碎片属性	
Volume Shader	体积着色器	
Volume Speed Attributes	体积速度属性	
Volume Type	体积类型	
Volumetric Materials	体积材质	
Volumetric Shader	体积着色器	
Volumize	体积膨胀	
Volumize A/B	体积保持 A/B	
Volumize Dist	体积保持距离	
Volumize Falloff	衰减体积碰撞	
Volumize Offset	体积保持偏移	
Volumize Puff	体积保持扩张	
Volume Light	体积光	
Volume Primitives	体积基本几何体	
Vortex	漩涡力重	
VRAM Management	VRAM 管理	
W		
Warning	警告	
Warp Image	扭曲图像	
Water	水	
Water Attributes	水属性	
Wave	波浪	
Wax Tool	上蜡工具	
Wedge	楔形	
Wedge Face	楔入面	
Weight	权重	
Weight Change Percentage	权重改变比率	
Weight Color	权重颜色	
Weight Distribution	最大影响	

(续)

英文命令	中文命令	备 注
Weight Hammer	权重锤	
Weight Locking	权重锁定	
Weight Slider	权重滑块	
Weight Start/Weight Mid/Weight End	权重起始端/中部/末端	
Weight Threshold	权重阈值	
Weight Deformer	权重变形	
Weighted/Non-Weighted Tangents	加权/非加权切线	
Weighting	权重	
Weighting on Edge	设置边的权重	
Weights	权重	
Weld Adjacent Borders	焊接邻近边界约束	
When Clicking on the ViewCube	在视图导航上单击时	
When Dragging on the ViewCube	在视图导航器上拖拽鼠标时	
Whiteness	白色	
Width	宽度	
Width Brush	宽度笔刷	
Width/Height	宽度/高度	
Width divisions	沿宽度方向细分	
Width Focus	宽度焦距	
Width Spread	宽度扩展	
Width/Height	宽度/高度	
Width/Height divisions	宽度/高度细分	
Width/Height/Depth	宽度/高度/深度	
Width/Height/Depth divisions	宽度/高度/深度细分	
Width/Length/Height	宽度/长度/高度	
Wind Dir X/Y/Z	风向 X/Y/Z	
Wind Field Generation	产生风场	
Wind Noise	风波动	
Wind Noise Dirty	风波动模糊	
Wind Noise Scale	风的波动缩放	
Wind Speed	风速	
Wind Strength	风强度	
Window	窗口	
Wire	线	
Wire Color	线框颜色	
Wire Dropoff Locator	线衰减定位器	
Wire Node	线节点	
Wire Tool	线工具	

（续）

英文命令	中文命令	备注
Wire frame	线框	
Wireframe Color	线框颜色	
With Connections to Network	用连接关系链接到网络	
With Curve	关联曲线	
Without Network	无网络	
Wood	木纹	
Work Area	工作区域	
Working Units	工作单位	
Workspaces	工作区	
Word	世界	
World Coordinate System	世界坐标系	
World Space Coordinates	世界空间坐标系	
World up Object	世界上方对象	
World up Type	世界上方类型	
World up Vector	世界上方向量	
Wrap	包裹	
Wrap Amplitude	包裹振幅	
Wrap U/Wrap V	U/V 包裹	
Wrinkle A/B	褶皱 A/B	
Wrinkle Spread	褶皱扩展	
Wrinkle Strength	褶皱强度	
Wrinkle Tool	褶皱工具	
Write Materials	读入材质	
X		
X Factor/Z Factor	X/Z 因数	
Xgen Library	Xgen 库	
X/Y resolution	X/Y 轴向的分辨率	
X/Y/Z values	X/Y/Z 值	
X-Ray	X 射线	
X-Ray Joint Display	X 射线关节显示	
X-Ray Mode	X 射线模式	
X-Ray on/off	启用/关闭 X 射线	
Z		
Zero Key Layer	零值关键帧层	
Zero Length Handling	零长度操作	

(续)

英文命令	中文命令	备注
Zero Weight and Key Layer	零权重并对层设置关键帧	
Zeroth Scanline	零扫描线	
Zoom Factor	缩放因子	
数字		
3D Cut and sew UV Tool	3D 切割和缝合 UV 工具	

附录 A　Maya 常用命令操作快捷键

快捷键	快捷键功能	快捷键	快捷键功能
A	在当前视图框显示所有对象	Shift+A	在所有视图框显所有对象
B	修改笔刷上限半径	Shift+B	修改笔刷下限半径
C	捕捉到曲线	Shift+D	用变换复制
E	旋转工具	Shift+E	设置旋转关键帧
F	在当前视图框显示选择对象	Shift+h	显示选择物体
G	重复上一次操作	Shift+I	查看选择
H	UI 模式标记菜单	Shift+M	面板菜单条切换
I	插入关键帧工具激活	Shift+P	解除父物体
L	锁定曲线长度	Shift+Q	选择元素工具标记菜单
N	修改绘画值	Shift+R	设置缩放关键帧
O	多边性笔刷标记菜单	Shift+S+左键	关键帧切线标记菜单
P	建立父物体	Shift+W	设置平移关键帧
Q	选择工具	Shift+Z	重做
R	缩放工具	Ctrl+A	显示属性编辑器或通道框
S	设置关键帧	Ctrl+Alt+S	递增并保存
T	显示操纵	Ctrl+B	倒角
U	绘制操作标记菜单	Ctrl+C	复制
V	捕捉	Ctrl+D	关联复制
W	移动工具	Ctrl+Del	删除边和顶点
X	捕捉到栅格	Ctrl+E	挤出
Z	撤销	Ctrl+G	成组
>	增长	Ctrl+H	隐藏选择物体
<	收缩	Ctrl+M	切换主菜单条
Ctrl+N	新建场景	Alt+D	取消全部选择
Ctrl+O	打开场景	Alt+F	覆盖表面（绘制权重时）
Ctrl+Q	退出	Alt+H	隐藏未选对象
Ctrl+R	创建引用…	Alt+R	折射切换
Ctrl+S	保存场景	Alt+V	向后播放
Ctrtl+V	粘贴	Alt+Shift+A	扩展选择区域
Ctrl+X	剪切	Alt+Shift+S	收缩选择区域
Ctrl+Y	重做	Alt+Shift+V	回到最小
Ctrl+Z	撤销	Alt+F9	选择顶点面组件
Ctrl+>	沿循环方向扩大	Alt+,	前一帧
Ctrl+<	沿循环方向收缩	Alt+.	后一帧

附录 A　Maya 常用命令操作快捷键

快捷键	快捷键功能	快捷键	快捷键功能
Ctrl+Alt+Left	反转自旋边	Alt+↑	逐像素上移
Ctrl+Alt+Right	正向自旋边	Alt+↓	逐像素下移
Ctrl+F9	转换选择到顶点	Alt+←	逐像素左移
Ctrl+F10	转换选择到边	Alt+→	逐像素右移
Ctrl+F11	转换选择到面	F1	帮助
Ctrl+F12	转换选择到 UV	F2	显示动画
Ctrl+Shift+`	平滑代理	F3	显示建模
Ctrl+Shift+~	平滑代理选项	F4	显示动力学
Ctrl+Shift+A	全部	F5	显示渲染
Ctrl+Shift+D	特殊复制	F8	选择对象模式切换
Ctrl+Shift+I	反转	F9	选择顶点组件
Ctrl+Shift+Q	四边形绘制	F10	选择边组件
Ctrl+Shift+S	场景另存为	F11	选择面组件
Ctrl+Shift+X	多切割	F12	选择 UV 组件
Alt+B	循环背景颜色	0	默认画质显示
2	中等画质显示	Shift+}	后一个视图布局
3	高画质显示	Shift+<	收缩多边形选择区域
4	显示线框	Shift+<	扩展多边形选择区域
5	显示材质	Shift+D	复制并变换
6	显示材质和纹理	↑	将当前选择的 NURBS 的 CV 或多边形顶点向上拾取
7	显示灯光	↓	将当前选择的 NURBS 的 CV 或多边形顶点向下拾取
8	绘画效果面板	←	将当前选择的 NURBS 的 CV 或多边形顶点向左拾取
"-"	减小操纵器大小	→	将当前选择的 NURBS 的 CV 或多边形顶点向右拾取
"+"	增加操作器大小	Page_Up	增加光滑代理模型的细分层级
[	撤销视图变换	Page_Down	减少光滑代理模型的细分层级
]	重做视图变换	Insert	进入中枢轴编辑模式
,	前一个关键帧	Enter	完成当前操作
.	后一个关键帧	Space	在单一视图和四视图间切换
Shift+{	前一个视图布局	—	—

附录 B 本书教学视频链接网址和密码

01_基础模块：https://pan.baidu.com/s/19N3DFvzUUogpUtpn8l-cOw 密码：uexs
02_建模模块：https://pan.baidu.com/s/1lzwrpK-Q7qxCQX4ySAvUcQ 密码：q973
03_装备模块：https://pan.baidu.com/s/1OxU59f5ouIzXGL3qygMYRQ 密码：smal
04_动画模块：https://pan.baidu.com/s/1vLhhJxXGGlWCQXLY1vCnpA 密码：lhg6
05_FX 模块：https://pan.baidu.com/s/1zO1EmA9B6GzYHPpW94dPKg 密码：311w
06_渲染模块：https://pan.baidu.com/s/1zPMMTDMMnJg31fuwcgo-qQ 密码：qcfr
07_视图窗口菜单组：https://pan.baidu.com/s/1ToZVbzz7ONcxYu7RTFLLIQ 密码：kvrk
08_对象：https://pan.baidu.com/s/1A0LneoqZ_mfZtbiaZiNOJg 密码：vsmn
09_对象：https://pan.baidu.com/s/1-apDUGkqtTXkHlud7C26rA 密码：lqj5
10_显示：https://pan.baidu.com/s/1010a0FVKhzuvUUDjevmB6Q 密码：ot8d
11_渲染动画：https://pan.baidu.com/s/1grg05_jbeIfWhQ59RCF1bw 密码：adi4
12_渲染：https://pan.baidu.com/s/14dNoJu7eaRvc0LDySjZ-Iw 密码：mxdx